Creationism in Europe

MEDICINE, SCIENCE, AND RELIGION IN HISTORICAL CONTEXT
Ronald L. Numbers, *Consulting Editor*

Creationism in Europe

EDITED BY

Stefaan Blancke, Hans Henrik Hjermitslev,

AND Peter C. Kjærgaard

Foreword by Ronald L. Numbers

Johns Hopkins University Press

Baltimore

 Published 2014
Printed in the United States of America on acid-free paper
2 4 6 8 9 7 5 3 1

Johns Hopkins University Press
2715 North Charles Street
Baltimore, Maryland 21218-4363
www.press.jhu.edu

Library of Congress Cataloging-in-Publication Data

Creationism in Europe / edited by Stefaan Blancke, Hans Henrik Hjermitslev, and Peter C. Kjærgaard.
pages cm. — (Medicine, science, and religion in historical context)
Includes bibliographical references and index.
ISBN 978-1-4214-1562-8 (hardcover : alk. paper) — ISBN 978-1-4214-1563-5 (electronic) — ISBN 1-4214-1562-3 (hardcover : alk. paper) — ISBN 1-4214-1563-1 (electronic) 1. Creationism—Europe. I. Blancke, Stefaan, 1976– editor.
BS651.C75 2014
231.7'652094—dc23 2014010423

A catalog record for this book is available from the British Library.

*Special discounts are available for bulk purchases of this book.
For more information, please contact Special Sales at
410-516-6936 or specialsales@press.jhu.edu.*

Johns Hopkins University Press uses environmentally friendly book materials, including recycled text paper that is composed of at least 30 percent post-consumer waste, whenever possible.

Contents

Foreword, by Ronald L. Numbers *vii*

Acknowledgments *xvii*

Introduction: Creationism in Europe or European Creationism? 1

STEFAAN BLANCKE, HANS HENRIK HJERMITSLEV, AND PETER C. KJÆRGAARD

1 France 15

THOMAS LEPELTIER

2 Spain and Portugal 31

JESÚS I. CATALÁ-GORGUES

3 United Kingdom 50

JOACHIM ALLGAIER

4 The Low Countries 65

STEFAAN BLANCKE, ABRAHAM C. FLIPSE, AND JOHAN BRAECKMAN

5 Scandinavia 85

HANS HENRIK HJERMITSLEV AND PETER C. KJÆRGAARD

6 Germany 105

ULRICH KUTSCHERA

7 Poland 125

BARTOSZ BORCZYK

8 Greece 144

EFTHYMIOS NICOLAIDIS

9 Russia and Its Neighbors 162

INGA LEVIT, GEORGY S. LEVIT, UWE HOSSFELD, AND LENNART OLSSON

10 Turkey 180

MARTIN RIEXINGER

11 Catholicism 199
RAFAEL A. MARTÍNEZ AND THOMAS F. GLICK

12 Intelligent Design 214
BARBARA FORREST

13 The Rise of Anti-creationism in Europe 228
PETER C. KJÆRGAARD

Afterword: Reclaiming Science for Creationism 242
NICOLAAS A. RUPKE

A Note on Sources, by Stefaan Blancke 251
List of Contributors 257
Index 265

Foreword

RONALD L. NUMBERS

Until fairly recently the notion of a history of creationism in Europe would have struck many readers as preposterous. Although the United States had become notorious for spawning creationism, the rest of the world seemed relatively immune. As the late Stephen Jay Gould assured non-Americans in 2000, creationism, though "insidious," had remained confined to the United States; indeed, in his opinion, it had remained largely "a local, indigenous, American bizarrity." His Harvard colleague Richard C. Lewontin, a prominent evolutionary biologist, shared this view, writing that "creationism is an American institution, and it is not only American but specifically southern and southwestern."[1]

But even as Gould and Lewontin assured the rest of the world's population that it was safe from this uniquely American concept, creationism was spreading beyond the confines of the United States. Despite its "Made in America" label, it had begun to emerge among conservative Protestants in various parts of the world. At first, beginning in the late 1960s, the leading proponents were men such as Henry M. Morris and Duane Gish from the Institute for Creation Research (ICR) near San Diego, California, which eventually published creationist tracts in some two dozen languages. After the mid-1990s, leadership passed increasingly to Kenneth A. Ham at Answers in Genesis (AiG), a Kentucky-based operation located just south of Cincinnati. In less than a decade Ham and his AiG colleagues had created a network of international branches and distributed books in languages ranging from Afrikaans and Albanian to Romanian and Russian.[2] During that time scientific creationism and its younger cousin, intelligent design, also spread from their evangelical Protestant bases to Catholicism, Eastern Orthodoxy, Islam, Judaism, and beyond.[3]

As the last sentence suggests, the term *creationism* spans a spectrum of meanings, from relatively rare supernatural interventions in the natural order to the creation of the entire universe no more than ten thousand years ago. Although much of the world's population rejected (and continues to reject) naturalistic evolution, organized antievolutionism did not appear until the early 1920s, and then in the United States among conservative Protestants who believed in an inerrant Bible. With few exceptions, however, even the most literalistic Bible believers accepted the antiquity of life on earth as revealed in the paleontological record. They typically did so either by interpreting the days of Genesis 1 as vast geologic ages (the day-age theory) or by inserting a series of catastrophes and recreations or ruins and restorations into an imagined gap between the first two verses of the Bible (the gap theory). By the close of the nineteenth century, virtually the only Christians writing in defense of the recent appearance of life on earth and attributing the fossil record to the action of Noah's flood were Seventh-day Adventists, a fundamentalist group then numbering fewer than 100,000 members.

The leading popularizer of this uncompromisingly literal reading of Genesis was the Canadian-born Adventist "geologist" George McCready Price, who dubbed his version of creationism *flood geology*. Price's idiosyncratic interpretation of earth history remained on the periphery of fundamentalism until after 1961, when John C. Whitcomb Jr., an Old Testament scholar, and Henry M. Morris, a hydraulic engineer, brought out a book called *The Genesis Flood*. Whitcomb and Morris presented Price's flood geology as the only acceptable interpretation of the first chapters of Genesis. Their insistence on beginning with a literal reading of the Bible and then trying to fit science into that context, rather than constantly accommodating the Bible to the findings of science, struck a responsive chord with many concerned Christians. In substantial, though undetermined, numbers they abandoned the once-favored day-age and gap theories, which allowed for the antiquity of life on earth, accepting instead the strict young-earth creationism urged by Whitcomb and Morris.

In 1963 Morris and nine other like-minded creationists banded together to form the Creation Research Society (CRS). About 1970, in an effort to make their Bible-based creationism sufficiently secular to be taught in state schools, leaders of the CRS rechristened flood geology "creation science" or "scientific creationism." In the context of creation science, the sequence and timing of key events, such as a recent special creation and subsequent worldwide flood, remained the same, but all direct references to biblical characters and places, such as Adam and Eve, the Garden of Eden, and Noah and his ark, disappeared from the stripped-down narrative.

The young-earth creationists associated with the CRS and Morris's Institute for Creation Research, established near San Diego in 1972, proved highly effective in promoting Price's version of creationism among conservative Christians. Within a decade or two, the tireless proselytizers for scientific creationism had virtually co-opted the generic creationist label for their hyperliteralist views, which only a half century earlier had languished on the margins of American fundamentalism. After the 1970s people who called themselves creationists typically assumed that most listeners would identify them as believers in a young earth.

In the early 1980s various state legislatures debated the wisdom of mandating the teaching of "creation science" in public-school classrooms whenever "evolution science" was taught. Two states, Arkansas and Louisiana, passed such "balanced treatment" laws. In 1987, however, the US Supreme Court ruled that these laws violated the First Amendment of the Constitution requiring the separation of church and state. This created an opening for a new species of anti-evolutionism: intelligent design theory (ID). Although some critics dismissed it as "the same old creationist bullshit dressed up in new clothes" and disparaged it as merely the latest "creationist alias," promoters of ID stressed the differences between young-earth creationism and their own position. The authors of *Pandas and People: The Central Question of Biological Origins* (1989), the first book explicitly to promote intelligent design, insisted that their text implied "absolutely nothing about beliefs normally associated with Christian fundamentalism, such as a young earth, a global flood, or even the existence of the Christian God." Hoping to distance themselves from the intellectually marginal creation scientists and to avoid endless niggling over the meaning of the Mosaic story of creation, design theorists carefully avoided any mention of Genesis or God, although, as one of them confessed to some fellow Christians, referring to an intelligent designer was merely a "politically correct way to refer to God." In *Darwin on Trial* (1991), another early book advocating ID, the Berkeley law professor Phillip E. Johnson wrote, "This book is 'creationist' only in the sense that it juxtaposes a paradigm of 'intelligent design,' with the dominant paradigm of (naturalistic) evolution, and makes the case for the former. It does not rely upon the authority of the Bible."[4]

Most controversially, the ID theorists tried "to reclaim science in the name of God" by rewriting the rules of science. Although they targeted evolution, they identified the real enemy as scientific naturalism, which ruled god-talk out of science. Most troubling of all was the widespread acceptance of so-called methodological naturalism among even evangelical Christians, who, despite their beliefs, left God at the laboratory door and refrained from invoking the supernatural when trying to explain the workings of nature. In contrast to "metaphysical naturalism," which denied the existence of a transcendent God, methodological naturalism implied nothing about God's existence and activities. But to critics such as Johnson, it smacked of thinly veiled atheism.

Young-earth creationists generally wanted as little to do with ID as the ID theorists wanted to do with them. Although some creation scientists applauded the effort to discover evidence of God in nature, the leaders of the movement never warmed up to ID theory as such. They especially disliked the ID theorists' marginalization of biblical concerns in the interest of mounting a united attack against Darwinism. When the "Goliath" of naturalistic evolution "has been tumbled," reasoned the proponents of intelligent design, "there will be time to work out more details of how creation really did occur." Henry Morris, the grand old man of scientific creationism, expressed admiration for the efforts of ID theorists to refute Darwinism but deplored their apparent lack of concern for theological niceties. He feared that many Christians would embrace ID as a way to avoid "having to confront the Genesis record of a young earth and global flood." He dismissed as "nonsense" the claim that the intelligent designer need not be God—or even "a deity." He correctly predicted that, despite having compromised on the plain meaning of the Bible, the proselytizers for ID theory would find no more favor with naturalistic evolutionists than he himself had.

The low point for the ID movement came late in 2005, when a federal district court judge in Pennsylvania ruled on a case in which the Dover Area School District had promoted ID and urged students to read *Pandas and People*. In his decision the judge excoriated the school board for its actions, which he described as a "breathtaking inanity." He ruled that ID was "not science" because it invoked "supernatural causation" and failed "to meet the essential ground rules that limit science to testable, natural explanations." The board's promotion of it thus violated the establishment clause of the First Amendment to the US Constitution. Since that judicial setback, the promoters of ID have urged educators to teach "the strengths and weaknesses of evolution."

Over the past three decades, beginning in 1982, the Gallup Organization has surveyed American attitudes toward creation and evolution. During this period the number of those reporting that "God created humans in present form" remains fairly stable, fluctuating between 43 and 47 percent (with 46% in 2012). The next most popular view, ranging from 32 to 40 percent, affirmed that "humans evolved, with God guiding." Only 9 to 16 percent (15% in 2012) believed that "Humans evolved, but God had no part in process."[5]

Creationism Goes Global

Few countries outside the United States initially gave creation science a warmer reception than Australia. A visit from Morris in 1973 sparked interest in flood geology. Within five years young-earth creationists, led by Ham, then a schoolteacher, and Carl Wieland, a physician, had organized the Creation Science Foundation (CSF), which from its headquarters in Brisbane quickly became the center of antievolutionism in the South Pacific. A 2013 survey of scientific literacy in Australia, commissioned by the Australian Academy of Science, found that more than a quarter of those surveyed believed that humans and dinosaurs lived together on earth and nearly one in ten rejected evolution outright. Similar developments occurred in New Zealand, only more slowly and with less fanfare. In the mid-1990s a popular magazine announced, to the surprise of many readers, that "God and Darwin are still battling it out in New Zealand schools." In contrast to the common image of a thoroughly secular educational system, the popular magazine revealed that "specialists with science degrees" had been promulgating creationism in the country's classrooms—with great success, especially among the Maori and Pacific Island peoples.[6]

In Asia the Koreans emerged as *the* creationist powerhouse, propagating the message at home and abroad. Since its founding in the winter of 1980–81, the Korea Association of Creation Research, warmly supported by many Christians in the country, has flourished. Within fifteen years of its founding the association had spawned sixteen branches in Korea, recruited several hundred members with doctorates of one kind or another, and published dozens of creationist books and a bimonthly magazine, *Creation*, with a circulation of four thousand. By 2000 the membership stood at 1,365, giving Korea claim to being the creationist capital of the world, in density if not in influence. That same year the Korea Association of Creation Science made history by dispatching the first creation-science missionary, Kwang Ho Jun, formerly with the US National Institutes of Health, to Muslim Indonesia, where the association had been sending lecturers for some time. During the late twentieth century the association established several branch chapters along the West Coast of North America. A survey conducted in 2012 discovered that nearly a third of South Koreans rejected evolution. The following year a breakaway group from the Korea Association of Creation Science ignited a national controversy by pressuring publishers to delete the "error" of evolution from science textbooks by removing examples of how evolution occurred. However, a government-appointed advisory committee urged the publishers to resist the creationists' efforts to sanitize the texts.[7]

Elsewhere in East and Southeast Asia creationism took hold in Hong Kong, Taiwan, the Philippines, and Japan, where, despite having a relatively low percentage of creationists, Answers in Genesis in 1998 established AiG/Japan, with offices in the Tokyo suburb of Nagaoka. Because of its hostility to Christian proselytizers, the People's Republic of China posed a huge challenge to creationists.

From time to time *People's Daily*, the official newspaper of the Communist Party of China, covered the creation-evolution controversy in the United States; in the late 1990s the party-owned Central Compilation and Translation Press in Beijing published Chinese translations of two major books advocating ID.[8]

South Asia proved to be almost as impenetrable as China, although many Muslims in India, Pakistan, and Bangladesh rejected evolution. By the 1990s creationist missionaries, such as the peripatetic Gish, were visiting India, but the customary reports of throngs of eager listeners did not follow. As late as 2005 Jyoti P. Chakravartty, founder of the small Calcutta-based Creation Science Association of India, could not identify any other "creationist group in South Asia." Nevertheless, a poll conducted in 2011 suggests that a third of the population of India could be classified as creationists.[9]

Organized creationism has remained relatively weak in sub-Saharan Africa, largely because conservative Christianity had grown so strong that creationists have rarely found an evolutionist establishment to attack. Not surprisingly, a poll of adolescents in Kenya showed that 68 percent of them believed "that Christianity is necessarily creationist." According to one observer, "within many home and church environments in Kenya there is no recognition that there are any views of origins which are consistent with Christian commitment other than that of creationism."[10]

The comparatively prosperous Republic of South Africa welcomed creationism the most enthusiastically of any sub-Saharan country. In 1948 the pro-apartheid National Party inaugurated a program of Christian National Education acceptable to the Reformed church. As a matter of course, South African students, including those in teacher-training schools, learned only the Genesis story of creation. After Nelson Mandela and the African National Congress came to power in 1994, the restrictions against the teaching of evolution eased to the point that South African creationists were soon complaining that it had become "almost impossible to get attention for a creationist point of view in present day South Africa, which has a communist government after the last election." If anything, however, organized creationism became more visible than ever, with 56 percent of South Africans identifying as creationists in 2011.[11]

Writing in 2000, one observer claimed that "there are possibly more creationists per capita in Canada than in any other Western country apart from the US." Though much has been written lately about the de-Christianization of Canadian culture, the claim may have been true. In 1993 *Maclean's*, "Canada's Weekly Newsmagazine," published a surprising public-opinion poll showing that "even though less than a third of Canadians attend a religious service regularly . . . 53% of all adults reject the theory of scientific evolution." This figure may no longer reflect Canadian opinions. In 2010 a poll showed that 61 percent of Canadians accepted human evolution, including evolution directed by God, while 24 percent favored special creation "within the last 10,000 years." The prairie provinces of Manitoba and Saskatchewan led the country in support for creationism.[12]

After a very slow start in Latin America, creationists witnessed "an explosion" of interest in the late 1990s, paralleling that of evangelical Christianity generally. Nowhere in South America did antievolutionists make deeper inroads than in Brazil, where, according to a survey in 2004, 31 percent of the population believed that "the first humans were created no more than 10,000 years ago" and the overwhelming majority favored teaching creationism. In 2004 the evangelical governor of the state of Rio de Janeiro, Rosinha Mateus, announced that public schools would be teaching creationism. "I do not believe in the evolution of species," she declared. "It's just a theory." Evolutionists tried to mount a protest, but they failed to generate much interest. The Catholic majority in the country, explained a dispirited scientist, had become confused and overwhelmed by the aggressive evangelical Protestants who "imported creationism from the U.S." A survey of public opinion a short time later revealed that 89 percent of Brazilians agreed with the governor on teaching creationism, with 75 percent of the respondents saying that only creationism should be taught in the public schools. Thirty-one percent favored young-earth creationism; 54 percent, old-earth creationism; nine percent, naturalistic evolution.[13]

Creationism in Modern Europe

In all comparative surveys, Europe collectively ranks at or near the top (with China and Japan) in embracing evolution. But, as recent surveys have shown, creationism is far from absent on the continent, with about one in five Europeans rejecting human evolution. In Turkey, which straddles Europe and Asia, more than 50 percent of the population repudiates evolution. Cyprus comes in second with 36 percent, followed by Greece (32%); Lithuania (30%); Slovakia (29%); Austria, Switzerland, and Croatia (28%); Finland, the Czech Republic, Latvia, and Poland (27%); and Malta, Romania, and Slovenia (25%). The most receptive to human evolution are Iceland (with a mere 7% embracing creationism); France (12%); Denmark, Sweden, and the United Kingdom (13%), Spain (16%); Norway (18%); Estonia (19%); Italy (20%); Belgium, Bulgaria, Hungary,

Ireland, and Portugal (21%); and Germany, Luxembourg, and the Netherlands (23%). The most surprising pattern, observed by one analyst, is the generally rising rate of creationist sentiment as one moves east, into the former communist (and officially atheistic) countries of the Eastern bloc.[14]

Opinion polls, though preferable to mere impressions, can also confuse or mislead. In the above survey, for instance, the United Kingdom appears to be solidly pro-evolution. However, another survey, conducted in Britain in 2009, in connection with Charles Darwin's two-hundredth birthday, showed that "four out of 10 people in the UK think that religious alternatives to Darwin's theory of evolution should be taught as science in schools" and that only 48 percent of Britons believed that the theory of evolution "best described their view of the origin and development of life"; 22 percent said that "creationism" best described their views, 17 percent favored "intelligent design," and 13 percent remained undecided. In other words, as *The Guardian* announced in a headline, "Half of Britons Do Not Believe in Evolution."[15]

Except for such polls and repeated warnings from European scientists about the invasion of American-style creationism, we know virtually nothing about the cultural origins and meanings of antievolution sentiment in Europe. But now the historiographical desert disappears with the appearance of this extraordinary collection of original essays, *Creationism in Europe*, edited by Stefaan Blancke, Hans Henrik Hjermitslev, and Peter C. Kjærgaard. I welcome this pioneering book with the greatest of enthusiasm.

NOTES

1. Portions of this foreword appeared in an earlier form in Ronald L. Numbers, *The Creationists: From Scientific Creationism to Intelligent Design*, expanded ed. (Cambridge, MA: Harvard University Press, 2006). Unless otherwise noted, documentation can be found there.

2. In a bitter split in 2005 Ken Ham and AiG broke with their former mates in Australasia, who took the name Creation Ministries International. See Jim Lippard, "Trouble in Paradise," *Reports of the National Center for Science Education* 26 (2006): 4–7.

3. On the exceedingly high levels of creationism among Muslims, see, e.g., Salman Hameed, "Bracing for Islamic Creationism," *Science* 322 (2008): 1637–1638; Ipsos Global @dvisory, "Supreme Being(s), the Afterlife and Evolution," released Apr. 25, 2011, www.ipsos-na.com/news-polls/pressrelease.aspx?id=5217. Among the twenty-four countries surveyed, Saudi Arabia ranks as first in creationist sentiment (75% of the population), Turkey as second (60%), and Indonesia as third (57%). For a full listing of the survey's findings, see www.ipsos-na.com/download/pr.aspx?id=10670.

4. On the history of ID, see Numbers, *The Creationists*, chap. 17, "Intelligent Design."

5. Gallup, "Evolution, Creationism, Intelligent Design: Gallup Historical Trends," accessed Nov. 1, 2013, www.gallup.com/poll/21814/evolution-creationism-intelligent-design.aspx. In one of the few comparisons between creationism and intelligent design, Gallup in 2005 asked respondents to identify the beliefs they found "Definitely/Probably True"; 58% said creationism, 55% evolution, 31% intelligent design. For a somewhat more positive survey of belief in evolution, see Ben Henderson, "Belief in Evolution Up since 2004," July 22, 2013, http://today.yougov.com/news/2013/07/22/belief-in-evolution-up-since-2004.

6. On global creationism, see Numbers, *The Creationists*, chap. 18, "Creationism Goes Global." Cameron Williams, "Australia's Science Literacy Falls: Survey," June 17, 2013, www.science.org.au/news/media/17july13.html. The results of the survey are available at www.science.org.au/reports/literacysurvey. See also Ipsos Global @dvisory, "Supreme Being(s), the Afterlife and Evolution," where 15% of Australians identified as creationists.

7. Soo Bin Park, "South Korea Surrenders to Creationist Demands," June 5, 2012, www.nature.com/news/south-korea-surrenders-to-creationist-demands-1.10773; Soo Bin Park, "Science Wins over Creationism in South Korea," Sept. 6, 2012, www.nature.com/news/science-wins-over-creationism-in-south-korea-1.11377; Ipsos Global @dvisory, "Supreme Being(s), the Afterlife and Evolution," reports that 24% of South Koreans were creationists.

8. Ipsos Global @dvisory, "Supreme Being(s), the Afterlife and Evolution," reports that only 10% of Japanese identify as creationists. A similarly low fraction is also shown in Jon D. Miller, Eugenie C. Scott, and Shinji Okamoto, "Public Acceptance of Evolution," *Science* 313 (2006): 765–766. Despite the official communist ideology, 11% of Chinese view themselves as creationists; see Ipsos Global @dvisory, "Supreme Being(s), the Afterlife and Evolution."

9. Ibid. On Hindu critiques of evolution, see C. Mackenzie Brown, *Hindu Perspectives on Evolution: Darwin, Dharma, and Design* (London: Routledge, 2012).

10. Peter Fulljames and Leslie Francis, "Creationism among Young People in Kenya and Britain," in *Cultures of Creationism: Anti-evolutionism in English-Speaking Countries*, ed. Simon Coleman and Leslie Carlin (Aldershot: Ashgate, 2004), 165–173.

11. Ipsos Global @dvisory, "Supreme Being(s), the Afterlife and Evolution."

12. National Center for Science Education, "Polling Evolution in Three Countries," July 16, 2010, http://ncse.com/news/2010/07/polling-evolution-three-countries-005708. See also National Center for Science Education, "Polling Creationism in Canada," Aug. 8, 2008, http://ncse.com/news/2008/08/polling-creationism-canada-001375; and Ipsos Global @dvisory, "Supreme Being(s), the Afterlife and Evolution," which estimates that 22% of Canadians are creationists.

13. Luisa Massarani, "Few in Brazil Accept Scientific View of Human Evolution," Jan. 28, 2005, www.scidev.net/global/news/few-in-brazil-accept-scientific-view-of-human-evol.html. Ipsos Global @dvisory, "Supreme Being(s), the Afterlife and Evolution" found that 47% of Brazilians identified as creationists, the highest percentage of the 23 countries surveyed outside of Saudi Arabia, Turkey, Indonesia, and South Africa. Mexico came in at 32%; Argentina at 26%.

14. The observation and the statistics come from Kari A. Tikkanen, "Evolution vs Creationism in Europe," Aug. 2006, www.Student.oulu.fi/~ktikane/EUevocre.html. See

also Ulrich Kutschera, "Darwinism and Intelligent Design: The New Anti-evolutionism Spreads in Europe," *NCSE Reports* 23 (2003): 17–18; and Miller, Scott, and Okamoto, "Public Acceptance of Evolution."

15. "Britons Unconvinced on Evolution," BBC News, Jan. 26, 2006, http://news.bbc.co.uk/2/hi/science/nature/4648598.stm; Riazat Butt, "Half of Britons Do Not Believe in Evolution," Feb. 1, 2009, www.theguardian.com/science/2009/feb/01/evolution-darwin-survey-creationism.

Acknowledgments

The editors acknowledge the financial support of Ghent University toward the organization of the workshop in November 2011 that marked the start of this project and toward the funding of Stefaan Blancke's research. We also acknowledge FWO Flanders for Stefaan Blancke's short research stay in May 2012 at Aarhus University. We thank the contributors to the workshop and the book for their enthusiasm and solid work. We would like to thank Deborah Bors and Katherine Curran from Johns Hopkins University Press for carefully guiding us through the publication process. We want to express special gratitude to our acquisitions editor at JHUP, Jacqueline Wehmueller, for her endless patience and concern, and to Ronald L. Numbers for supporting this project from the very start. We also thank the reviewers for their comments, Brian MacDonald for copyediting, and Deborah Tourtlotte for preparing the index.

Creationism in Europe

Creationism in Europe or European Creationism?

STEFAAN BLANCKE, HANS HENRIK HJERMITSLEV,
AND PETER C. KJÆRGAARD

On October 4, 2007, the Parliamentary Assembly of the Council of Europe passed Resolution 1580, which issued a serious warning. Creationists across the European continent were adopting a model originating in North America to target education at all levels. The council cautioned of "a real risk of serious confusion being introduced into our children's minds between what has to do with convictions, beliefs, ideals of all sorts and what has to do with science." "An 'all things are equal' attitude," the resolution concluded, "may seem appealing and tolerant, but is in fact dangerous." To counteract this potential risk, the council urged member states to "defend and promote scientific knowledge" and "firmly oppose the teaching of creationism as a scientific discipline on an equal footing with the theory of evolution."[1]

The resolution did not pass unanimously. The Vatican, the European Evangelical Alliance, and the Islamic Science Research Foundation (BAV) lobbied against the resolution and exerted considerable influence on the conservative members of the council. The first draft was sent back to committee, and only after toning down the language of the resolution and including a preamble stating that the resolution was not directed against faith as such, but only at faith posing as science, the council passed the resolution with the votes 48–25.[2]

Creationist lobbying also has been taking place within the offices of the European Union. In October 2006 the Polish Catholic creationist Maciej Giertych organized a seminar entitled "Teaching Evolutionary Theory in Europe: Is Your Child Being Indoctrinated in the Classroom?" for the members of the European Parliament in Brussels. Outside the political sphere, European creationists have established cross-border networks and have met at European creationist congresses since 1984. These congresses have been dominated by Protestant creationists from northwestern Europe who invited American creationists and translated their works to European languages.[3]

These activities demonstrate that creationism is on the rise in Europe. The central aim of the creationist groups is to influence educational policies and introduce alternatives to the theory of evolution in biology classes. Several European politicians, including ministers of education of several European countries, have supported the introduction of alternative theories such as scientific creationism and intelligent design in biology classes. Also, at local levels, both public and religious schools have supplemented the national curriculums with creationist textbooks.[4]

Even though a general pattern of increased creationist activism is detectable throughout the European continent, spanning from Sweden to Turkey and from the United Kingdom to Russia, we should be cautious about grand narratives and quick conclusions. Different cultural, educational, religious, and political contexts result in a varied and complex picture. It is possible to identify local, or native, creationist traditions in some countries, while other countries have imported an American-style creationism and sometimes adapted it to local needs.

By investigating creationism in seven major European countries, in three European regions, and within the framework of Catholicism, intelligent design, and anti-creationism, this volume bears out these points. The main focus of the volume lies with recent and contemporary history from the 1970s until the present, but central actors, publications, and episodes from earlier European history are included in order to demonstrate continuities and changes in the religiously motivated opposition to the theory of evolution in Europe.

The North American Roots of Creationism

In order to understand the history of creationism, we have to turn to North America, where an organized opposition against the teaching of evolution first occurred as part of the fundamentalist movement among evangelical Protestants in the aftermath of World War I.

When Charles Darwin published *On the Origin of Species* in 1859, religious people responded to evolution and evolutionary theory in a variety of ways. Some accepted evolution, albeit often conditionally, whereas others rejected it straightforwardly. However, organized religious opposition to evolution did not emerge until the 1910s when a series of events in the United States resulted in a genuine antievolutionary movement. The movement was led by William Jennings Bryan, a politician who by the turn of the century had run for president three times as the Democratic nominee. He employed his political and rhetorical skills in combating evolution, which he considered to be a "dogma of darkness and death" that could threaten children's faith and thus their salvation. By the end of 1928, five states had adopted laws and regulations that were directed against the teaching of evolution. Moreover, in order to secure their sales, publishers deleted evolution from schoolbooks that were used across the entire country. As a consequence, during the 1930s, evolutionary theory largely disappeared from American science education.[5]

This situation lasted well into the 1950s and changed only when the Soviet Union succeeded in launching the first satellite, the Sputnik, in orbit around earth in 1957. American politicians felt that they were losing the space race and a large part of the blame for this embarrassment was put on the lamentable state of American science education. Soon thereafter, in 1958, the US government passed the National Defense Education Act, through which it entrusted expert committees with the task of rewriting the science textbooks, including the ones used in biology classes. With the integration of Darwin's theory of evolution by natural selection and Mendelian genetics in the 1920s and 1930s, known as the modern synthesis, and the discovery of the structure of DNA in 1953 by James Watson and Francis Crick, evolutionary theory had become one of the cornerstones of modern biology and, consequently, was reintroduced into the American science curriculum. Creationists would soon rise to oppose this revived encroachment on the souls of the American children.[6]

During the 1960s, however, the meaning of creationism shifted drastically. In the 1920s, most fundamentalists, except for a minority of Seventh-day Adventists, accepted an old earth, reconciling a literal reading of the book of Genesis with the evidence for an old earth. They either interpreted the days of creation as long ages or posited a gap of time between the first and second verses of Genesis. From the 1930s, creationism became an emblem of a variety of fundamentalist interpretations of scripture and its relation to the geologic and biological sciences. In 1961, however, Henry Morris, a hydraulics engineer, and John C. Whitcomb, a conservative theologian, published a book called *The Genesis Flood*, in which they propounded the young-earth creationist beliefs of the Seventh-day Adventists. In the course of the 1960s, these views became so popular among evangelicals and fundamentalists that by the beginning of the 1970s, the term *creationism* was understood to refer only to the young-earth variant.[7]

In 1968 the creationists suffered a major blow. After more than forty years, the US Supreme Court ruled the antievolution law of Arkansas unconstitutional on the argument that such laws violated the separation of church and state stated in the Establishment Clause of the First Amendment, which reads: "Congress shall make no law respecting an establishment of religion." In the same vein, the verdict proscribed the teaching of biblical creationism in public schools. In response, the creationists adopted the term *scientific creationism* or *creation science* in support of a new strategy by which they would strive for what they called *equal time* or *balanced treatment*. The switch to a more scientific image initially worked to the benefit of the creationist movement. Throughout the 1970s, the popularity of the movement reached unprecedented heights. By the beginning of the 1980s, two states, Arkansas and Louisiana, passed equal-time laws decreeing a balanced treatment of evolution and creation in public schools. The success was, however, brief. In 1982, in *McLean v. Arkansas*, a US district court declared these laws to be unconstitutional. Five years later, in *Edwards v. Aguillard*, the US Supreme Court ruled the Louisiana law unconstitutional as a violation of the First Amendment of the Constitution. Although this verdict put an end to the strategy of equal time and balanced treatment, evangelical fundamentalists continued to teach creationism at home or in private Christian schools and to influence local authorities or advisory committees on education.[8]

In the 1990s a new form of creationism emerged under the label *intelligent design*. The proponents purported to have scientific evidence of intentional design in nature, however, without identifying the designer. Emphasizing their scientific ambitions and not mentioning the god of the Christian Bible, intelligent design supporters pushed their views as a scientific alternative to evolutionary theory. In support of their plans they established the Center for the Renewal of Science and Culture (later renamed the Center for Science and Culture), a subsidiary of the Discovery Institute, a Seattle-based conservative think tank. For a while, they were quite successful in drawing popular and academic attention and influencing school boards. In 2005 a school board in Dover, Pennsylvania, adopted a policy requiring biology teachers to read a statement saying that students should be aware that there existed alternatives to evolutionary theory, including intelligent design, and inviting them to consult copies of the intelligent design textbook *Of Pandas and People* in the school library. A group of parents who were concerned about their children's constitutional rights filed a lawsuit against the policy. The verdict of *Kitzmiller v. Dover* put an end to the plans of intelligent design adherents to get their ideas taught in science classes. The judge ruled that intelligent design constituted a religious view and that teaching or promoting intelligent design as a valid alternative to evolutionary theory consequently violated the constitutional separation between church and state. Following the *Kitzmiller v. Dover* verdict, the Discovery Institute changed tactics and began promoting laws permitting teachers to have their students think critically about or discuss the strengths and weaknesses of scientific theories, including evolutionary theory. This is the most recent attempt to find ways to allow teaching creationist views in schools. This new strategy has so far had some success, as Louisiana in 2008 and Tennessee in 2012 have adopted such a law.[9]

As the history of modern creationism demonstrates, the term *creationism* can have several meanings, including old-earth creationism, young-earth creationism, and intelligent design. In this book we adopt an inclusive definition of creationism referring to religious belief systems that oppose established evolutionary science. As such, creationism includes intelligent design, as well as young- and old-earth, scientific and biblical, and indigenous and imported creationism. Moreover, the religious opposition to the theory of evolution has been driven by various concerns. In nineteenth-century and early twentieth-century Europe, critiques of Darwinian evolution usually supported a general attack on philosophical materialism. By contrast, the antievolution movement in North America focused primarily on biblical and moral concerns, attracting the interest of evangelical laypeople who regarded evolution as a threat to their traditional interpretation of scripture and humans' relationship with apes as a moral degradation. As the chapters in this book demonstrate, moral and scriptural reservations to evolution have now become significant issues among fundamentalist Christians and Muslims in Europe, just as they have been since the 1910s among groups in North America.

Global Creationism

Organized creationism originated in a specific North American context. During the twentieth century, however, small factions of creationists began to operate in other parts of the world. In the 1920s the leading Scottish anatomist and anthropologist, Arthur Keith, warned against the spread of *Daytonism* (after the trial in Dayton, Tennessee) in Britain, and in the early 1930s the Evolution Protest Movement was founded in England. It was later renamed the Creation Science Movement and still boasts to be the oldest creationist movement worldwide. Others followed, and by the end of the twentieth century Christian creationism had expanded to work on a global scale. Ideas, arguments, and evidence were appropriated by various religious groups to fit specific national and local contexts. Creationism was no longer restricted to Christians and now appeared in a multitude of forms and denominations. In the twenty-first century, creationism has become a highly flexible repository for legitimating religious group identity, highlighting cultural barriers, and providing a stage for antagonists. Muslims, Hindus, and Jews adopted strategies from Christian creationists, who in turn embraced other successful schemes for pushing their agendas around the world, including Europe.[10]

The global spread of creationism has generated a large anti-Darwin enterprise with events, books, pamphlets, websites, and videos covering all continents and all main religions. Creationists all over the world have been successful in exploiting every media and any opportunity to promote their antievolution messages, not always playing with open cards. As preparation for the Darwin anniversary year in 2009, the Australian–New Zealand creationist documentary *The Voyage That Shook the World* duped three leading Darwin scholars into participating in what they were told was an educational drama documentary for the Australian broadcasting television. Only afterward did it appear that the organization behind it was Creation Ministries International, a young-earth creationist organization based in Australia, with autonomous ministries in Africa, Europe, and North America, that had commissioned the documentary to make a technologically cutting-edge production to persuade the general public to think critically about evolution. In order not to antagonize the main audience, the documentary was not overtly against Darwin as a person and was thus in stark contrast to the American documentary *Expelled: No Intelligence Allowed* from 2008 that linked the theory of evolution to fascism, Nazism, the Holocaust, atheism, abortion, and eugenics.[11]

The very strong antievolution line playing on communism and terrorism and making Darwin the evil face of evolution ultimately responsible for some of the worst crimes of the twentieth and twenty-first centuries has been taken up by the well-funded Muslim creationist organization, based in Turkey and fronted by Adnan Oktar. He became known across Europe in 2007 by sending out tens of thousands of copies of the massive and lavishly produced *Atlas of Creation* in Turkish, English, and French translations written under the pen name Harun Yahya. Initially, the Muslim creationists were trained by conservative Christian American creationists, and to support the program Christian creationist texts were translated and converted to fit a new Muslim creationism. Ties were close, and allies made between different religious groups as strategies to oppose the evolutionary adversaries were shared. This produced a series of unexpected alliances across the globe. On the other hand, it also increased the competition between creationist groups. Europe is a good example of the complexity of rivalry, cooperation, and competition among creationists of any denomination.[12]

Polling Creationism

While historical analysis investigates the main actors and episodes including the political, cultural, and religious context of creationism, quantitative studies provide data to disclose more intuitive creationist beliefs that do not translate into activist creationism. In the United States, opinion polls have repeatedly been conducted to investigate what Americans believe concerning evolution,

TABLE 1
Responses to European Commission 2005 Survey on Creationism and Evolution (in percentages)

"Here is a little quiz. For each of the following statements, please tell me if it is true or false. If you don't know, say so, and we will go on to the next one. Human beings, as we know them today, developed from earlier species of animals."

	True	False	Don't know
Belgium	74	21	5
Denmark	83	13	4
France	80	12	8
Germany	69	23	8
Greece	55	32	14
The Netherlands	68	23	9
Norway	74	18	8
Poland	59	27	14
Portugal	64	21	15
Spain	73	16	11
Sweden	82	13	5
Turkey	27	51	22
United Kingdom	79	13	8

creationism, and the origin of the human species. Even though it is difficult to compare polls with different wording, the results indicate that Americans are divided on the question with equal numbers supporting creationism and evolutionism, and that these numbers have remained remarkably stable across time.[13]

Unfortunately, for Europe, we do not have this kind of data. However, several surveys give us at least some idea of how popular creationist beliefs are in Europe. They have clearly indicated that human evolution is more readily accepted in European countries than in the United States, with the exception of Turkey.[14]

It is, however, highly problematic to draw general conclusions about Europe from surveys like this. A closer look at the data reveals significant differences between the European countries. The European Commission's Eurobarometer survey from 2005 (Table 1), which includes thirteen countries presented in this book, demonstrates that in highly secularized countries such as Sweden, Denmark, and France, around 80 percent of the population accepts human evolution and about 10 percent rejects it. Generally, European countries have an acceptance rate of around 70 percent, with 20 percent of the population rejecting it. However, in Poland and in Greece the level of acceptance of human evolution is below 60 percent. With 27 percent of the population accepting human evolution, Turkey is the only European country discussed in this book with a lower level of acceptance of human evolution than that of the United States.[15]

TABLE 2

Responses to Ipsos Global @dvisory 2011 Survey on Creationism and Evolution in 24 Countries (in percentages)

"There has been some debate recently about the origins of human beings. Please tell me which of the following is closer to your own point of view."

A. Some people are referred to as "evolutionists" and believe that human beings were in fact created over a long period of time of evolution growing into fully formed human beings they are today from lower species such as apes.
B. Some people are referred to as "creationists" and believe that human beings were in fact created by a spiritual force such as the God they believe in and do not believe that the origin of man came from evolving from other species such as apes.
C. Some people simply don't know what to believe and sometimes agree or disagree with theories and ideas put forward by both creationists and evolutionists.

	A	B	C
Belgium	61	8	31
France	55	9	36
Germany	65	12	23
Great Britain	55	12	34
Poland	38	25	37
Russia	26	34	40
Spain	53	11	37
Sweden	68	10	21
Turkey	19	60	21
United States	28	40	32
All 24 countries	41	28	31

According to a 2011 poll by Ipsos Global @dvisory of twenty-four countries (Table 2), including nine of the European countries discussed in this book, 41 percent of the global population prefers an evolutionary account of human origins, whereas 28 percent favors a creationist explanation; 31 percent remained indecisive. Sweden (68%), Germany (65%), and Belgium (61%) had the highest levels of acceptance of human evolution. In these countries, only 8 to 12 percent identified as creationists. However, in Italy, Poland, and Russia, the percentage of creationists rose to 21, 25, and 34 percent respectively. Of all European countries, the highest percentage of creationists lived in Turkey (60%).[16]

Clearly, creationist beliefs are not equally popular in each and every European country. In Eastern European countries and Turkey, creationism seems to have a substantial following, whereas in Northern Europe most people accept human evolution. However, with the exception of Turkey, evolution is much more accepted in Europe than in the United States. These differences at international, national, and regional levels definitely require an explanation. This volume is the first attempt to take up this challenge.

European Creationism

The study of individual European countries or regions demonstrates some similarities but also, more importantly, great local and national differences with respect to the history, organization, and spread of creationism. As the chapters in this book document, one cannot talk about European creationism. Creationism in Europe is so many different things to different populations for different reasons. A brief summary highlights some of these differences.

Details can be found in the following chapters.

France

In 1905 church and state were formally separated in France. The combination of a strong anticlerical tradition in a secular country and an ambivalent relationship to the United States leading to strong anti-American sentiments across the political spectrum make French creationism a special case. One of the main challenges to French creationists has consequently been to fashion a strategy that dissociated creationism from America. Except for recent attempts to introduce Muslim creationism, the school system has not been targeted. Instead creationists in France have taken the debate to public forums, using media, magazines, and books. There is some organized creationism in France, mostly with marginal influence. In recent years, intelligent design has been used to promote a creationist agenda that has been met with a strong anticreationist reaction.

Spain and Portugal

In the predominately Catholic countries on the Iberian Peninsula, the theory of evolution was received in different ways around 1900. In Spain Jesuit theologians were hostile toward the theory, which they regarded as an example of atheist materialism and left-wing politics. In Portugal, on the other hand, Catholic thinkers were more reluctant to criticize evolution. During the fascist regimes in mid-twentieth century, Catholic antievolutionism gained some ground at least in the Spanish school system, but Catholic voices open to evolution were also apparent. Since around 1980 evangelical members of the Protestant minorities in Spain and Portugal have imported American creationism through translations, a short-lived journal, and antievolutionary teaching at some evangelical schools. However, their influence is restricted to their own religious circles. In the first decade of the twenty-first century, antievolutionary views including the theory of intelligent design also entered mainstream media and gained open support by some conservative Catholic thinkers.

United Kingdom

British creationism has existed in a more or less organized form since the early 1930s but has had very little impact on public debates and the school system. This includes a creation museum and a creation zoological garden. During the 1990s and the first decade of the twenty-first century, a series of episodes including lecture series, events, campaigns, and school controversies have heightened public, religious, and academic awareness of a growing phenomenon. Polls have shown an increase in acceptance of creationist ideas and rejection of the theory of evolution in both rural and urban populations. This has generated a strong anti-creationism reaction, especially from academic and humanistic sides. The Church of England has officially denounced creationist ideas and openly endorsed evolution.

The Low Countries

Creationism has played a very different role in the Netherlands than in Belgium. In the Protestant tradition of the Netherlands, there have been debates about evolution between liberal and orthodox groups and within orthodox groups since the nineteenth century. In the 1970s, when most members of the mainstream churches had come to accept evolution, the orthodox found an ally in the evangelicals, forming an evangelical-Reformed movement, in which creationism was almost an article of faith. About three decades later, intelligent design began to be explored by prominent members of this movement as an alternative to young-earth creationism. This development was not appreciated by the rank and file, which evoked a new wave of creationism during the Darwin year 2009. In Belgium, a country with a Catholic tradition, creationism has never been an issue. Already in the early twentieth century, an influential group of liberal and moderate theologians actively promoted the reconciliation of evolution with Catholic faith, and later the theistic evolutionary views of Teilhard de Chardin became very popular. Creationism is prevalent in Islamic groups and is taught at some Protestant and orthodox Jewish schools.

Scandinavia

In the highly secularized Scandinavian countries, creationism remains a relatively marginalized phenomenon. Organized creationist movements did not appear until the 1980s, when creationist groups and individuals in Sweden, Norway, and Denmark began to cooperate. In 1983 evangelical antievolutionists from Norway and Denmark launched the journal *Origo*. Since then, Scandinavian creationist groups to the right of the Protestant theological spectrum have joined forces in translating books, organizing networks and conferences, and building websites. In 1996 a creationist museum was established in Umeå in northern Sweden, and in the Darwin anniversary year in 2009 Norwegian and Danish antievolutionists generated some media attention partly through polemic anti-Darwinian websites. During the past few years, Muslim old-earth creationists and Vedic intelligent design advocates have also entered the Scandinavian scene through websites, lectures, media appearances, and publications.

Germany

The rise of the modern German creationist movement is marked by the publication of Joachim Scheven's booklet *Data on the Teaching of Evolution in Biology Classes* in 1982. Scheven, a schoolteacher with a PhD in paleontology, had strong ties to a well-organized creationist group, the Study Community Word and Knowledge. This was the first of a series of creationist publications, among which a textbook appearing in several editions and translations became close to a European creationist best seller. The German creationists actively pursued an international network and have succeeded in publishing articles in respected scientific journals. Also, creationism is being taught at conservative evangelical private schools, which have been growing in number. There has been some local political support for introducing creationist ideas in the school curriculum. The minister of culture in the State of Hessen, for example, saw nothing wrong when it was disclosed that teachers in several schools were teaching biblical creation in biology lessons.

Poland

Most of the Polish population is Roman Catholic. Creationism became an issue only after the fall of the Iron Curtain in 1989. The communist regime had restricted the influence of religious groups considerably, and the few academic debates on evolution were confined to specialized philosophical and theological journals. Since 1989 this picture has changed. Organized creationism appeared in both Catholic and Protestant circles, while intelligent design has been promoted by a group of philosophers. The Polish Catholic Church has not yet taken a firm stance on evolution, thus allowing Catholics to propagate creationist views. Despite the support of some politicians, including outspoken remarks from government officials, modern Polish creationism has played only a minor role in mainstream politics.

Greece

The dominance of the Greek Orthodox Church provided the religious framework for creationist debates in Greece. By the end of the nineteenth century, it was tied to a politically charged discussion about the future of Greece as either a modern state or a traditional society. Creationism was linked to conservative values and Darwinism to socialist ideals. In the history of Greece in the twentieth century with dictatorships, civil war, and fierce battles between communists and anticommunists, creationism continued to be as much a political issue as a religious or scientific one. Creationism has been actively discussed in educational contexts with conservatives arguing for an exclusion of the teaching of evolution and changes in textbooks and school curricula.

Russia and Its Neighbors

Decades of atheist propaganda and anticlerical repression under the communist regime have had a strong influence on religion in Russia. A general admiration for Western goods and values and a certain degree of theological inexperience rendered the population highly receptive to religious proselytizing. Under the banner of religious freedom, various Protestant churches actively promoted creationism and found an audience that associated evolutionary theory with Soviet ideology. As such, antievolutionism became highly popular in Russia. Until the end of the 1990s, in the Russian Orthodox Church creationism was being discussed only within the church. This changed when certain events attracted the attention of the media. Today the Russian Orthodox Church stands divided over the issue of evolution and creation between a liberal and a radical wing, the latter being the most influential. In Belarus and Ukraine, countries that are still closely connected with Russia, Protestant creationism appears to be on the rise.

Turkey

Early in the twentieth century, Turkish Islamic intellectuals associated the theory of evolution with materialism and a denial of God's agency. However, opposition to the theory of evolution was not a major issue before the 1970s, when members of the conservative Nurcu movement attacked the theory of evolution in order to undermine leftist materialism. This was part of a political agenda, not a religious one. For this purpose, Turkish conservatives adopted auxiliary arguments from Christian creationists. Since the late 1990s, Turkish creationism has predominantly been an Islamic issue owing to a highly active organization run by Adnan Oktar. The new Turkish creationism has been very effective in setting an agenda for Muslim creationism around the world.

NOTES

1. Parliamentary Assembly of the Council of Europe, "The Dangers of Creationism in Education," Resolution 1580, Oct. 4, 2007, §§7, 19.1 and 19.4, http://assembly.coe.int/main.asp?link=/documents/adoptedtext/ta07/eres1580.htm#1.

2. Ibid.; Andrew Curry, "Creationist Beliefs Persist in Europe," *Science* 323 (2009): 1159; Birthe B. Pedersen, "Hårdt opgør i vente i Europarådet om kreationisme og Darwin," *Kristeligt Dagblad*, Oct. 3, 2007, www.kristeligt-dagblad.dk/artikel/263342:Udland—Haardt-opgoer-i-vente-i-Europaraadet-om-kreationisme-og-Darwin; Stefaan Blancke, Hans Henrik Hjermitslev, Johan Braeckman, and Peter C. Kjærgaard, "Creationism in Europe: Facts, Gaps and Prospects," *Journal of the American Academy of Religion* 84:4 (2013): 996–1028.

3. Ulrich Kutchera, "Devolution and Dinosaurs: The Anti-evolution Seminar in the European Parliament," *Reports of the National Center for Science Education* 26:5 (2006): 10–11; Ronald L. Numbers, *The Creationists: From Scientific Creationism to Intelligent Design*, expanded ed. (Cambridge, MA: Harvard University Press, 2006), 367–368, 405–416.

4. Blancke, Hjermitslev, Braeckman, and Kjærgaard, "Creationism in Europe."

5. Edward J. Larson, *Trial and Error: The American Controversy over Creation and Evolution* (New York: Oxford University Press, 2003); Bryan quotation in Christian C. Young and Mark A. Largent, *Evolution and Creationism: A Documentary and Reference Guide* (Westport, CT: Greenwood Press, 2007), 165; Peter J. Bowler, *Monkey Trials and Gorilla Sermons: Evolution and Christianity from Darwin to Intelligent Design* (Cambridge, MA: Harvard University Press, 2007); Ronald L. Numbers, "Darwinism, Creationism and Intelligent Design," in *Scientists Confront Intelligent Design and Creationism*, ed. A. J. Petto and L. R. Godfrey (NewYork: W. W. Norton, 2007), 31–58; Adam R. Shapiro, *Trying Biology: The Scopes Trial, Textbooks, and the Antievolution Movement in American Schools* (Chicago: University of Chicago Press, 2013).

6. Bowler, *Monkey Trials and Gorilla Sermons*; Larson, *Trial and Error*; Numbers, *The Creationists*; Numbers, "Darwinism, Creationism and Intelligent Design"; Eugenie C. Scott, *Evolution vs. Creationism: An Introduction* (Berkeley: University of California Press, 2004).

7. Ronald L. Numbers, "Creating Creationism: Meanings and Uses since the Age of Agassiz," in *Darwinism Comes to America* (Cambridge, MA: Harvard University Press, 1998), 49–57.

8. Larson, *Trial and Error*; Numbers, *The Creationists*; Numbers, "Darwinism, Creationism and Intelligent Design"; Scott, *Evolution vs. Creationism*; Eugenie C. Scott, "The Once and Future Intelligent Design," in *Not in Our Classrooms: Why Intelligent Design Is Wrong for Our Schools*, ed. Eugenie C. Scott (Boston: Beacon Press, 2006), 1–27.

9. Numbers, "Darwinism, Creationism and Intelligent Design"; Eugenie C. Scott, "Creation Science Lite. 'Intelligent Design' as the New Anti-evolutionism," in *Scientists Confront Intelligent Design and Creationism*, ed. A. J. Petto and L. R. Godfrey (NewYork: W. W. Norton, 2007), 59–109; Forrest, chapter 12 in this volume.

10. Glenn Branch, "Creationism as a Global Phenomenon," in *Darwin and the Bible: The Cultural Confrontation*, ed. Richard H. Robbins and Mark N. Cohen (Boston: Penguin, 2009), 137–151; Peter C. Kjærgaard, "The Darwin Enterprise: From Scientific Icon to Global Product," *History of Science* 48:1 (2010): 105–122; Numbers, "Darwinism, Creationism and Intelligent Design"; Ronald L. Numbers, "Myth 24. That Creationism Is a Uniquely American Phenomenon," in *Galileo Goes to Jail and Other Myths about Science and Religion*, ed. Ronald L. Numbers (Cambridge, MA: Harvard University Press, 2009), 215–223.

11. Kjærgaard, "The Darwin Enterprise."

12. Riexinger, chapter 10 in this volume.

13. Bronislaw Szerszynski, "Understanding Creationism and Evolution," in *Science and Religion: New Historical Perspectives*, ed. Thomas Dixon, Geoffrey Cantor, and Stephen Pumfrey (Cambridge: Cambridge University Press, 2010), 153–174; *National Center for Science Education*, "Polls and Surveys," accessed Nov. 1, 2013, http://ncse.com/creationism/polls-surveys.

14. European Commission, *Social Values, Science and Technology*, Special Eurobarometer 225, Wave 63.1 (Brussels: European Commission, 2005), accessed Nov. 1, 2013, http://ec.europa.eu/public_opinion/archives/ebs/ebs_225_report_en.pdf; Jon D. Miller, Eugenie C. Scott, and Shinji Okamoto, "Public Acceptance of Evolution," *Science* 313 (2006): 765–766; Angus Reid Public Opinion, *Origin of Humans: Americans Are Creationists; Britons and Canadians Side with Evolution*, released July 15, 2010, www.visioncritical.com/wp-content/uploads/2010/07/2010.07.15_Origin.pdf; Angus Reid Public Opinion, *Britons and Canadians More Likely to Endorse Evolution than Americans*, released Sept. 5, 2012, www.angus-reid.com/wp-content/uploads/2012/09/2012.09.05_CreEvo.pdf; Ipsos Global @dvisory, "Supreme Being(s), the Afterlife and Evolution," released Apr. 25, 2011, www.ipsos-na.com/news-polls/pressrelease.aspx?id=5217.

15. Szerszynski, "Understanding Creationism and Evolution"; European Commission, *Social Values, Science and Technology*.

16. Ipsos Global @dvisory, "Supreme Being(s), the Afterlife and Evolution."

CHAPTER ONE

France

THOMAS LEPELTIER

In their book *The Creationists: A Threat to French Society?*, Cyrille Baudouin and Olivier Brosseau argue that, in France, even when you are a creationist, "it is better not to appear as such."[1] Outspoken critics of creationism, Baudouin and Brosseau denounce as biblical creationists or supporters of intelligent design certain individuals who deny that they are creationists. If it is true that French creationists are not openly so, we need to look into the reasons why. One reason these two critics offer is that in the French media creationists are identified with American Protestant fundamentalists who try to introduce creationism into the school curriculum.[2] In France, a secular country since the 1905 law separating church and state, there has been a strong anticlerical and anti-American tradition. Consequently, being identified as an American Protestant fundamentalist has negative connotations. For example, a Catholic author such as Dominique Tassot, who, as we shall see, clearly defends a creationist conception of the world, says: "I never present myself as a creationist, because that would make me a sitting target. . . . I am anti-Darwinian, I am anti-evolutionist."[3] The unwillingness to carry the creationist label is not universal. There are a few authors or researchers who explicitly state that they are creationists, but they remain a minority.[4] The wariness of the term *creationist* and the tendency to avoid the label even among people who in any other context would be identified as such demonstrate an inherent difficulty for any study of creationism in France. How should we define French creationism when many of the very people who seem to be creationists deny being such? Which criteria can we use to designate as creationists people who say they are not?

Another feature of French creationism is its almost total lack of a connection with the education system. In the United States, debates about creationism have centered extensively around education. American creationists have consistently worked to have their doctrine taught in schools. The legal battles that followed, as reported in the media, have been instrumental in giving visibility to creationism on the European side of the Atlantic. This has served as an inspiration for creationists in some European countries but not in France. The centralized organization of education makes it very difficult for creationism to infiltrate the curriculum and textbooks. These are not set locally, as in the United States, but by the Ministry of Education, so that it is difficult for any lobby group to change their content. Even if many scholars took up creationism, it would be difficult to introduce it in schools because the Ministry of Education, which decides the content of the curriculum, could easily reject it in the name of the 1905 national law separating church and state. Creationist strikes against the educational system are therefore few. One exception is the mass mailing in January 2007 of the Turkish creationist Adnan Oktar's *Atlas of Creation*. But the immediate reaction of public authorities who successfully ordered the book to be removed from school libraries confirms instead the practical barriers that prevent creationism from infiltrating the state education system.[5]

Consequently, in France the "battle" over creationism takes place outside school, mainly in the media, magazines, and books. Of course, the major strands of this French creationism, or at least of the French contestations of Darwinism, do not arise from nowhere. There have deep historical roots.

Historical Background

France has a Catholic tradition. The French reception of Darwinism must be understood in the light of this religious heritage. But there is a French specificity related to the persistence of a line of thought that is called, sometimes wrongly, *neo-Lamarckian*, in reference to the ideas of Jean-Baptiste Lamarck.[6] Charles Darwin admitted two mechanisms of evolution: the inheritance of acquired characteristics and the natural selection of characteristics that have varied randomly.[7] The first mechanism is often associated with Lamarck and his earlier formulation of a theory of evolution. Darwin certainly made reference to it but did not put it at the heart of his concept of evolution, nor was he the first to use it to explain the transformation of living forms. Traditionally, Lamarckism can be defined by the idea that what causes change in individual living organisms also causes evolution. Accordingly, individual variations determine the path of evolution. The strong influence of Lamarckism in France helps to explain why French researchers were initially opposed to neo-Darwinism and Mendel's genetics. This also explains why they did not participate in the development of the modern synthesis of genetics and the theory of evolution in the years 1920–50.[8]

However, the French neo-Lamarckism was not homogeneous. It can be separated into two main forms that were chronologically distinct: a materialist trend, followed by a vitalist trend. Adherents of the former, dominant from the 1880s to the 1930s, sought to develop a mechanical and deterministic theory of evolution.

With this approach, the causes of individual variations were to be found outside organisms, in their biological and physical environment. In short, the variations of an organism corresponded to direct adjustments to its environment. The second form of neo-Lamarckism was highly influential in the years 1930–70. During this period there was a new emphasis on the idea of the autonomy of life, so that the causes of individual variations were to be founded within organisms. Internal laws of development and structural constraints would guide individual transformations and therefore evolution.

The important figures of the vitalist tradition of neo-Lamarckism are two zoologists, Albert Vandel and Pierre-Paul Grassé. Both had an important influence on French biologists, especially Grassé, who, as the holder of a chair of zoology at the prestigious Sorbonne University, became a kind of mandarin in the French university system. Believing that random mutations and natural selection are not in themselves sufficient to explain the apparent purpose of evolutionary processes and the increasing complexity of living organisms over time, they concluded that the prime mover of evolution had to be a vital force. This meant that organisms evolved because they were pushed to do so by an internal force. Such internal and structural determinism implied that it was impossible to totally explain living organisms in mechanistic terms—hence, the criticism of materialistic conceptions of life and an openness to a form of teleology. For Vandel and Grassé there was no doubt that evolution was oriented. This approach opened the door to spiritualist considerations about the meaning of life, which fitted well with Grassé's Catholic belief in God.

The influence of this interpretation of the Lamarckist tradition in French biology did not completely end with the death of these two imposing figures, Vandel in 1980 and Grassé in 1985. Among the best known who followed in their footsteps was the entomologist Rémy Chauvin, who was also a professor at the Sorbonne.[9] In the French academic world, this tradition has now become marginalized. However, outside scholarly circles it remains to be a source of inspiration to critics of Darwinism, creationists among them.

Biblical Creationism

As in many Western countries, French religious practice is weakening. Nevertheless, according to recent studies, French citizens who still consider themselves Catholics are estimated to be between 50 and 65 percent. Those who consider themselves to be Muslims are between 3 and 5 percent.[10] Declared atheists or agnostics are between 30 and 40 percent.[11] So, despite a significant decline in religious practice in the sense that Christians attend church much less frequently, France remains a mainly Catholic country. While the Catholic Church accepts the basic idea of evolution, for a long time it defended a creationist worldview of biblical inspiration. Some religious minority groups like Jehovah's Witnesses, around 0.2 percent of the population, continue to believe in a young earth and a separate creation of living species. In France, there are three organizations in particular uniting those who adopt such a creationist approach.

The Cercle d'Études Scientifique et Historique (CESHE, Circle of Scientific and Historical Studies) is an association founded in 1971 that "aims to reconcile science and faith in heart and mind." In particular, it "professes the scientific and historical inerrancy of the Bible."[12] It draws on the work of Fernand Crombette, the self-taught Catholic author of voluminous works that deal with geography, ancient history, prehistory, and astronomy. His books endeavor to show the accuracy of the biblical story and defend a vision of the cosmos that places Jerusalem at the center of the universe. More than just geocentrism, Crombette endorses a strong Christocentrism.

While Crombette worked primarily on the reinterpretation of ancient texts, the CESHE also addresses the natural sciences. Its vice-president, Guy Berthault, an amateur geologist, has questioned the methods of dating geologic strata and the standard scenario of their formation. In the 1980s he published two articles in the *Comptes rendus de l'Académie des sciences* (Proceedings of the Academy of Sciences), a prestigious French scientific journal. In these articles, he described experiments in stratigraphy that would give credence to the idea that sedimentary layers of the planet were deposited following a major flood on the terrestrial scale. In Berthault's mind, this flood is of course none other than the Flood of the Bible. These publications have since been widely criticized by the community of geologists. But they continue to be highlighted on many websites worldwide related to the young-earth creationist movement.

The other active association in the defense of creationism is the Centre d'Études et de Prospective sur la science (CEP, Center for Studies and Foresight on Science). Founded in 1997, it proposes "to coordinate the various sciences . . . in the light of revealed truths." The Bible is an obligatory starting point for scientific practice, according to the association: "Just as the engineer is most familiar with the operation of the object he designed, who is most likely to tell us about the universe, its contents, and the law that governs it, if not its creator? By insisting on studying nature, societies, and their history as if God did not exist, a misunderstood secularism has deprived researchers of the highest, and thus ultimately the most useful, insight. While extravagant hypotheses are given a hearing, the precious information left by the Holy Spirit in the most revered text in human history is rejected out of hand."[13] To members of CEP, the Bible serves as the principal basis for the study of nature.

CEP is chaired by Dominique Tassot, author of several books, including *Evolution: A Challenge for Science, a Danger to the Faith*, and editor of the quarterly journal of the Association, entitled *Le CEP*.[14] In Tassot's writings one finds multiple attempts to return to a short chronology for life on earth, consistent with a literal interpretation of the Bible. Despite Tassot's arguments to the contrary, there is a clear connection to the young-earth creationist movement. Though CEP has appealed to some French Catholic circles, its general impact on society has been marginal.

Besides these two associations, created by Catholic figures, the association In the Beginning (Au commencement), founded in 1998 by the Protestant André Eggen, has also attracted attention. Like the others, it belongs to the young-earth creationist movement. But the academic status of its founder gives it a special profile: Eggen was a geneticist at a prestigious public research center, Institut National de la Recherche Agronomique (INRA, French National Institute for Agricultural Research), where he became a research director in 2004. There he participated in the international research program that decoded the genome of the cow, which led him to be interviewed by the mainstream press for his work, with subsequent articles appearing in *Le Monde*, *L'Express*, and *Libération*. At the time, this gave him wide publicity. He is, as a result, one of very few—perhaps the only—recognized French scientists who have adopted a creationist view of biblical inspiration. This has caused confusion within the INRA, and Eggen has been accused of using "his status as a recognized scientist to mystify those who are 'in search of meaning,' in particular by mixing the scientific approach with his personal beliefs."[15] In 2009, however, Eggen stopped working for this public research organization and joined a private company, Illumina (based in the United States), dealing with genetic analysis sequencing. This change of status has calmed the controversy raised about his dual activity. Despite the heated publicity, the association is by no means a powerful organization. In 2009 it had just thirty members.[16] Like the two Catholic associations, its impact on French society remains marginal.

While these organizations and individuals clearly fall into the category of creationists, another association that officially denies a creationist leaning has nonetheless caused confusion among opponents of creationism. This is Université Interdisciplinaire de Paris (UIP, Interdisciplinary University of Paris).

The UIP and Criticism of Darwinism

Despite its name, UIP is not a university but rather an organization dedicated to finding a common ground for dialogue between different academic disciplines through public meetings and courses. Unlike CEP, CESH, and Au commencement, UIP has managed to be more visible through many high-profile activities, including conferences with the participation of several Nobel Prize winners. Founded in 1995 and partly funded by the Templeton Foundation, this association aims to encourage a return to dialogue "between the realm of facts and the realm of values in order to better understand the relationship between scientific research and the quest for meaning."[17] It does not explicitly profess a creationist doctrine but is regularly accused of conveying ideas of intelligent design.[18]

Among the factors that raise suspicion in the eyes of its critics is the vocation of the UIP to open science up to a quest for meaning and to consider the compatibility of the facts established by science and the tenets of major religions. According to Guillaume Lecointre, professor at the Museum of Natural History, Paris, who is at the forefront of the fight against creationism in France, the UIP "commits one of the errors of creationists with respect to scientific investigation: the denial of methodological naturalism."[19] According to Lecointre, methodological naturalism is a necessary condition of science. It means that a scientist can apprehend in the real world only what is material. Trying to open science to a spiritual dimension is going against the very nature of science. For this reason, the UIP is accused of promoting a form of "philosophical creationism."

In a heated exchange with Jean Staune, the general secretary of the UIP, Lecointre wrote that those affiliated with the UIP

> seem to be promoting a "soft" creationism, a kind of scientific providence, in the sense that it takes science out of its legitimate domain by trying to marry it by force with the spiritual quest. As scientists, we restrict science to the purpose it has had since the eighteenth century: to explain nature with only the resources of nature, without recourse to spiritualist, supernatural, or transcendent entities (as a methodological principle) and without the influence of religions (as a political principle). This is a simple humility or lucidity. Jean Staune and the UIP are working in precisely the opposite direction in enlarging the field of scientific investigation beyond its own definition and legitimate domain. Not that it is forbidden—fortunately!—to discuss philosophically the "quest for meaning," but it would be a scientific usurpation of words to continue to call this metaphysical quest science.[20]

In addition to combining science and a quest for meaning, the UIP has also been accused of having close links to certain individuals considered by some to be creationists or at least endorsing a kind of creationism. One example is the biochemist Michael Denton, who is associated with the Discovery Institute and has played an important role in promoting intelligent design. To defend the UIP from this accusation, Jean Staune has argued that because of the different views in Denton's publications from the 1980s and the 1990s, one could see him as two different authors. There is the author of *Evolution: A Theory in Crisis* (1985), who could indeed be a target for anti-creationist criticism. But there is also the author of *Nature's Destiny: How the Laws of Biology Reveal Purpose in the Universe* (1998), who would apparently, at least according to Staune, escape this kind of criticism. It is on this second Denton that Staune draws. Another person who by some has been considered a creationist is Pierre Perrier, member of the Scientific Council of the UIP. A specialist in fluid mechanics and a corresponding member of the French Académie des sciences, Perrier wrote a positive preface to the French translation of the intelligent design proponent Michael Behe's *Darwin's Black Box* (1996), published in 2009.[21] This suggests that at least one member of the UIP defends the approach of one of the fathers of the intelligent design movement. However, Perrier is also a signatory of an open letter published by the most prestigious French newspaper *Le Monde*, in which some scientists denounced the confusion often committed in France between researchers studying the philosophical implications of science and proponents of intelligent design. As the signatories of the letter write, "Trying to use the existence of an intelligent design movement to discredit scientists who claim, on the basis of evidence, that scientific discoveries provide not proof, but some credit, to a nonmaterialistic worldview, is to make, intentionally or not, a confusion that should be denounced."[22] Signing a petition expressing an explicit distrust in intelligent design makes it difficult to determine whether Pierre Perrier identifies himself with the intelligent design movement.

The same ambiguity appears regarding the paleoanthropologist Anne Dambricourt-Malassé, who was a member of the Scientific Council of the UIP from 1997 to 2005. She wrote the preface to the French edition of the leading intelligent design advocate Phillip Johnson's *Darwinism on Trial* (1991), published in 1997.[23] In 2000 she published *The Cursed Legend of the Twentieth Century: The Darwinian Error*, in which she established a link between her Catholic faith and her scientific work.[24] Finally, in 2005 she signed a petition launched by the Discovery Institute gathering scientists skeptical about the claim that it is possible to explain the complexity of life by random mutations and natural selection. All this would be enough to make one suspect her of favoring intelligent design.

The main reason for the accusations of creationist leaning is her very particular conception of the evolution of hominin skulls.

From her work in paleoanthropology, Dambricourt-Malassé believes that human bipedalism and upright stature is due to a process of internal evolution, whose main actor is a relatively small bone in the skull, the sphenoid. This evolution would therefore not come, as argued by evolutionary biologists, from adaptation to the environment through the selection of the most successful mutations. In other words, the upright position of the body did not develop in hominins because it brought a strategic advantage, such as the ability to see above the tall grass of the savannah, but because this development was embedded in the structure of this bone. To imagine that there is an internal mechanism of this kind is one thing. But why would the changes of the sphenoid, which "directs" human stature, bipedism, and the shape of the skull and jaw, always tend toward the same direction? Dambricourt-Malassé does not have a specific answer to give, but she sees it as a sign of a teleology echoing her Christian worldview.

Dambricourt-Malassé's religious take on human evolution has generated much controversy in France. A television documentary dedicated to a discussion of her theory was scheduled by the national TV channel, ARTE, in October 2005. But when the program was announced Guillaume Lecointre and some of his colleagues organized a campaign to put pressure on ARTE, accusing the documentary of pushing a creationist agenda in disguise. Part of the press joined the protest, highlighting the "scandal" it would be to broadcast a creationist documentary on a public channel in a secular state. The broadcast was not canceled, but to accommodate the critical voices, it was decided that the channel should host a "debate" immediately following the program. However, there was little debate. Rather than getting critics and defenders of the Dambricourt-Malassé thesis together, only outspoken critics were invited. They could easily, with no opponents, discredit the documentary's scientific claims. More than anything, this episode highlights the tensions provoked by any questioning of Darwinism in France. The question remains whether the Dambricourt-Malassé thesis and, more generally, the conceptions of UIP belong to the intelligent design movement.

To many, their implicit association is evident. For example, in January 2006, under the headline "The Bible against Darwin," several articles in a special issue of a major weekly magazine, *Le Nouvel Observateur,* described the theses defended by UIP as a French version of intelligent design. This charge was also brought in an article in *Le Monde* in 2006 entitled "French Neo-creationism in Disguise."[25] Other media have been equally critical of UIP. To all these attacks, Jean Staune has responded consistently. First, he reiterates with great force that he defends the theory of evolution (the fact that all living beings have a common ancestor) and denounces traditional creationism (the idea that living things were created separately). Second, contradicting people who criticize him, he says that members of UIP do "not have recourse a priori to spiritualism, the supernatural, or the transcendental. The whole approach of the UIP, and of the researchers who are involved in its activities, is based on the exploration of reality without bias, or at least with as few a priori ideas as possible—no one can claim to have none at all. Philosophical conclusions are drawn from this exploration, even if they might strengthen a nonmaterialistic conception of the world."[26]

In two books, *Has Our Existence a Meaning?*, and especially *Beyond Darwin*, Staune argues that some research results from various disciplines—paleontology, genetics, and biochemistry—show that the structure of living organisms does not seem to come from a series of adjustments but is inscribed in the very laws of nature.[27] He sees living organisms not as randomly assembled machines of evolution but as having their own internal logic. It is this logic that, in the long term, drives the evolution of species, not selection induced by changes in the environment. For Staune, this implies that there is little contingency in evolution, contrary to what the neo-Darwinians argue. It appears to be this conception, which explicitly echoes Pierre-Paul Grassé's, of an evolution impelled by the physical and biological laws of nature that makes Staune, in the eyes of his critics, a proponent of intelligent design. Yet Staune spares no criticism of intelligent design.

For proponents of intelligent design, there exists a series of complex systems in the living world that cannot be formed step by step through a Darwinian process of trial and error. They argue that the emergence of these systems would have required the coordination of the variations of each components. To explain this apparent coordination, proponents of intelligent design invoke an intelligent designer. This designer intervened directly in the evolutionary process to make these complex systems. But, according to Staune, proponents of intelligent design and neo-Darwinians commit exactly the same error—that is, to believe that we know all the laws of nature that govern the evolution of living beings. When the former consider that Darwinian processes cannot explain the formation of complex structures, they resort to the action of a designer. The position of the neo-Darwinians comes, Staune claims, from the same kind of error, since they are convinced that all the laws governing living forms are known, and they refuse to consider that current laws may be insufficient to explain the evolution of living forms. Thus, they postulate that the formation of complex structures is possible through the Darwinian process, the only one they can accept. But, for Staune, it is possible to overcome this opposition. On the one hand, Staune agrees with proponents of intelligent design that Darwinian processes cannot explain the formation of *all* complex structures. But he rejects recourse to an intelligent designer to explain a given phase of evolution, including those showing so-called irreducible complexity. He believes that the evolution of life must be explained through natural means. On the other hand, he rejects the neo-Darwinians when they argue that chance and natural selection explain everything. For him, this postulate does not take into account that living forms are constrained, by their physical and biological structure, to evolve in a particular direction. Thus, evolution is channeled in one direction by already existing biological forms and not by the intervention of a supernatural being.

The film *Avatar,* released in December 2009, made some of these arguments resurface. In January 2010 in *Le Monde,* the biologist Thomas Heams accused the film's director, James Cameron, of not being Darwinian and of promoting intelligent design.[28] First, Cameron takes a position "strongly in favour of life as an inevitable consequence of cosmic evolution," which is already dubious, according to Heams. But there is an even more serious problem. As an evolutionist, Heams was struck by "the remarkable similarity between the structure of the biosphere on Earth and on Pandora, where life consists of carbon-based plants and animals, some of which are mammals. Now, according to the [Darwinian] theory, nothing strictly guarantees that extraterrestrial life would be like this, neither that it would take the form of cell-based life, a basic unit for us, from bacteria to redwood, and a sign of their common ancestry, nor that it would be DNA-based, which is essential for combining the hero Jake Sully's DNA with that of a Na'vi (resident of Pandora)." For Heams, Cameron's vision undermines Darwinism: "The mechanism of natural selection, that is to say, the local blind game of chance and selection proposed by Charles Darwin 150 years ago, and which to date remains a formidable tool to explain the whole evolution of life, takes a real knock. If similar structures appeared independently on distant planets, that would mean we have given too much importance to chance on earth. It would be urgent to consider other mechanisms that might explain shared trends, such as the transition to the multicellular state and the distinction between plants and animals, let alone the evolution of some living organisms to a form of culture and civilization." Although Heams sees Cameron as an evolutionist, he adds that the film director is "not very Darwinian in the modern sense. Taken to the extreme, [his] position can lead to all the excesses of 'Intelligent Design.' " Thus *Avatar* was accused of conveying a creationist conception of life.

To Staune, Heams's article perfectly reflects the error committed by Darwinians. In the same newspaper, *Le Monde,* he responded to Heams, not so much to defend Cameron or intelligent design, but to promote his own vision of a guided evolution.[29] In this response, Staune argued that "Thomas Heams's article is based on the standard Darwinian position. Each path leading from the common ancestor to the various living beings now existing on earth was only possible thanks to the accumulation of a large number of contingent events, which suggests that the probability of seeing the same course taking place, not only on earth but also elsewhere in the universe, and obtaining the same result, is almost zero." But, according to Staune, this is not a valid argument. On any other planet with similar characteristics, he estimated that one would find animals very similar to mammals, and mammals closely related to apes. Not identical, but similar, perhaps very similar. The reason was that if "contingency and chance play a role in evolution," they were nonetheless " channeled in such a way that evolution would be largely reproducible and predictable." He contrasted "an evolution that would take place mainly through natural selection" with an "evolution that would take place mainly under the influence of laws of nature. This explains that evolution could well be repeated on different planets if they have similar physical and chemical conditions."

Despite all his clarifications and attempts to show that his conception of evolution is distinct from that of the intelligent design movement, Staune has failed to wash away the accusation of creationism, or to exonerate the UIP of collusion with this movement. For many opponents of creationism, he still appears to be a creationist in disguise and his association is still seen as a Trojan horse for American creationism. What seems to have fueled suspicion is that Staune makes no secret of being a believer in God and, specifically, a Catholic. For some of his opponents in secularized France, it seems that the mere fact of believing in a God that laid the foundation of the world is enough to make someone a creationist.

Darwinian Catholicism

As a clearly anti-Darwinian organization, UIP has been heavily criticized by opponents of creationism. Still, neither the more or less covert creationist organizations nor the UIP are representative of the majority of believers in France. French Catholics generally believe in a Creator. But they do not necessarily ascribe to him an explicit role in the unfolding of life on earth. However, according to some critical of Catholicism, doctrines are creationists if, at "any point in their argument, they involve a transcendent being, outside the natural world, endowed with intentionality and will, . . . creator and organizer of the arrangement of the universe, its components and its living entities."[30]

French theologians are often criticized by French anti-creationists. One of the reasons is that they frequently draw on the work of Pierre Teilhard de Chardin. This Jesuit theologian and paleontologist developed a Christian evolutionism in which, although chance does play a role, evolution is directed by an internal logic. In 1955, the year of his death, his best-known work *The Phenomenon of Man* was published.[31] He had finished it in 1930, but initially the Vatican prohibited its publication. Contrary to the Darwinian consensus that life evolved through random mutations and natural selection, Teilhard de Chardin describes in this book evolution with direction from single-cell organisms to man and beyond. At a certain level of evolution, the individual consciousness had to appear; at a further level, a planetary consciousness (the "noosphere") should appear and lead to a reconciliation with God. Thus, the universe would evolve from the point "Alpha," corresponding to its creation, to the point "Omega," corresponding to a happy end of time. The question is, Does this theory qualifies Teilhard de Chardin and the followers of his teleological conception of evolution as creationists?

At the beginning of the twenty-first century, most theologians say they reject any creationist doctrine and accept the Darwinian theory in its entirety. Among the examples are three well-known Catholic thinkers who have all dealt explicitly with the issue of creationism: Jacques Arnould, Francois Euvé, and Jean-Michel Maldamé. Arnould, a Dominican, is project manager on ethical issues at Centre National d'Études Spatiales (CNES, National Center of Space Research). His many publications include, for example, *The Creationists* and *God versus Darwin: Will Creationists Triumph over Science?*[32] Euvé is a Jesuit and dean of the Faculty of Theology of the Centre Sèvres, where he holds the Teilhard de Chardin Chair, and is author of *Darwin and Christianity: True and False Debates.*[33] Maldamé is also a Dominican, for a long time dean of the Faculty of Theology at the Catholic Institute of Toulouse and a member of the Pontifical Academy of Sciences. He is the author of *Creation through Evolution: Science, Philosophy and Theology.*[34] All three are Catholics who can be taken as representative of the stance on creationism for a large part of the Catholic Church in France.

All three theologians explicitly reject creationism, whether it is creationism inspired by the Bible or intelligent design, and claim to be Darwinians. They do, however, believe that God created the universe and has maintained a special relationship with man. They have tried to justify a compatibility with Darwinian evolution in various ways. For example, Euvé published an article in the journal *Études* arguing for a God that does not intrude into his creation.[35] According to this view of Euvé, after initiating the process, this divine being allows ongoing creation to take place autonomously. Euvé states that God's creation "is not only a single action that brings being out of nothingness, but an ongoing process, a 'continuous creation,' that makes new forms appear." With regard to the randomness of the Darwinian approach, Euvé argued that it did not contradict the notions of purpose, plan, or intent conveyed by his vision of creation. The element of chance in the evolutionary process fits well with the idea that the history of life in Euvé's Christian conception was not scheduled in advance, and consequently the divine plan with regard to humanity was not predetermined. As a result, the relationship between the divine and humanity has not followed a predetermined fate. God let life develop by itself. Long after the initial spark, he decided to establish a special relationship with humans because he saw them as the pinnacle of his creation. Through this conception of the divine, Euvé believed it was possible to perceive an echo of the evolving vision of life peculiar to Darwinism in the Christian biblical view.

Despite Euvé's attempts to advocate a concilliatory view of Catholic faith and Darwinian evolution, he has not managed to convince some of the more outspoken opponents of creationism, who suspect all theologians who claim to be Darwinian of being creationist in disguise.[36] Accordingly, whatever the believer does to distance herself from creationist theories faithful to the book of Genesis, one cannot be a true Darwinian if he or she considers that evolution follows a divine plan, with or without chance playing a role in this process. That would make one a creationist. According to this "equation," every believer is a creationist.

This argument has effectively been asserted by the publisher Marc Silberstein, who has been very active in the fight for materialism and against creationism. He does not hesitate to cast a wide net in his condemnations: "The French biologist Pierre-Paul Grassé, national leader of anti-Darwinism during the years 1940–80, was a convinced evolutionist, neo-Lamarckian, just as he was a believer, convinced that God had initiated the evolutionary process. Strictly speaking, Grassé was, according to the typology I try to explain here, a creationist, albeit in a particularly sophisticated way. For similar reasons, it is the same with the supporters of UIP or intelligent design, among others."[37] Silberstein further explains his thinking: "Whatever the degree of evolutionism—or, if I may, of methodological naturalism inherent to modern science—that is included in a theory of life, to decree, or infer (whichever applies), a first principle, an initial cause, that is ontologically divine, is to slip into one of the many forms of creationism." In other words, to this extreme form of anti-creationism, every believer is a creationist. This radical definition highlights a crucial difference between the French intellectual debate about creationism and the attitude in the United States, where more scientists define themselves as believers. Silberstein is not alone in holding this position, but neither is his argument accepted unanimously among philosophers of biology.[38]

To answer the question whether France is threatened by creationism or not depends on what is meant by creationism. Biblical creationism is very marginal and, given the way the French educational system is organized, it is unlikely to be taught to schoolchildren in the near future. With regard to intelligent design and for the same reason, it is very unlikely that it will take root in France. This doctrine seems more likely to attract a portion of the population that still believes in God but is detached from any literal interpretation of the Bible. However, most of the people accused of promoting intelligent design in France tend to keep a distance from it. So it is difficult to see how a doctrine that is denigrated by those that are called its supporters could gain ground. The question is less clear-cut when we turn, on the one hand, to criticisms of Darwinism from people close to UIP and, on the other hand, to theological interpretations of Darwinism. The diffusion of the latter depends on the fate of Christianity or of a similar theology. The spread of anti-Darwinism advocated by people like Anne Dambricourt-Malassé and Jean Staune depends on the evolution of Darwinism itself in the scientific community. Characteristic of the French debate about creationism is the fact that very few are willing to defend a creationist position. As such, recent history of creationism in France is strongly linked to the evolution of French anti-creationism.

ACKNOWLEDGMENTS

I thank Danny and Jennifer Yee for checking and improving my English.

NOTES

1. Baudouin and Brosseau, *Les Créationnistes. Une menace pour la société française* (Paris: Éditions Syllepse, 2008), 100.

2. Ibid., 90.

3. Cited in ibid.

4. Among others, one could include the journalist Laurent Glauzy, *Le mystère de la race des géants à la lumière du créationnisme* (L. Glauzy, 2006). This book attempts to show that scientists hide the existence of a now extinct race of giants, which would contradict the Darwinian theory of evolution. Under the name Laurent Blancy, the same journalist also wrote several articles between 2003 and 2006 in which he explicitly defends biblical

creationism, in *Le Libre journal de la France courtoise* (available at www.france-courtoise .info/theme/creation.php).

5. This does not mean that some pupils are not influenced by creationist ideas. But this influence coming from outside school does not affect the content of formal teaching. The principal question it raises is how teachers should respond to pupils' creationist remarks or beliefs in class. On this subject, see Corinne Fortin, *L'évolution à l'école. Créationnisme contre darwinisme?* (Paris: Armand Colin, 2009).

6. On this French neo-Lamarckism, see Laurent Loison, "French Roots of French Neo-Lamarckism, 1879–1985," *Journal of the History of Biology* 44 (2011): 713–744; and Laurent Loison, *Qu'est-ce que le néolamarckisme? Les biologistes français et la question de l'évolution des espèces, 1870–1940* (Paris: Éditions Vuibert, 2010).

7. See, e.g., Peter Bowler, *Evolution: The History of an Idea* (Berkeley and Los Angeles: University of California Press, 2009).

8. Denis Buican, *Histoire de la génétique et de l'évolutionnisme en France* (Paris: Presses Universitaires de France, 1984).

9. See, e.g., his book with an explicitly anti-Darwinian title, *Le Darwinisme ou la fin d'un mythe* (Monaco: Éditions du Rocher, 1997).

10. Within the Muslim population in France, it is possible to observe the circulation of creationist ideas coming from Muslim authors. In addition to Adnan Oktar's books coming from Turkey but often translated into French, one could mention the book of Mohammed Keskas, *La Théorie de Darwin. Le hasard impossible. La théorie de l'évolution des êtres vivants analysée par un croyant* (Darwin's theory: The impossible chance; the evolutionist theory of living beings analyzed by a believer) (Gennevilliers: Le Figuier, 2002). While this type of publication seems to have an impact in parts of the French Muslim population, it does not appear to have any influence beyond. Moreover, the creationist ideas addressed to the French Muslim population do not seem to reflect any French specificity but are related to creationism as it can be found in traditional Muslim countries. However, it is worth observing that the challenge to Darwinism coming from the socially disadvantaged Muslim minority sometimes corresponds to a political demand for recognition (see, e.g., Weronika Zarachowicz, "Mais que vient faire Darwin dans le 9–3?," *Télérama*, Hors-série "Charles Darwin" [2009]: 73–75).

11. These figures are approximate and cut across various surveys conducted in France in recent years (see, e.g., http://atheisme.free.fr/Religion/Statistiques_religieuses.htm).

12. See the website of the association: www.ceshe.fr/, accessed Oct. 9, 2011.

13. See the website http://le-cep.org/, accessed Oct. 10, 2011.

14. Dominique Tassot, *L'évolution. Une difficulté pour la science, un danger pour la foi* (Paris: Pierre Tequi Éditions, 2009).

15. Olivier Brosseau and Cyrille Baudouin, "L'arbre qui cache la forêt? Un créationniste français à l'honneur," *Agora Vox*, May 6, 2009. Text available at www.agoravox.fr/actualites/societe/article/l-arbre-qui-cache-la-foret-un-55449.

16. See "André Eggen, un scientifique au service de la Bible," *Charlie Hebdo*, May 13, 2009.

17. According to its website: http://uip.edu/presentation, accessed Sept. 17, 2011.

18. For example, Baudouin and Brosseau's book, *Les Créationnistes*, discusses the UIP at length. The association is also accused by Association française pour l'information scientifique (Afis, French Association for Scientific Information), which works explicitly against creationist ideas. UIP is also mentioned by the French Member of European Parliament

Guy Lengagne in his Report to the "Committee on Culture, Science and Education" of the Council of Europe about "The Dangers of Creationism in Education" (June 8, 2007).

19. Website of the CNRS: www.cnrs.fr/cw/dossiers/dosevol/decouv/articles/chap1/lecointre4.html, accessed Oct. 9, 2011.

20. Guillaume Lecointre, "Guillaume Lecointre répond à Jean Staune au sujet de l'UIP et du sens des actions à mener contre elle," www.assomat.info/Guillaume-Lecointre-encore-une, accessed Oct. 12, 2011.

21. Michael J. Behe, *La boîte noire de Darwin. L'intelligent design* (Paris: Presses de la Renaissance, 2009).

22. Jacques Arsac et al., "Pour une science sans *a priori*," *Le Monde*, Feb. 23, 2006.

23. Phillip E. Johnson, *Le Darwinisme en question. Science ou méthaphysique?* (Paris: Éditions Exergue, 1997).

24. Anne-Dambricourt Malassé, *La Légende maudite du vingtième siècle. L'erreur darwinienne* (Strasbourg: Éditions Nuée Bleue, 2000).

25. Michel Alberganti, "Le jeu de masque du néo-créationnisme français," *Le Monde*, Sept. 2, 2006.

26. Text published on Jean Staune's website, www.staune.fr/Theorie-et-pratique-de-l.html, accessed Oct. 12, 2011.

27. Jean Staune, *Notre existence a-t-elle un sens?* (Paris: Presses de la Renaissance, 2007) and *Au-delà de Darwin* (Paris: Éditions Jacqueline Chambon, 2009).

28. Thomas Heams, "James Cameron, encore un effort pour être darwinien!," *Le Monde*, Jan. 16, 2010.

29. Jean Staune, "Le non-darwinisme visionnaire de James Cameron," *Le Monde*, Jan. 21, 2010.

30. Olivier Brosseau and Marc Silberstein, "Évolutionnism(s) et créationnisme(s)," in *Les mondes darwiniens*, ed. Thomas Heams et al. (Paris: Éditions Syllpese, 2009), 1027.

31. Pierre Teilhard de Chardin, *Le phénomène humain* (Paris: Éditions du Seuil, 1955).

32. Jacques Arnould, *Les créationnistes* (Paris: Le Cerf, 1996) and *Dieu versus Darwin. Les créationnistes vont-ils triompher de la science?* (Paris: Albin Michel, 2007).

33. François Euvé, *Darwin et le christianisme. Vrais et faux débats* (Paris: Buchet Chastel, 2009).

34. Jean-Michel Maldamé, *Création par évolution. Science, philosophie et théologie* (Paris: Cerf, 2011).

35. François Euvé, "Le quadruple défi darwinien," *Études*, Nov. 2009, 485–496. In another article, "L'hérésie du dieu programmateur" (*Nouvel Observateur*, Hors-Série 61, "La Bible contre Darwin" [2005], 48), François Euvé criticizes the concept of a programmer God and even tries to demonstrate the compatibility of Darwinism with Teilhard de Chardin's conception of evolution.

36. See, e.g., the accusation against Jacques Arnould in Pierre Deleporte and Jean-Sébastien Pierre, "Jacques Arnould et le recul élastique du dogme," in *Les matérialismes (et leurs détracteurs)*, ed. Jean Dubessy et al. (Paris: Édtions Syllepses, 2004), 545–553.

37. Marc Silberstein, "L'unité des créationnistes," *L'idée Libre* 279 (Dec. 2007); text available on the website www.assomat.info/L-unite-des-creationnismes-Par, accessed Oct. 12, 2011.

38. See, e.g., Michael Ruse, *Can a Darwinian Be a Christian?* (Cambridge: Cambridge University Press, 2001).

CHAPTER TWO

Spain and Portugal

JESÚS I. CATALÁ-GORGUES

Spain and Portugal share several social and historical traits. Both countries are chiefly Catholic, and matters about the political significance of the church have marked their contemporary history. In Western Europe, only the Iberian countries have been governed by fascist regimes after World War II, and both established their present democratic systems almost simultaneously in the 1970s. Apart from that, troubles and difficulties have been common in their efforts to consolidate an institutionalized and normalized practice of science. Nowadays, the situation of religion in both countries is quite similar. According to research conducted by the Spanish Center of Sociological Research in November 2011, 74.1 percent of Spaniards reported to be Catholic; 14.0 percent, nonbelievers; 8.4 percent, atheists; and 1.6 percent had other faiths. In Portugal data from a 2001 census showed that about 85 percent of the Portuguese claimed to be Catholic, 4 percent had no religion, and 3 percent had other faiths. Among the relatively few citizens in Spain and Portugal with other faiths than Catholic, about 1 million people in Spain and 100,000 in Portugal belong to Protestant denominations. Despite these data, Spain and Portugal are relatively secularized countries. Of Spanish believers, only 15.8 percent attends religious services regularly, whereas the Portuguese Catholic Church admitted that only 20 percent of believers attended Sunday mass in 2001.[1]

Certainly, Portugal and Spain also show major differences. Portugal is a one-nation state, whereas Spain is a conglomerate of diverse national sensibilities, some demanding a higher degree of autonomy or complete independence. Therefore, Spain is characterized by linguistic plurality, while Portugal has only one language. For the specific purposes of this chapter, there are important dissimilarities in the historical background of evolutionism and creationism in Portugal and Spain that need to be taken into consideration. However, this study shows a Spanish bias because there is more literature on the subject available in Spain than in Portugal.

Polarizations around Evolutionism in Spain

The historiography about the reception and diffusion of the theories of evolution in nineteenth-century Spain is rich in detail. Since the seminal efforts of authors like Thomas Glick and Diego Núñez, an increasing number of works have improved our knowledge on the question. During the last third of the nineteenth century, Spain, a Catholic country with an incomplete bourgeois revolution, was extremely polarized about the state model, the role of the church in social life, and ideological freedom. In that context, Darwinian disputes achieved an unusual prominence in a country with important shortages of scientific education and research practice. Charles Darwin was waved as a flag representing the subversion of the ancient intellectual and moral order by freethinkers and left-wingers. Anti-Darwinians embodied the cause for the traditional values that Catholicism represented. Certainly, reality was more complicated. For example, some Spanish Catholics already strove to conciliate the evolutionary worldview with their own religious faith. But the deafening racket promoted by journalists, columnists, agitators, and vociferous preachers did not make room for a silent assimilation.[2]

Only at the turn of the century were positions mollified enough for a civil discourse on evolution—and on science in general—to arise. The tone of discussion was moderated, and it became possible to be a Catholic and an evolutionist at the same time. Nevertheless, the old manners sometimes reappeared. The scandal of the homage to Darwin at the University of Valencia in 1909 provides a good example. Some students organized a meeting with professors and intellectuals for commemorating the centennial of Darwin's birth. The conservative press reacted against them and described the event as "a homage to the devil."[3]

Some Catholic figures remained openly and actively hostile to evolutionism. The most renowned ones were a handful of Jesuit priests devoted to scientific research. The Jesuit secondary schools were a powerful force in the ecclesiastic attempt to maintain a room in the realm of science. In parallel, the Society of Jesus founded multiple scientific institutions in the beginning of the twentieth century and in 1913 launched a sophisticated popular science journal. *Ibérica* provided Jesuit naturalists and biologists a platform for spreading their antievolutionary ideas. The most prominent scholar among them was the embryologist Father Jaume Pujiula, who disseminated his antievolutionism in both books and talks. The main target of Pujiula and his brothers in religion was human evolution as a foundation of a materialistic anthropology. The Jesuits were deeply worried about the negative influence that could be exerted on the youth by such materialism. Coherently, they developed a kind of rhetoric that explicitly warned of the corrosive educational effect of the theory of evolution.[4]

The brief democratic period of the Second Republic (1931–36) meant the official secularization of the Spanish state and the expulsion of Jesuits. The teaching of the theory of evolution was firmly established after a weak attempt in 1927. But the victory of General Francisco Franco and his army in the Civil War (1936–39) abolished those regulations and restored Catholicism as the official religion of the state, establishing a genuine national-Catholic regime. During the 1940s, the evolutionary discourse was banned, and antievolutionism thrived. Once again, the Jesuits stood out because of their antievolutionary zeal. In Valencia, Father Ignacio Sala taught biology presupposing "a Supreme Artisan, God, Creator of the Universe." In his turn, Father Alejandro Roldán propounded a nonevolutionary solution for anthropogenesis and expressed agreement with Douglas Dewar's theories. A cofounder of Britain's first creationist society, the Evolution Protest Movement, in 1932, Dewar used the scarcity of transitional forms in the fossil record and the alleged survival of several fossil genera to impugn evolution. Roldán utilized these paleontological arguments at length.[5]

From *Humani Generis* to the Death Throes of Censorship

The encyclical *Humani Generis*, promulgated by Pope Pius XII in 1950, was an important landmark in the relationship between Catholicism and science, including evolutionary theories. Although human evolution was discussed very superficially in that pontifical document, it was the first official declaration of the Magisterium that declared some kinds of evolutionism as compatible with Catholic faith. The response in Spain was relatively quick: a meeting held in the Pontifical University of Salamanca in 1954 devoted to the philosophical and theological implications of evolutionism. Ten contributions, mainly written by Dominican friars, but also by Franciscan, Jesuit, and lay theologians, were published two years later. The road to a more conciliatory approach lay open. Meanwhile, the stimulating ideas of the recently deceased Pierre Teilhard de Chardin were studied by some Spanish theologians.[6]

However, Teilhard's doctrines went beyond theology and penetrated the natural sciences—paleontology chiefly. The Catalan paleontologist Miquel Crusafont, an internationally renowned specialist on fossil mammals and a friend of the U.S. paleontologist George G. Simpson, was the main champion of Teilhard in Spain. A fervent Catholic, Crusafont attempted to synthesize neo-Darwinism with a Teilhardian, finalistic worldview. The paleontologist and Jesuit priest Emiliano Aguirre, who worked in Madrid, was also attracted by Teilhard's doctrines. Crusafont, Aguirre, and Bermudo Mélendez, professor of paleontology at the University of Madrid, edited a thick volume entitled *La Evolución* in 1966. Spanish biologists, geologists, physicians, theologians, and philosophers worked together on this book, which was published by the most influential Catholic publishing company in the country, La Editorial Católica, in its most prestigious collection, Biblioteca de Autores Cristianos. Perhaps it is ironic, but the first Spanish general book on evolution was published by a Catholic firm, blessed by the ecclesiastical hierarchy, and coedited by a member of the Society of Jesus.[7]

Despite this liberal atmosphere, skepticism against evolution reappeared from time to time. In 1971 a letter to the director of the Catholic magazine *Roca Viva* expressed unease because of a TV program devoted to popular science, *Planeta Azul*, which had included evolutionary themes. The annoyed viewer considered the evolutionary theories fictitious, untenable, and atheistic, hence his complaint. The letter was read by the religious assessor of the Ministry of Information and Tourism, who urged the director of the program, the physician and filmmaker Félix Rodríguez de la Fuente, to make it clear that the theory of evolution was not a reliable scientific postulate. Rodríguez, who attained great fame because of his documentaries and books on wildlife, obtained the support of Crusafont, but it was in vain. *Planeta Azul* was pushed into an unfavorable time slot, and its scripts were censored. Certainly, the most reactionary elements of Franco's regime could admit a room for evolution in selected circles—for example, the book edited by Crusafont and his collaborators. But an open diffusion in the mass media was something completely different.[8]

The most liberal groups in the government thought differently. The new education law, promulgated in 1970 and developed in the next years, reintroduced the theory of evolution in the syllabus of secondary education. Thus, the death of the dictator in 1975 and the beginning of the democratic transition coincided with a reintroduction of evolutionary teaching. At the same time, religious influence on Spanish society was diminishing.[9]

Around the Centenary of Darwin's Death

Throughout the 1980s, the Spanish public was informed about American creationism. The Spanish translations of Stephen Jay Gould's series of *Reflections in Natural History* attained great success and familiarized their readers with that issue, the news of the 1982 creation trial in Little Rock, Arkansas, appeared on the pages of national newspapers, and Darwin returned to the front pages at the centenary of his death in 1982.[10]

Several books in Spanish and Catalan, monographic issues in journals, and articles in newspapers from different cities and regions recalled the figure of the famous English naturalist. Although a laudatory tone was predominant, less enthusiastic approaches also appeared. There were no openly creationist contributions, but some would correspond to a mitigated creationism. A representative of this line of thought was the famous lawyer and diplomat Antonio Garrigues Díaz-Cañabate, who was the only participant in an international meeting, held in Caracas, on the social implications of the theory of evolution who talked restrictively with regard to human evolution. Though a fervent Catholic, Garrigues was conciliatory in political and religious aspects. However, he was adamant in his refusal of natural selection and skeptical about the conciliation of creation and evolution. He expressed these ideas in several articles in important newspapers.[11]

Garrigues's stance was not typical of Spanish Catholics at that time. The daily *ABC*, very close to the church hierarchy, did not hesitate to dedicate an issue to Darwin, including his portrait on the front page. Its leader was explicit: "Around 1850, a part of the Catholic intelligentsia refused Darwinism because of its incompatibility with respect to the biblical revelation. Nowadays, the theory of evolution has been accepted and incorporated to the Christian explanation of man's origin. . . . The theory of natural selection has been turned into normal science, according to Thomas Kuhn's expression, and the critical awareness must be on the alert, not because it fears heterodoxy, but dogmatism." There is a marked contrast between this tone and the style of those Catholic dailies that in 1909 roared against the "homage to the devil" at the University of Valencia. The young Spanish democracy seemed to have sent Darwin back to the public sphere without serious controversies.[12]

Several Catholic intellectuals took a receptive attitude to evolution during the 1980s, coinciding with a renewed interest in theology of creation. There were some Spanish theologians willing to offer audacious approaches to the creation-evolution debate. In this regard, the most prominent author was probably Juan Luis Ruiz de la Peña, professor at the Pontifical University of Salamanca and at the High Institute for Theological Studies of Oviedo. Ruiz explicitly admitted that the evolutionary view of the world was the current one. So the classical definition of creation as *productio rei ex nihilo* was clearly unsatisfactory once the static conception had been overcome. Ruiz found a new sense for divine creation in the autotranscendence of beings, which can improve because God's causality allows them to go beyond their ontological threshold. He made a harsh appraisal on the historical reception of Darwinian theses: "The reaction of the majority of Christian theologians to these theses was, from the outset, very severe, and it constitutes one of the least glorious pages in the history of theological thinking."[13]

Ruiz also wrote a merciless attack against "scientific creationism." Typically optimistic and Eurocentric, he underlined its irrelevance in Europe. At the same time, he warned that it was necessary to learn from the past, so theologians and believers, for their own and their faith's good, ought to avoid certain closed worldviews.[14]

A more conservative approach appeared in Mariano Artigas's work. A professor at the University of Navarra, a prestigious institution belonging to Opus Dei, Artigas was a Catholic priest, philosopher, physicist, and historian of science. As a writer of popular science, he published *Las fronteras del evolucionismo* (Borders of Evolutionism) in 1985. This book was a great success and has been reprinted six times and translated into Portuguese and Italian. Artigas admitted the compatibility between creation and evolution for the believer. But he also assumed that it is not possible to assert the factuality of evolution because it is neither directly observable nor strictly demonstrable, so evolution is only a hypothesis, not a fact. There would be underlying ideological purposes for presenting evolutionism as the only option, Artigas claimed.[15]

Neo-Darwinism in general and natural selection in particular were the targets of Artigas's criticism. The main issue was the role of chance. Artigas insisted on the existence of a natural finality for reducing natural selection to a mere complementary mechanism. But Artigas also used more idiosyncratic means for combating neo-Darwinism. For example, he cited Niles Eldredge and Gould's theory of punctuated equilibrium as evidence of the crisis of the synthetic theory. However, he refused the scientific character of Gould's views by arguing that they were irrefutable and indemonstrable at the same time. There are so many pages criticizing Gould in the book that the absence of invectives against more explicit and dogmatic materialists—for example, Richard Dawkins, whose books have been translated into Spanish since the 1970s—is surprising. Probably, this was a reaction against the fact that Gould was in vogue in Spain during the 1980s. Carl Sagan, whose TV series *Cosmos* was broadcast in the country during that period, was criticized by Artigas too and epitomized as a typical materialist who slipped his ideological bias into a popular science discourse. Thus, the impact of materialism on mass culture was still a major concern for religious writers at the end of the twentieth century.[16]

Artigas also tackled the question of American scientific creationism, which he classified as "ultra-creationism." He stated that its partisans did not defend the Christian tradition, because they were forcing and inventing certain interpretations of the Bible in an entirely untraditional way. Coherently, he rejected their proposals. But when he was asking for the social background of that phenomenon, he found an important reason in the reactive attitude against ultra-evolutionists. This appraisal of the causes of scientific creationism, emphasizing the external circumstances, contrasts vividly with the self-criticism of Ruiz's assessment.[17]

In sum, Spanish Catholic writers at the end of twentieth century were diverse in their appraisals of the factuality of evolution but unanimous regarding its full compatibility with divine creation. However, invectives against materialism were still a commonplace, so the cause for concern was not evolution itself but the uses of evolution for an atheistic worldview.

The Arrival of American Creationism in Spain

Despite the claims of Ruiz about scientific creationism describing it as an exclusive American phenomenon, some creationist books were already being circulated in Spain when he wrote his works. The transition to democracy favored this circumstance, after several Protestant believers had suffered reprisals at the hands of Franco's regime.[18]

CLIE, a Protestant publishing house located in Catalonia, had distributed a collection of Spanish translations of brief works on the question since 1979 under the general series title Creación y ciencia (Creation and Science). The first book published in that collection was an anthology of articles about evolution and the fossil record written by significant creationists such as Duane T. Gish and Bolton Davidheiser. The volume also included a brief contribution by a Spanish author, Santiago Escuain, centered on the discontinuities in the fossil record. Escuain, who was also the translator of the other chapters, proclaimed that "the fossil record doesn't just provide any support to evolutionism, but it is openly hostile to it." Escuain, a chemist and activist of Protestant evangelicalism, is also a promoter of Servicio Evangélico de Documentación e Información (SEDIN, Evangelical Service for Documentation and Information), a platform for diffusing texts and news related to that denomination.[19]

During the following years, the Creación y ciencia series added new titles, almost all of them translated exclusively by Escuain. Creationist works by Henry M. Morris, Willem Ouweneel, Harold S. Slusher, Thomas G. Barnes, John C. Whitcomb, and other creationist authors were thus offered to the Spanish-speaking public, together with Whitcomb and Morris's classic *The Genesis Flood*.[20]

Escuain's translations reached their apex with the appearance in Spanish of Phillip E. Johnson's *Darwin on Trial*, which was published by Portavoz—a subsidiary of Kregel Publications, a veteran (founded in 1949) U.S. evangelical publishing house. In practice, these products were restricted to evangelical circles, as their commercial distribution was limited to Protestant bookshops. This circumstance, in the absence of an indigenous literature, marginalized the impact of this type of creationism in Spain and explains the minimal attention it received in academic circles during the last decades of the twentieth century.[21]

Escuain's activism motivated him to found and run a creationist journal, *Génesis*, which published four issues between 1993 and 1995. Its basic contents consisted of translations and adaptations of articles previously published in American creationist media. Only the editorial articles, some brief comments accompanying articles, and a few book reviews were written by Spanish authors. One of those editorial pieces summed up the sessions of the Sixth European Creationist Congress in Soesterberg, the Netherlands, in August 1995, which focused on biblical geology and was attended by Escuain. Another piece was devoted to combating those who were attempting to harmonize Christian faith and evolution.[22]

Not all Spanish evangelical believers and writers were antievolutionists associated with current views of creation science. For example, one prominent leader, Enrique Mota, a professor of mathematics at the University of Valencia, wrote a brochure addressed to undergraduates, in which he stated: "The theory of evolution, with its consequences, is the *bête-noire*, the sworn enemy, for some Christians. And they look for an alternative theory that could agree with (their interpretation of) biblical teachings while they try to set out the scientific weaknesses of the theory of evolution." The words in parenthesis indicate that Mota distanced himself from creationist interpretations of the Bible. In addition, the books that Mota recommended were more balanced and conciliatory than any creationist would admit.[23]

Entering the New Century

Despite the efforts of people like Escuain or evangelical publishing houses like CLIE and El Peregrino, creation science had not aroused a great deal of interest in Spain at the beginning of twenty-first century. However, the most recent years are characterized by an increase in creation-evolution debates in the media and a less peripheral presence of creationist works on the bookshelves. The emergence of intelligent design has been received by some people as a respectable alternative to evolutionism without the crude literalism of the traditional forms of creationism.[24]

Significantly, some nonconfessional, general-interest publishing companies have launched intelligent design books on the Spanish market. The Spanish version of Michael Behe's *Darwin's Black Box* was published in 2000 by Editorial Andrés Bello, a company that also publishes works by Martha Nussbaum, Pablo Neruda, and François Mitterrand. No charges of ideological narrowness would be possible in this case. A different case is that of Homo Legens, whose specific aim is "to rehabilitate Western Christian culture values." This firm published a Spanish translation of William A. Dembski's *Intelligent Design*, but this book coexists in its catalog with other ones sympathetic to an evolutionary worldview. An interesting example of publishing policies linked to ideological interests is Amit Goswami's *Creative Evolution*, a plea for reconciling intelligent design and Darwinism through quantum physics, which is Goswami's specialty. The Spanish translation was published by La Esfera de los Libros, a major firm in the publishing industry that belongs to a media consortium led by the daily newspaper *El Mundo*, the second-largest print newspaper in Spain whose editorial policy is opposed to the Socialist Party and other left-wing political forces.[25]

The connection between right-wing media and more or less vague antievolutionism is increasingly common in Spain nowadays. The Spanish press treatment of the bicentenary of Darwin's birth in 2009 is illuminating in this context. *El Mundo* stood out by the high quality of the numerous articles devoted to the commemoration and "offered an obvious defense of the theory of evolution and was critical of the opposing positions, which were branded as 'fundamentalist.'" The daily *ABC*, in its turn, "treated evolution as a valid theoretical model deserving a serious consideration." But the contents of its religious—Catholic—supplement were somewhat different. Some articles insisted on the compatibility between evolution and creation, while one showed "unquestionable support for creationism." Other right-wing newspapers like *La Razón* and *Diario Ya* did not devote "extensive coverage to the bicentenary." However, they included some articles with certain creationist sympathies under sensationalist headlines. These different stances illustrate the different audiences that each newspaper strives to gain. In short, *El Mundo* represents a center-right ideology. *ABC* has always maintained a prestige among educated Catholic people. *La Razón* flirts with sensationalism and circles close to the extreme right, while *Diario Ya* is a confessional Catholic, conservative, on-line daily.[26]

However, it is difficult to correlate such general religious and political attitudes with concrete stances about creationism or evolutionism. This is easier for individual columnists and journalism personalities who maintain an opinion-forming discourse. Juan Manuel de Prada, a successful novelist and columnist, has come to the defense of creationists. In 2008 he argued that media usually caricature the creationists and their theories, presenting them as a sort of freaks. But he took this farther by saying: "A creationist . . . is just a person who refuses . . . the ideological fodder with which they try to control us." That "ideological fodder" is the idea of continuity in the evolutionary process applied to man. Another piece by Prada clarified his antievolutionism. The article dealt with creationist paleoanthropology, which Prada esteems as evidence against the idea of man's animal descent. He thus discredits human evolution: "What we find in such caverns . . . refutes those evolutionary hypotheses that confuse and complicate everything so we cannot understand the truth." Both articles appeared in a popular weekly magazine published by the consortium that supports *ABC*, the daily paper in which Prada writes a regular column. The provocative tone is usual in Prada's works. According to Prada, who declares himself to be a Catholic, one day the attacks against the church prompted a revitalizing experience of his faith, which turned him into a champion of Catholicism.[27]

The famous journalist César Vidal is another example of public adhesion to antievolutionary theses. Vidal is a former lecturer in history who has obtained fame through his frequent appearances on radio and TV programs and in the press. He is a prolific essayist, too. His ultraconservative ideas permeate his works, including revisionist books on contemporary history of Spain. He is also a prominent evangelical Christian. Vidal refuses to accept the scientific nature of Darwinism, denigrates paleontology as openly "novelistic," and claims the respectability of intelligent design theory. His discourse expresses several suspicions about scientific theory and its practice not only in the biological and geologic realm but also in climatology, cosmology, and theoretical physics. However, he declares that no religious concerns hinder him from accepting evolution—only strictly scientific ones.[28]

It is difficult to separate what is a true conviction from a mere political stance in this kind of public statement. Maybe more Catholic believers are sympathetic to creationism by mere ideological vogue than Protestant followers, probably more self-convinced. In any case, there is a clear link between conservative ideology and creationist ideas in Spain that goes beyond the concrete religious options. A good example of the pluralism within the evangelical sphere is a volume sponsored by the Spanish Evangelical Alliance that brings together four contributions about controversial issues. One essay, written by César Vidal, is devoted to the historical evidences of Jesus of Nazareth, while another includes a devastating argument against intelligent design and in favor of the theory of evolution. Its author is the chemist Pablo de Felipe, who teaches science and faith at the Evangelical United Seminary of Theology in Madrid. De Felipe has been critical toward other Spanish Protestants because of their antiscientific attitudes.[29]

American-style creationism linked to conservative evangelicalism continues to be promoted by writers such as the indefatigable Santiago Escuain, whose translation of Thomas Woodward's *Darwin Strikes Back* was published by the American evangelical press Portavoz in 2007. The documentation service that Escuain and his collaborators maintain offers a growing selection of creationist items. Some publishing companies like El Peregrino offer more recent creationist literature in Spanish. For example, Antonio Cruz, a biologist and evangelical minister in Terrassa in Catalonia, wrote *Darwin no mató a Dios* (Darwin Did Not Kill God), which was published in the United States and constitutes one of the few examples of original Spanish creationist essays. The first paragraph immediately reveals Cruz's approach: "Since the publication of *On the Origin of Species* . . . Darwin's ideas have killed God within the conscience of several people. This event happens usually during the early years of school formation. When a teenager faces the biology teacher's forceful statements about evolution as a proven scientific fact . . . , it is not difficult to understand that the pupil starts questioning several things and even loses his or her elders' faith."[30]

In any case, Spanish creationism continues to be connected to American initiatives. Recently, Escuain was involved as a translator of a lecture series that an international anti-Darwinian organization, Physicians and Surgeons for Scientific Integrity (PPSI), sponsored in Spain in January 2008. The speakers included Thomas Woodward, an evangelical theologian from Florida and author of *Darwin Strikes Back*; Geoffrey Simmons, a physician and fellow member of the Discovery Institute; Isaac Lorencez, a Swiss engineer; and Antonio Martínez, a Spanish ophthalmologist and PSSI member. The lecture series was planned to visit the cities of Madrid, Barcelona, Málaga, León, and Vigo, giving a total of ten lectures (two per city), but in the last two cities it was possible to offer only one talk, because the local universities refused to provide a venue. These educational institutions had been warned by the Sociedad Española de Biología Evolutiva (SESBE, Spanish Society for Evolutionary Biology) about the creationist orientation of PPSI.[31]

The year before, SESBE had already warned a university establishment in Gijón against sponsoring a talk by the Cuban American astronomer Guillermo González, senior fellow of the Discovery Institute. The veto in León and Vigo provoked a debate in the media about freedom of expression and the limits of science and scientific teaching. In some respects, there were certain resemblances to earlier controversies on evolutionism in Spain. The tone of the president of

SESBE, Manuel Soler, is particularly significant. With regard to PSSI and its lecture series, he wrote: "This invasion is a very serious attempt to win [the battle of ideas] that will surely continue. We have to get ready for the defense . . . Now we are at war against the invaders."[32]

No doubt, these military metaphors reveal a grave concern. They are a response to a situation that Soler and other evolutionary scientists consider alarming. The rhetoric of warfare is encouraged by some scientists who not only declare themselves as atheists but also scorn the religious attitudes of some of their colleagues. For example, evolutionary ecologist Santiago Merino used a review of Richard Dawkins's *The God Delusion* for expressing his surprise when he finds scientists defending the compatibility between science and religion. Similarly, geneticist Arcadi Navarro assesses Francis Collins's *The Language of God* as a "vehement but impossible attempt to reconcile God . . . with the scientific advances to which he himself has contributed" and claims that Collins deceives himself.[33]

The public visibility of intelligent design and other creationist currents has increased during the first decade of the twenty-first century. Probably several Spanish evolutionary scientists have underestimated the previous presence of creationism in their country. The presentation of certain favorable opinions on intelligent design from Catholic voices such as Cardinal Schönborn has caused a special concern. In fact, the Jesuit paleontologist Leandro Sequeiros has written a book with the purpose of challenging intelligent design. In other publications, Sequeiros notes that he is seriously worried about the penetration of creationist discourses into Catholic circles.[34]

However, creationists are still a relatively small minority in Spain. According to a 2005 Eurobarometer survey of Spaniards, 73 percent accepts human evolution and 16 percent rejects it. Likewise, a 2011 Ipsos survey revealed that 53 percent of the Spanish population identified as evolutionists and only 11 percent as creationists, while those claiming to be undecided on the question totaled 37 percent.[35]

Indeed, the presence of evolutionary theory in the secondary-school curriculum has declined after the reform of the education system in the 1990s. Some voices demand a greater effort in this aspect to prevent the spreading of creationism. But the increasingly polarized Spanish society, with its proclivity to pack together politics, education, science, and religion, while in the midst of a deep crisis (and not only the economic one), could reissue old polemics in new formats in the future.[36]

Debating Evolution and Creationism in Portugal

Unlike Spain, Portugal did not experience a controversial reception of Darwinism. At least, this is the image pictured by some authoritative accounts, which stress the open-minded attitude of several university professors and do not identify a broad opposition to evolutionary ideas. In contrast to their Spanish counterparts, Portuguese Jesuits did not challenge the theory of evolution. Since 1902 in Portugal, the Society of Jesus had been publishing the scientific journal *Brotéria* aimed at promoting natural science. From 1907 until 1924, *Brotéria* had a specific series devoted to popular science, which, however, avoided discussing evolutionary theories. This reluctance vividly contrasted with the combative line held by the aforementioned Spanish Jesuit journal *Ibérica*. The nonpolemical line of the Portuguese Jesuits might be explained by the strong anticlericalism apparent around the time that the republic was proclaimed in 1910.[37]

There is an important historiographical gap with regard to evolutionary thinking in Portugal after 1914, which limits the knowledge about the situation during the so-called Estado Novo period—the authoritarian regime established by António de Oliveira Salazar in 1926. During its forty-eight years of existence, the Estado Novo was supported by Catholic authorities. Some historians define it as a clerical-fascist regime, whereas other scholars prefer to describe it as an example of national Catholicism. However, Catholicism was never declared the official religion in Portugal during that period, unlike what occurred in Spain during Franco's dictatorship. In fact, the Catholic Church remained formally separated from the Portuguese state. Elucidating the fate of evolutionism during those five decades is an important challenge for historians of Portuguese science.[38]

Nowadays, creationism and antievolutionism are present in Portugal in different ways. Some Catholic thinkers, such as Carmelite father Armindo dos Santos Vaz, are very critical of creationist movements, considering them dangerous for faith and pernicious for theological prestige, while other Catholic authors uphold more ambiguous stances. The Franciscan theologian Joaquim Carreira das Neves, retired professor at the Portuguese Catholic University, doubts the truthfulness of Darwinism and is sympathetic to ideas suggesting that human biological nature is not evolving.[39]

Some Protestant groups are promoting creationist initiatives on Portuguese ground. The most spectacular one is a creationist museum in Mafra, a town near Lisbon. The initiative had the support of the Discovery Institute and was announced in 2006. However, it has not come to fruition yet. More modestly, several Protestant denominations are spreading critical reports on evolutionist teaching in the Portuguese education system. In fact, creationism is taught in public schools within the syllabus of an optional subject, "Moral and Religious Evangelical Education," offered to a few thousand Protestant pupils.[40]

The most renowned champion of creationism in Portugal, Jónatas Machado, is a professor of law at the University of Coimbra. In 2004 Machado wrote a creationist paper, published in an academic journal, in which he refuses "an epistemological dichotomy" between the Bible and science and rejects any kind of reconciliation between Christian faith and evolutionism.[41]

Creationists in Portugal live on relatively fertile ground compared to those in Spain. Thus, according to the 2005 Eurobarometer survey, the Portuguese acceptance of human evolution is 9 percent lower than it is in Spain. In Portugal 64 percent accepts human evolution, 21 percent rejects it, and 15 percent claims to be undecided. In spite of this relatively sympathetic atmosphere, no organized creationist societies and journals have been established in Portugal.[42] Some surveys specifically centered on teachers' perceptions show 17 percent of biology teachers declaring that the theory of evolution contradicts their own beliefs, compared to a significant 26 percent of nonbiology teachers. Among Portuguese teachers, 45.5 percent refer to God as being active in the origin of mankind, while there is a broad acceptance of explanations combining evolutionary processes with divine action.[43]

Since around 1980, evangelical members of small Protestant minorities in Spain and Portugal have imported American creationism through translations and a short-lived journal and introduced creationism at some evangelical schools. However, the influence of Protestant creationists is restricted to their own religious circles. Among conservative Catholics, antievolutionary voices are heard, but they have neither initiated organized campaigns against evolution nor received strong financial or political support. In 2005, 16 percent of Spaniards and 21 percent of Portuguese rejected human evolution, placing the Iberian countries in between the pro-evolutionary countries in northwestern Europe and the more antievolutionary countries of Eastern Europe.

ACKNOWLEDGMENTS

The research for this chapter was funded by HAR2010-21333-C03-01 (MICINN, Government of Spain) and PRCEU-UCH40/10 (Universidad Cardenal Herrera-CEU).

NOTES

1. Centro de Investigaciones Sociológicas, *Estudio 2917. Barómetro de noviembre de 2011* (Madrid: CIS, 2011), accessed Jan. 23, 2012, www.cis.es/cis/opencms/-Archivos/Marginales/2900_2919/2917/Es2917_mapa.htm; Instituto Nacional de Estatística, *XIV Recensamento Geral da Populaçao* (Lisboa: INE, 2001), accessed Jan. 23, 2012, http://censos.ine.pt/xportal/xmain?xpid=INEandxpgid=censos_historia_pt_2001; "Portugueses menos católicos," *Paróquias de Portugal* (2001), accessed Jan. 23, 2012, www.paroquias.org/noticias.php?n=939.

2. Thomas F. Glick, "Spain," in *The Comparative Reception of Darwinism*, ed. Thomas F. Glick (Austin: University of Texas Press, 1972), 307–345; Thomas F. Glick, *Darwin en España* (Barcelona: Península, 1982); Diego Núñez, *El darwinismo en España* (Madrid: Castalia, 1977); Diego Núñez, "Darwinisme espagnol," in *Dictionnaire du darwinisme et de l'évolution*, ed. Patrick Tort (Paris: Presses Universitaires de France, 1996), 896–900; Jesús I. Catalá-Gorgues, "Cuatro décadas de historiografía del evolucionismo en España," *Asclepio* 61 (2009): 9–66; Rafael Sanus, "Algunos aspectos de la apologética española en la segunda mitad del siglo XIX," *Almena* 2 (1962): 11–32; Francisco Pelayo, *Ciencia y creencia en España durante el siglo XIX* (Madrid: CSIC, 1999); Diego Núñez, "El darwinismo en España: un test significativo de nuestra situación cultural," *Revista de Hispanismo Filosófico* 2 (1997): 31–36.

3. Thomas F. Glick, "Ciencia, política y discurso civil en la España de Alfonso XIII," *Espacio, Tiempo y Forma. Serie V. Historia Contemporánea* 6 (1993): 81–98; Thomas F. Glick, "The Valencian Homage to Darwin in the Centennial Date of His Birth," in *III Congreso Nacional de Historia de la Medicina. Actas*, vol. 2 (Madrid and Valencia: SEHM), 577–601.

4. Mercè Durfort, "Jaume Pujiula i Dilmé, S.I (Besalú, Garrotxa, 1869–Barcelona, 1958). La morfologia microscòpica," in *Ciència i Tècnica als Països Catalans: una aproximació biogràfica*, ed. Josep Maria Camarasa and Antoni Roca (Barcelona: FCR, 1995), 827–858; Francisco Teixidó-Gómez, "El jesuita Jaime Pujiula Dilmé, científico clave de la investigación biológica barcelonesa," *Llull* 33 (2010): 355–380; Jesús I. Catalá-Gorgues, "Los jesuitas españoles ante el evolucionismo durante el período restauracionista (1875–1922)," in *Darwinismo, biología y sociedad*, ed. Rosaura Ruiz, Miguel Ángel Puig-Samper, and Graciela Zamudio (Aranjuez: Doce Calles, 2013), 211–233.

5. Francisco Blázquez-Paniagua, "La evolución biológica en los cuestionarios oficiales de bachillerato en España (1927–1978)," *eVOLUCIÓN* 6:1 (2011): 39–44; Francisco Blázquez-Paniagua, "A Dios por la ciencia. Teología natural en el franquismo," *Asclepio* 63 (2011): 453–476; Ignacio Sala de Castellarnau, *Zoología* (Valencia: Bello, 1948), 7 (quotation); Alejandro Roldán, *El problema de la evolución y de la antropogénesis* (Barcelona: Atlántida, 1950), 210–213; Ronald L. Numbers, *The Creationists: From Scientific Creationism to Intelligent Design*, expanded ed. (Cambridge, MA: Harvard University Press, 2006), 166–177.

6. Francisco J. Ayala, *Darwin y el diseño inteligente*, trans. Miguel Ángel Coll (Madrid: Alianza, 2007), 172; Mariano Artigas, Thomas F. Glick, and Rafael A. Martínez, *Seis católicos evolucionistas. El Vaticano frente a la evolución (1877–1902)* (Madrid: BAC, 2010), xiii; R. Scott Appleby, "Exposing Darwin's 'Hidden Agenda': Roman Catholic Responses to Evolution, 1875–1925," in *Disseminating Darwinism: The Role of Place, Race, Religion, and Gender*, ed. Ronald L. Numbers and John Stenhouse (Cambridge: Cambridge University Press, 1999), 173–207; Pawel Kapusta, "Darwinism from *Humani generis* to the Present,"

in *Darwin and Catholicism*, ed. Louis Caruana (London: T&T Clark, 2009), 27–42; *El evolucionismo en filosofía y en teología* (Barcelona: Juan Flors, 1956); J. Guerra, "El evolucionismo de Teilhard de Chardin," *Compostellanum* 2 (1957): 501–520; Manuel Benzo, "Evolución y dogma," *Revista de la Universidad de Madrid* 8 (1959): 533–559.

7. M. Mañosa, "Miquel Crusafont i Pairó. Sabadell, 1910–1983. L'escola paleontològica de Sabadell," in Camarasa and Roca, *Ciència i Tècnica als Països Catalans*, 1443–1472; Thomas F. Glick, *Darwin en España*, expanded ed. (Valencia: Universitat de València, 2010), 113–133; Emiliano Aguirre, "Aspectos filosóficos y teológicos de la evolución," *Revista de la Universidad de Madrid* 8 (1959): 445–531; Miguel Crusafont, Bermudo Meléndez, and Emiliano Aguirre, eds., *La Evolución* (Madrid: La Editorial Católica, 1966).

8. Carlos Acosta-Rizo, "La teoría de la evolución y la censura en TVE. Entre el fijismo, el finalismo ¿y el neodarwinismo?," *Actes d'Història de la Ciència i de la Tècnica* 1 (2008): 271–277; Sara del Cerro, Rodrigo Megía, Juan Rivero de Aguilar, Josué Martínez de la Puente, and Santiago Merino, "Félix Rodríguez de la Fuente y la teoría de la evolución por selección natural," *eVOLUCIÓN* 5:1 (2010): 21–24.

9. Óscar Barberá, Beatriz Zanón, and José-Francisco Pérez-Plá, "Biology Curriculum in Twentieth-Century Spain," *Science Education* 83 (1999): 97–111; Blázquez-Paniagua, "Evolución biológica."

10. Stephen Jay Gould, *Dientes de gallina y dedos de caballo: más reflexiones acerca de la historia natural*, trans. Antonio Resines (Madrid: Hermann Blume, 1984); "Un tribunal americano deberá decidir si el relato bíblico de la creación es científico," *El País*, Dec. 10, 1981.

11. Catalá-Gorgues, "Cuatro décadas," 28–34; "Bibliografía hispánica sobre Darwin y el darwinismo," *Anthropos* 16–17 (1982): 15–54; Ramón Tamames, "Evolucionismo y crisis / 1," *El País*, Dec. 6, 1982; Joaquín Bardavío, "Antonio Garrigues Díaz-Cañabate, paradigma del reformismo," *El Mundo*, Feb. 26, 2004; Antonio Garrigues Díaz-Cañabate, "La dignidad del origen de la vida: evolución o creación," *El País*, Jan. 22, 1980; Antonio Garrigues Díaz-Cañabate, "La vida," *El País*, June 24, 1983; Antonio Garrigues Díaz-Cañabate, "El Génesis y Darwin. La evolución del evolucionismo y la presencia de Dios," *La Vanguardia*, May 23, 1982.

12. "Cien años. Por qué Darwin," *ABC* (Madrid), Mar. 28, 1982.

13. Carlos Díaz, "Juan Luis Ruiz de la Peña, *in memoriam*," *Acontecimiento* 41, suppl. (1996): 1–8; Juan Luis Ruiz de la Peña, *Teología de la creación* (Santander: Sal Terrae, 1986), 119–124; Ruiz de la Peña, *Imagen de Dios. Antropología teológica fundamental* (Santander: Sal Terrae, 1988), 249–250 (quotation).

14. Ruiz de la Peña, *Imagen de Dios*, 261.

15. Santiago Collado-González, "Mariano Artigas (1938–2006) *in memoriam*," *Anuario de Historia de la Iglesia* 17 (2008): 418–425; Mariano Artigas, *Las fronteras del evolucionismo* (Madrid: Palabra, 1991), 78–84, 152–155.

16. Artigas, *Fronteras*, 141–142.

17. Ibid., 140–144.

18. Manuel López-Rodríguez, *La España protestante. Crónica de una minoría marginada (1937–1975)* (Madrid: Sedmay, 1976).

19. Duane T. Gish et al., *Creación, evolución y el registro fósil*, trans. Santiago Escuain (Terrassa: CLIE, 1988); Duane T. Gish, *Especulaciones y experimentos relacionados con teorías del origen*, trans. Santiago Escuain (Barcelona: Portavoz Evangélico, 1978); Santiago

Escuain, "Las discontinuidades del registro fósil. ¿Fósiles perdidos, o engendros mentales que nunca existieron?," in Gish et al., *Creación*, 91–99 (quotation); Servicio Evangélico de Documentación e Información, accessed Oct. 29, 2011, www.sedin.org/propesp/X0001_Qu.htm.

20. SEDIN, "Libros Creación y Ciencia," accessed Oct. 29, 2011, www.sedin.org/libroscc.html; John C. Whitcomb and Henry M. Morris, *El diluvio del Génesis*, trans. Dante M. Roso (Terrassa: CLIE, 1982); David C. C. Watson, *Mitos y milagros* (Terrassa: CLIE, 1980); David C. C. Watson, *El gran fraude intelectual* (Terrassa: CLIE, 1981); Malcolm Bowden, *Los hombres simios: ¿realidad o ficción?* (Terrassa: CLIE, 1984); John N. Moore y otros, *Vida, herencia y desarrollo* (Terrassa: CLIE, 1985), all trans. Santiago Escuain.

21. Phillip E. Johnson, *Proceso a Darwin*, trans. Santiago Escuain (Grand Rapids, MI: Portavoz, 1995); Kregel Publications, "About Us," accessed Dec. 17, 2011, http://kregel.com/ME2/dirmod.asp?sid=A12DB34B70B34EA28EA748A96CD5AEFEandtype=genandmod=Core+Pagesandgid=A615A3BFB3EA496C83F0F4CFDCBE387D.

22. "Geología—interpretación, Revelación y reconstrucción," *Génesis* 2 (1994–95): 1, 6; "El universo inteligente, el dios de los evolucionistas y la Revelación," *Génesis* 2 (1994–95): 2–5.

23. Enrique Mota, *Ciencia y fe ¿en conflicto?* (Barcelona: Andamio, 1995), 52.

24. Edgard H. Andrews, *De la nada a la naturaleza*, trans. Demetrio Cánovas (Moral de Calatrava: El Peregrino, 1988).

25. Michael Behe, *La caja negra de Darwin: el reto de la bioquímica a la evolución*, trans. Carlos Gardini (Barcelona: Andrés Bello, 2000); Homo Legens, accessed Oct. 30, 2011, www.homolegens.com; William A. Dembski, *Diseño inteligente*, trans. Carmen García-Trevijano (Madrid: Homo Legens, 2006); Diego Martínez-Caro, *Génesis: el origen del universo, la vida y el hombre* (Madrid: Homo Legens, 2008); Manuel Guerra-Gómez, *La evolución del universo, de la vida y del hombre* (Madrid: Homo Legens, 2009); Amit Goswami, *Evolución creativa: la física cuántica reconcilia el darwinismo y el diseño inteligente* (Madrid: La Esfera de los Libros, 2009).

26. Esther Díez, Anna Mateu, and Martí Domínguez, "Darwin in the Press: What the Spanish Dailies Said about the 200th Anniversary of Charles Darwin's Birth," *Contributions to Science* 5 (2009): 193–198.

27. Juan Manuel de Prada, "Creacionismo," *XL Semanal* 1092 (2008), accessed Oct. 31, 2011, http://xlsemanal.finanzas.com/web/firma.php?id_firma=7091andid_edicion=3487; Juan Manuel de Prada, "La firma del hombre," *XL Semanal* 978 (2006), accessed Oct. 31, 2011, www.xlsemanal.com/web/firma.php?id_edicion=1207andid_firma=2851; Jesús G. Sánchez-Colomer, "Entrevista a Juan Manuel de Prada," *Época*, Oct. 22, 2006, accessed Nov. 1, 2011, www.interrogantes.net/Entrevista-a-Juan-Manuel-de-Prada-Revista-Epoca-220X006/menu-id-29.html.

28. Iglesia Evangélica Bautista de Córdoba, "El mito de la evolución—César Vidal," recorded talks, June 21 and 27, 2009, www.iebcordoba.es/2009/06/el-mito-de-la-evolucion-cesar-vidal.html; "Es la noche de César," TV program, Sept. 24, 2011, http://videos.libertaddigital.tv/2011-09-24/tertulia-de-cesar-vidal-rubalcaba-y-tarda-aGdjtruMazM.html; "Diálogo con César Vidal," *Libertad Digital.es*, Dec. 18, 2008, www.libertaddigital.com/opinion/chat-chat/del-30-de-abril-con-9089/.

29. César Vidal, "Evidencias históricas del cristianismo: Jesús en fuentes históricas no cristianas," in *Temas apologéticos de hoy* (Barcelona: Andamio, 2010), 77–86; Pablo de

Felipe, "El diseño inteligente y las alternativas apologéticas en las ciencias," in *Temas*, 43–75; Pablo de Felipe, "Orangutanes y Dios," *Protestante Digital*, June 9, 2007, www.protestantedigital.com/ES/Magacin/articulo/2581/Orangutanes-y-dios.

30. Thomas Woodward, *Darwin contraataca*, trans. Santiago Escuain (Grand Rapids, MI: Portavoz, 2007); John Blanchard, *La evolución, ¿realidad o ficción?*, trans. Demetrio Cánovas and David Cánovas (Moral de Calatrava: El Peregrino, 2006); Andy McIntosh, *Génesis para hoy: la pertinencia del debate creación-evolución para la sociedad de hoy*, trans. Demetrio Cánovas and Ana Juliá (Moral de Calatrava: El Peregrino, 2009); Antonio Cruz, *Darwin no mató a Dios* (Miami: Vida, 2004), 5.

31. PSSI, "Lo que Darwin no sabía, 17–25 de enero de 2008," schedule of talks, accessed Nov. 6, 2011, www.loquedarwinnosabia.com/; Geoffrey Simmons, "A Successful PSSI Lecture Tour in Spain," *Evolution News and Views*, Mar. 11, 2008, www.evolutionnews.org/2008/03/a_successful_pssi_lecture_tour004978.html; Manuel Soler, "Ya están aquí los creacionistas," *eVOLUCIÓN* 3:1 (2008): 3–4.

32. Manuel Soler, "Problemas de la Teoría Evolutiva en España en la actualidad," *eVOLUCIÓN* 2:2 (2007): 3–4; Javier Rico, "El creacionismo llega a España," *El País*, Jan. 10, 2008; Manuel Soler, "¿Desciende el hombre del mono?: los creacionistas y la teoría evolutiva," *El País*, Feb. 23, 2008; Diego Barcala, "Los anti-Darwin no tienen sitio en la Universidad," *Público*, Jan. 17, 2008; "Lo que Darwin no sabía," *Los fallos de Darwin*, blog, Oct. 10, 2010, http://los-fallos-de-darwin.blogspot.com/2010/10/lo-que-darwin-no-sabia.html; Soler, "Ya están," 4.

33. Santiago Merino, "*El espejismo de Dios* de Richard Dawkins. 2007. Ciencia y religión son incompatibles," *eVOLUCIÓN* 3:1 (2007): 65–66; Arcadi Navarro, "Un llibre impotent i dos d'impossibles," *El Temps*, Feb. 13, 2007, 20.

34. José Antonio Díaz-Rojo, "El caso Schönborn: un cambio retórico en la postura católica ante la evolución biológica," *Ilu. Revista de Ciencias de las Religiones* 14 (2009): 33–58; Juli Peretó, "Entre la bioquímica obsoleta y la seudociencia," *SEBMM* 153 (2007): 18–22; Leandro Sequeiros, *El diseño chapucero* (Madrid: Khaf, 2009); Leandro Sequeiros, "Cuando hablamos de 'evolución biológica,' ¿de qué evolución estamos hablando? Implicaciones teológicas," *eVOLUCIÓN* 4:1 (2009): 43–54.

35. European Commission, *Social Values, Science and Technology*, Special Eurobarometer 225, Wave 63.1 (Brussels: European Commission, 2005), accessed Nov. 1, 2013, http://ec.europa.eu/public_opinion/archives/ebs/ebs_225_report_en.pdf; Jon D. Miller, Eugenie C. Scott, and Shinji Okamoto, "Public Acceptance of Evolution," *Science* 313 (2006): 765–766; Ipsos Global @dvisory, "Supreme Being(s), the Afterlife and Evolution," released Apr. 25, 2011, www.ipsos-na.com/news-polls/pressrelease.aspx?id=5217.

36. Laureano Castro-Nogueira, "Docencia y evolución: la biología evolutiva en la enseñanza secundaria," *eVOLUCIÓN* 2:1 (2007): 63–66; Laureano Castro-Nogueira, "La evolución y el mundo educativo," *eVOLUCIÓN* 3:1 (2008): 55–57.

37. Carlos Almaça, *Evolutionism in Portugal* (Lisboa: Museu Bocage, 1993); Carlos Almaça, *O Darwinismo na Universidade Portuguesa (1865–1890)* (Lisboa: Museu Bocage, 1999); Ana Leonor Pereira, *Darwin em Portugal (1865–1914)* (Coimbra: Almedina, 2001); Marta Mendoça, "*Brotéria* e a difusão da ciencia em Portugal: ideário e conteúdos. Uma análise da Série de Vulgarização Científica (1907–1924)," in *Fê, ciência, cultura:* Brotêria—*100 anos*, ed. Hermínio Rico and José Eduardo Franco (Lisboa: Gradiva, 2003), 255–276.

38. Manuel Braga da Cruz, *O Estado Novo e a Igreja Católica* (Lisboa: Bizâncio, 1998).

39. Augusta Gaspar, Teresa Avelar, and Octávio Mateus, "Criacionismo e sociedade no séc. XX," in *Evolução e criacionismo: uma relação impossível*, ed. Augusta Gaspar (Quasi: Vilanova de Famalicão, 2007), 133–160; Joaquim Carreira das Neves, "Darwin—Ciência e Fé," in *Ainda Darwin: evolução, ética e direitos humanos*, ed. Cassiano Reimão (Lisboa: Universidade Lusíada, 2011), 23–30.

40. Gaspar, Avelar, and Mateus, "Criacionismo," 155–160.

41. Jónatas Machado, "Criacionismo Bíblico: Súmula dos principais fundamentos teológicos e científicos," *Estudos NS* 2 (2004): 107–166.

42. European Commission, *Europeans, Science and Technology*.

43. Pierre Clément, Marie Pierre Quessada, Charline Laurent, and Graça S. Carvalho, "Science and Religion: Evolutionism and Creationism in Education. A Survey of Teachers' Conceptions in 14 Countries," in *XIII IOSTE Symposium*, Izmir (Turkey), Sept. 21–26, 2008, on "The Use of Science and Technology Education for Peace and Sustainable Development," accessed Sept. 10, 2013, http://repositorium.sdum.uminho.pt/bitstream/1822/8934/1/IOSTE_Evolution.pdf; Graça S. Carvalho, Rosa Branca Tracana, Paloma R. Silva, Elaine Araújo, and Ana Maria Caldeira, "The Influence of Religion on Portuguese and Brazilian Teachers' Conceptions about the Origin of Life," in *eBook Proceedings of the ESERA 2011 Conference—Science Learning and Citizenship. Part 11: Cultural, Social and Gender Issues*, ed. C. Bruguière, A. Tiberghien, and P. Clément (Lyon: ESERA, 2012), 6–11.

CHAPTER THREE

United Kingdom

JOACHIM ALLGAIER

The United Kingdom is a unitary state that consists of four countries: England, Scotland, Northern Ireland, and Wales. It is an active member of the European Union, but it is not part of the European Economic and Monetary Union. The United Kingdom is one of the five permanent members of the United Nations Security Council and a founding member of the North Atlantic Treaty Organization (NATO). It has a colonial past, which has resulted in an ethnically diverse population today. In the nineteenth century the British Empire stretched over one-fourth of the earth's surface. The population estimate for July 2011 is close to 63 million people. According to the 2001 national census, the population consists of the following ethnic groups: white, 92.1 percent; black, 2 percent; Indian, 1.8 percent; Pakistani, 1.3 percent; mixed, 1.2 percent; other, 1.6 percent. The breakdown among the white population is English, 83.6 percent; Scottish, 8.6 percent; Welsh, 4.9 percent; Northern Irish, 2.9 percent. The major religions are, according to the same census data from 2001, Christian (Anglican, Roman Catholic, Presbyterian, Methodist), 71.6 percent; Muslim, 2.7 percent; Hindu, 1 percent; other, 1.6 percent; unspecified or none, 23.1 percent.[1]

Historically Great Britain has played a leading role in developments concerning science, medicine, and technology. According to a recent report, the United Kingdom today spends 4 percent of the world's gross expenditure on research and development on 6 percent of the world's researchers, who are authors of 8 percent of the world's research articles and reviews. The report concludes that the United Kingdom's average research impact now surpasses that of the United States.[2] The United Kingdom is also home of Charles Robert Darwin and his theory of evolution. *On the Origin of Species by Means of Natural Selection* was published on November 24, 1859, by John Murray in London. The first edition was sold out to booksellers the day it was published.[3] The official view of Darwin's eminence as a scientist was demonstrated by a ceremonial funeral in Westminster Abbey, where he was buried close to another national scientific hero, Isaac Newton. Another sign that official Britain is paying deference to Darwin and his lifework is that he is depicted on the £10 note. However, it has been less clear what the British population knows of and thinks about Darwin and the theory of evolution. Leading up to the double Darwin anniversary in 2009 various opinion polls have helped to shed some light on the public attitude toward Darwin and evolution, and also on attitudes toward God, creation, and intelligent design.

What People in Britain Think about Evolution and Creation

In 2006 the BBC commissioned Ipsos Mori, a respected UK market research company, to poll 2,112 adults in Britain about their views on what should be taught in science classes: 69 percent said evolution, 44 percent said creationism, and 41 percent said intelligent design (the poll allowed more than one answer). Given the option of three explanations of the origin of life, 48 percent opted for evolution without God, 22 percent chose creationism, and 17 percent chose intelligent design. The results that less than half of the polled adults chose evolution without God as explanation for the origin of life and that four out of ten said science classes should include intelligent design were rather unexpected and a cause of concern for many scientists.[4]

A couple of months later OpinionPanel, an independent market research agency specializing in student views, asked more than one thousand students in the United Kingdom for their views on creationism and evolution. Here 12 percent preferred creationism, 19 percent favored intelligent design, while evolution was supported by 56 percent. Another interesting result was that nearly 20 percent of the students said they had been taught creationism as fact by their main school. The survey also documented that not only British Christians but also Muslims were challenging evolution: a third of those who said they were Muslims and more than a quarter of those who said they were Christians supported creationism.[5] A comparative poll around the same time found that the proportion of adults in Great Britain that thought that evolution is "absolutely false" was 15 percent.[6]

According to a 2008 poll by Teachers TV, a now defunct satellite TV channel for teachers, one in three teachers believed that creationism should have the same status in class as evolution.[7] For this survey twelve hundred registered viewers, mainly teachers, were questioned. Also in 2008 the public theology think tank Theos commissioned the polling company ComRes for a detailed quantitative research study on what the people in the United Kingdom knew about Darwin and what they believed in terms of human origins.[8] The results pointed to the fact that many people's opinions are not necessarily well formed or coherent and that they articulate opinions that seem inconsistent.[9] Responses depend on how precisely the questions are asked and what the options for answers are. It is also not always clearly stated what belief the term creationism actually refers to. This had been one of the main criticisms of earlier surveys in which the respondents sometimes had limited answering options from which to choose. Consequently, this kind of survey should generally be read with caution. The findings of the ComRes study are based on interviews with 2,060 British adults from a demographically representative sample. In this study, special care was taken that the respondents had a broad and detailed range of options to answer the questions.[10] Among the key results were that merely 37 percent of respondents agreed that Darwinian evolution was a theory so well established that it was "beyond reasonable doubt," with 19 percent believing it had little or no supporting evidence; 36 percent stated that the theory was still waiting to be proved or disproved. Of the 124 Muslims that participated in the survey 23 percent were young-earth creationists, 16 percent favored intelligent design, 6 percent theistic evolution, and 5 percent atheistic evolution.

Another interesting result was that only 10 percent considered science and religious belief to be entirely incompatible.[11] The data from this research also allowed regional differentiation of the results. For example, Northern Ireland had the highest proportion of people believing in intelligent design (16%) and creationism (25%).[12] In 2010 the British Centre for Science Education (BCSE), a group dedicated to defending science education in the United Kingdom against creationism, confirmed this view and reported that Northern Ireland is "the creationist capital of Europe" with some unique local creationist groups, such as Queen's University Belfast Creation Society, the only university creation society in Europe.[13]

The ComRes survey also revealed a relatively high proportion of people in London who believe in creationism. Paul Woolley, director of the Theos think tank, was quoted in an article by the *Guardian* saying: "Whereas the national average is 17% who believe that human beings were created by God in the last 10,000 years . . . in London, that figure is 20%. That may well be due to the growth of Pentecostal churches in London, which are growing at an extraordinary rate."[14] In general, creationist attitudes are often associated with rural areas and lifestyles. However, in the United Kingdom this does not seem to be the case. The high number of people in London subscribing to creationist positions is probably a result of the numerous African and Afro-Caribbean Pentecostalist churches that have many members in the London area.[15]

Another poll commissioned by the British Council and conducted by Ipsos Mori around the same time also found a national division of beliefs and confirmed that nearly a quarter (23%) of people living in London believed in creationism. That was more than the national average of 16 percent who believe in creationism.[16] Further results from the British Council survey, in which 973 Britons were asked, were that more than half (54%) of British adults thought that intelligent design and creationism should be taught alongside evolution in school. This proportion was slightly higher than in the United States, where, of 991 adults responding to the same survey, 51 percent agreed that evolution should be on the curriculum alongside other theories, such as intelligent design.

A qualitative study of different beliefs in British churches brings further information about specific preferences within denominations with regard to creationism but confirms the general result of the quantitative studies. It shows that a large majority (69% in total; 38% of the Anglican and Methodist churchgoers and 82% of the Evangelical, Pentecostal, and Charismatic) rejected that all life evolved from simple life forms and around 33 percent of all questioned churchgoers were young-earth creationists, similar to the results of the ComRes survey. The majority of this view is held among Evangelical, Pentecostal, and Charismatic churchgoers.[17]

A series of surveys on the attitudes of young people in Britain toward evolution and creationism in the late 1980s and 1990s found that significant proportions of young people in England, Scotland, and Northern Ireland hold creationist views or think that Christianity is necessarily creationist (or against evolution); however, these views often seem to change with age.[18]

Creationist Organizations in Britain

Religious groups and organizations opposing the idea that humans developed from simple life forms are not a recent phenomenon in Britain. In 1865 the Victoria Institute, or Philosophical Society of Great Britain, was founded with the objective to defend "the great truths revealed in Holy Scripture . . . against the opposition of Science falsely so called" and can be seen as part of a religious reaction to the publication of Darwin's theory of evolution.[19] However, it was a rather moderate organization that was not distinctively opposed to evolution. In 1932 Bernard Acworth proposed the formation of the Evolution Protest Movement as an explicit antievolution society, with a scientific rather than a religious angle, emerging from the Victoria Institute. In the first years the Evolution Protest Movement existed solely as a paper organization, but in February 1935 it was launched publicly at a meeting at Essex Hall in London, which was attended by more than six hundred people.[20]

The leaders of both organizations, the Victoria Institute and the early Evolution Protest Movement, espoused old-earth creationism. The goal of the Evolution Protest Movement was to counter the spread and teaching of evolution in schools and to bring the people back to the Bible, mainly for moral reasons. It claims to be the oldest creationist movement in the world. It used to have branches in Canada, Australia, New Zealand, and South Africa, but owing to the formation of rival creationist groups, none of them survived. The Evolution Protest Movement was renamed the Creation Science Movement in 1980 because both its leadership and its base had switched to young-earth creationism.[21] Today its headquarter is located in Portsmouth in the South of England. In 2000 the organization opened a museum supporting a literal interpretation of the Bible and promoting a young-earth perspective. Genesis Expo in Portsmouth is the biggest creationist exhibition in Britain. The number of followers still seems to be rather low judging from the fifteen hundred people on the Creation Science Movement's mailing list.[22] The Creation Science Movement is also connected with Biblical Creation Ministries, a charitable trust that supports creation speakers Paul Garner and Dr. Steve Lloyd.

There are at least another half a dozen active creationist groups in Britain. One is the Biblical Creation Society founded in 1977 by the Scottish Minister Nigel M. de S. Cameron. This organization was established as a counterweight to the Evolution Protest Movement and its "wholly negative" attitude. Leaders of the Biblical Creation Society refused to limit membership to young-earthers, and in choosing a name it followed John C. Whitcomb more than Henry M. Morris, who was by this time promoting scientific creationism.[23] The Biblical Creation Society publishes a journal called *Origins* (formerly *Biblical Creation*) and is based in Rugby, Warwickshire. The British Centre for Science Education assumes that it is the second-largest creationist organization in Britain.[24] Further, it is claimed that it has some highly educated members and associates in the educational establishment.

Another rather curious creationist institution is Noah's Ark Zoo Farm, a zoo advocating a young-earth perspective, located in Wraxall near Bristol. The Creation Resources Trust was founded in 1981 by the preacher Geoff Chapman and a small group of Christians. Their main aim is providing and sharing evidence for creation and highlighting flaws in the theory of evolution. This small organization is based in Yeovil, Somerset.

The former schoolteacher John Mackay from Queensland, Australia, has run an international creationist network called Creation Research and in that capacity often visited Britain on his lecture tours promoting a young-earth perspective and a literal interpretation of the Bible. Creation Research also has a branch in the United Kingdom run by Randall Hardy from an office in Ashton-under-Lyne near Manchester with a mailing list of two thousand individuals.[25]

The best-known creationist movement in Britain is Answers in Genesis.[26] Its headquarters is in the United States, in Hebron, Kentucky, but it also has a British office in Leicester. Answers in Genesis is one of the premier creationist organizations in the United States, where it also runs a creation museum near Cincinnati. Whereas many creationist organizations aim at an intellectual audience, *Answers in Genesis* reaches out to ordinary people using a multitude of media channels such as comic books, radio stations, video clips, and social media. The organization promotes a rigid literal interpretation of the book of Genesis. The members not only object to the scientific view that the earth is billions of years old and that life on earth has evolved from simple organisms but are also concerned with issues such as homosexuality, abortion, marriage, and race. The organization was funded by Ken Ham, a former science teacher from Queensland, Australia. In Australia, Ham was connected with John Mackay and the Creation Science Foundation. In the mid-1980s he joined the staff of the Institute for Creation Research in the United States. Ham established his own creationist organization in December 1993, which proved successful on its own and changed its name to Answers in Genesis in 1997.[27] This organization has good international connections, and Ham, a charismatic speaker, also went on regular international lecture tours to advocate his fundamentalist views. In 2008, for instance, he was on tour in Britain.[28]

The Emmanuel College Controversy

Ken Ham and Answers in Genesis were also connected to one of the biggest British creationism controversies in recent years, at least in terms of media coverage. In January 2002 it was reported in a newspaper specializing in education topics that educators at Emmanuel College in North England rented rooms to the creationist organization Answers in Genesis for a conference in March. Ken Ham was one of the speakers.[29] At the time of the conference a series of critical articles in the *Guardian* accused the school of teaching creationism in science classes.[30]

It was also claimed that some of Emmanuel College's senior staff were undermining the teaching of the theory of evolution. Nigel McQuoid, the head of the school, was quoted for saying that it was "fascist" to say that schools should not consider creationist theories. Several quotes by Emmanuel staff members, such as John Burn and Gary Wiecek, suggested that the educators at Emmanuel College see both evolutionary theory and the theory of creation as faith positions, and it was claimed that several of its staff members had urged teachers to "show the superiority" of creationist theories. Additionally, extracts of a lecture given at Emmanuel College by Stephen Layfield, head of science at Emmanuel College, in which he investigated ways how to present "the superiority of a creationist world-view against the prevailing orthodoxy of atheistic materialism and evolution in science" served as further evidence that creationism was taught at Emmanuel College.[31]

Emmanuel College was a very good school that consistently received excellent exam results. It was also a relatively new type of school: a city technology college. These are technically independent schools funded both by the state and by the private sector. Independently of the creationism controversy, the policy of bringing private funds into public education through the introduction of city technology colleges and city academies had already generated public debate. One of the controversial aspects was that city technology colleges could opt out of aspects of the national curriculum for England and Wales. Before the creationism debate began, the education inspectorate the Office for Standards in Education (OFSTED) had written a favorable report about the school, and it was regarded a beacon school by the government. The school's private sponsor was the Vardy Foundation, a charitable trust with a Christian ethos donating £2 million to the school. Behind the foundation stood Peter Vardy, a successful car dealer, who declared he wanted to give some of his wealth back to the local communities.

As the reports came in that Emmanuel College taught creationism in science classes, a member of the opposition addressed the issue in Parliament and asked the prime minister, Tony Blair, about his views on the issue. Blair backed the school for its good exam results, considered that the newspaper reports were exaggerated, and said that he welcomed diversity in education.[32] Hundreds of articles, opinion pieces, and letters have subsequently been written and published on creationism and Emmanuel College.[33] Various experts from different fields got involved in the public debate. The atheist and evolutionary biologist Richard Dawkins was among the staunchest critics, while Andrew McIntosh, a young-earth creationist and professor of thermodynamics and combustion theory at the University of Leeds, was supportive of the line at Emmanuel College.[34] Even though British creationists had been highly concerned with airing their opinions, in particular on the BBC, the case of Emmanuel College marks a departure from earlier strategies as the first public case of creationism in British schools.[35]

The national curriculum, introduced to schools in England and Wales in 1988 (Scotland has a different curriculum and education system), demands that evolution be taught in science classes. These classes are compulsory. Creation myths from various cultural backgrounds can be taught in religious education. In these cases students can opt out.[36] However, until 2006 the curriculum also stated that students should learn about how controversies can arise throughout scientific practice. The national curriculum for science described this in the following way: "(1) Pupils should be taught: . . . (b) how scientific controversies can arise from different ways of interpreting empirical evidence [for example, Darwin's theory of evolution]."[37] The fact that Darwin's theory of evolution was explicitly mentioned in the national curriculum as an example of a scientific controversy had important consequences for the Emmanuel College story, notably in terms of which controversies could or should be taught in school science classrooms. However, in the British context the theory of evolution was never under threat of being taken off the curriculum as in other countries. Effectively, the national curriculum determines what *has* to be taught. However, in practice it is possible that religious and other views are taught alongside the compulsory content of the science curriculum, which was possibly the teaching practice at Emmanuel College. However, the educators and the sponsor of Emmanuel College denied the accusations and stressed that all requirements of the national curriculum were followed in the school. This was later confirmed by education authorities, such as the Department for Education and Skills and OFSTED. Emmanuel College spokespeople generally emphasized the openness of the college to children of various cultural backgrounds and that schoolchildren were offered both, religious views on creation and the theory of evolution, so they could decide for themselves what they wanted to believe.

In the course of the controversy, various action groups got together to write letters and petitions; in response to further calls for action concerning the controversy, heterogeneous alliances emerged.[38] For instance, one group of scientists and philosophers had signed a petition that was aimed explicitly at the national curriculum for science and argued for a change of the formulation that referred to teaching scientific controversies. Beginning in September 2006 a new program was introduced for the study for science. Among other changes, the controversial paragraph 1b now read: "(1) Pupils should be taught: . . . (b) how interpretation of data, using creative thought, provides evidence to test ideas and develop theories."[39] The reference to Darwin's theory of evolution had been removed.

During and after the controversy opponents of the school consistently claimed that religious "indoctrination" and "brainwashing" would be taking place there and that the educators of the school and the Christian sponsor were pushing young-earth creationism in the school. This had consequences for other schools sponsored by the Vardy foundation. Later in the controversy the Vardy Foundation was renamed the Emmanuel School Foundation (ESF). The ESF website explicitly stated that this was the result of the very negative press coverage Peter Vardy and the Vardy Foundation received during the creationism debate. There have been protests every time the Vardy Foundation planned to set up a new school. Every time it was claimed that creationism would be taught at the school, which Vardy routinely denied. Currently, the Emmanuel School Foundation sponsors four schools in England: Emmanuel College in Gateshead, King's Academy in Middlesbrough, Trinity Academy in Doncaster, and Bede Academy in Blyth. In January 2011 it was reported that Peter Vardy had succeeded in libel action against the *Tribune*, a weekly newspaper, which had claimed that his foundation was imposing fundamentalist beliefs on children. The solicitor of Sir Peter Vardy stated that none of these allegations were correct and that Vardy had specifically requested that OFSTED inspectors look for creationism anywhere in the curriculum and that they found no evidence for creationist teaching.[40] Nonetheless, the debate whether private sponsors have an influence on what is taught in schools and what is happening also in free schools and faith schools in terms of creationism being taught has not been settled.[41]

Intelligent Design

In 2006 a new player entered the scene causing further controversy. The organization Truth in Science, founded in 2004, sent packs of teaching materials to five thousand secondary state schools in the United Kingdom in order to promote intelligent design as part of the science curriculum.[42] The packs contained lessons plans and two DVDs. The DVDs included contributions from members of the Discovery Institute, a leading Seattle-based intelligent design organization in the United States. Truth in Science was the first British organization that in a large-scale operation avoided explicit references to the Bible in its antievolutionist mission and promoted a neo-creationist intelligent design approach. There were close connections between people central to the Emmanuel College controversy and Truth in Science. Stephen Layfield, head of science at Emmanuel College, for example, was a member of the board until Peter Vardy told him to resign from the organization.[43] The Department for Education and Skills and the Qualifications and Curriculum Authority have both distanced themselves from the materials sent out by Truth in Science and clearly stated that neither creationism nor intelligent design is recognized as a scientific theory and should not be taught as such in school science lessons.[44] However, Truth in Science remains very active. In 2009 it distributed a free copy of the revised British edition of *Explore Evolution: The Arguments for and against Neo-Darwinism* to all British school libraries where biology is taught at the advanced level, bypassing science teachers altogether.[45] On the organization's website, there are also lesson plans in Welsh and specific materials that address issues particular to Scotland.

In 2010 Scotland got its own intelligent design organization, Centre for Intelligent Design, based in Glasgow.[46] The director, Dr. Alastair Noble, is a former school inspector and high school chemistry teacher with excellent connections in the Scottish education establishment. The purpose of the center is to advocate the intelligent design approach mainly in Scotland but with interests stretching to Northern Ireland. The organization has good relations with the Discovery Institute in the United States.[47]

Muslim Creationism

From 2006, evolution was also challenged by other movements. At Guy's Hospital site of King's College London, leaflets questioning Darwin's theory of evolution were distributed among students as part of the Islam Awareness Week.[48] This was organized by the university's Islamic Society. The leaflets were produced by the Al-Nasr Trust, a Slough-based charity set up in 1992 with the aim of improving the understanding of Islam. During the debate about Guy's Hospital, it emerged that Muslim students were failing their exams at other London campuses as they presented creationist ideas as scientific facts. In direct response, David Read, who served as vice-president of the Royal Society of London, asked the geneticist Steve Jones to deliver a public lecture on creationism and evolution.

This was not the only event putting Muslim creationism on the agenda in 2006. The lavishly produced antievolution book, *Atlas of Creation*, written by Adnan Oktar under the pen name Harun Yahya and sent to European schools and universities for free, also arrived in Britain.[49] Oktar, a Muslim creationist residing in Turkey, has published and sent out antievolutionist books before.[50] In 2008 Oktar Babuna and Ali Sadun, representing Harun Yahya, were giving a talk entitled "The Collapse of Evolution Theory" at University College London. The talk was arranged by the college's Islamic Society as part of Islam awareness week. It was scheduled to take place at the Darwin Lecture Theatre, but protests from senior academics led to a change of venues.[51] In April 2011 Harun Yahya's representatives Ahmet Oktar Babuna and Cihat Gündoğdu held conferences in Glasgow, Scotland. The events were part of a global Harun Yahya series of conferences and lectures.[52] After the British Humanist Association had launched a campaign in which red London buses were signed with the slogan "There is probably no God: Now stop worrying and enjoy your life," Oktar reacted by mocking the campaign. In 2011 posters advertising Harun Yahya's *Atlas of Creation* were put on London buses with the statement: "Modern Science Demonstrates that God Exists" in capital letters.

Not much is known about how widespread Muslim creationism is in Britain. One incident, however, demonstrates how serious the issue is taken by some when an imam of a mosque in East London, Dr. Usama Hasan, was subjected to death threats over his support for the theory of evolution.[53]

A Growing, Heterogeneous, and Fragmented Movement

In the revision of the national curriculum in 2006, focus shifted from teaching just the facts to teaching the processes of science and how science works. This created more room to introduce alternative views of life's origins and opened the possibility of talking about creationist issues in class. Studies have shown that younger groups of students in the age group from eleven to fourteen are more prone to accept creationism. Evolution is introduced quite late in formal education, around the age of fourteen.[54] Consequently, various experts have called for an introduction of evolution already in the primary curriculum and a ban of teaching creationism in science classes.[55]

Despite the critical voices and various attempts from creationist organizations, teaching of evolution has been fairly well protected in the British education system. The official stance of the Church of England is in support of evolution and goes against introducing creationist views in the classroom. Consequently, the religious opposition to evolution has come from an incoherent group of evolution skeptics, including young-earth creationists, Muslim creationists, and intelligent design advocates. These groups rarely connect and communicate with each other. Ironically, one of the unifying factors among evolution skeptics has been vehement atheists such as the Oxford biologist Richard Dawkins, who has publicly been outspoken against religion.[56] Creationists of any denomination claim that figures like Dawkins have a galvanizing effect on the variety of evolution skeptics and help them to recruit other religious believers.[57]

The effect of outspoken evolutionists on the creationism debate was also seen in a specific case involving the director of education of the Royal Society, Michael Reiss. In 2008 he had to resign after stating in a public lecture that science teachers should deal with creationism if a child raised the topic. In such a case, he argued, science teachers should explain why evolution was a scientific theory and creationism was not. The lecture was misrepresented in various media reports claiming that Reiss argued for the teaching of creationism in science classes. Michael Reiss was not only professor of science education in London with a doctorate in evolutionary biology but also an ordained minister in the Church of England. Some atheist members of the Royal Society wrote protest letters arguing that a priest could not represent a scientific institution such as the Royal Society, eventually resulting in Reiss's resignation.[58] This story alienated many religious people in Britain, and creationist organizations used this story as evidence to show that religious thinkers and scientists were excluded from science on grounds of their personal belief.

As in other parts of the world, British creationists have successfully embraced modern communication technologies. Particularly the use of the internet has helped to make contacts and keep in touch with people, to maintain international networks and collaboration, and to recruit new followers. All of the contemporary British creationist organizations maintain their own internet presentations, websites, or blogs, where they comment on current issues, announce lectures and conferences, and publish their materials. Many of them are also using social media, such as Facebook, YouTube, and Twitter, to engage in debates and reach out to new audiences.

Creationist attitudes and organizations in the United Kingdom are heterogeneous and fragmented. There seem to be no strong connections to other sociopolitical issues, unlike in other countries such as the United States or Turkey. However, every now and again diverse creationist branches have joined forces in order to convert more people to antievolutionist beliefs. For instance, in 2004 the American intelligent design theorist Phillip Johnson and the Australian young-earth creationist Andrew Snelling teamed up on a lecture tour against the theory of evolution. The tour comprised twenty-two seminars in eleven cities in the United Kingdom.[59] Disparate attempts to set a creationist agenda in Britain since the 1920s have been unsuccessful in terms of creating a powerful centralized creationist organization. Despite that fact, creationist views in both rural and urban populations have been on the rise in the early decades of the twenty-first century.

NOTES

1. The UK country profile data have been taken from the Central Intelligence Agency's Word Factbook, https://www.cia.gov/library/publications/the-world-factbook/geos/uk.html.

2. Jonathan Adams, *Global Research Report: United Kingdom* (Leeds: Evidence / Thomson Reuters, 2011).

3. Alvar Ellegård, *Darwin and the General Reader: The Reception of Darwin's Theory of Evolution in the British Periodical Press, 1859–1872* (1958; rpt., Chicago: University of Chicago Press, 1990).

4. James Randerson, "Four Out of 10 Say Science Classes Should Include Intelligent Design," *Guardian*, Jan. 26, 2006, 16.

5. Harriet Swan, "How Did We Get Here? Evolution Is on the Way Out—More than 30% of Students in the UK Say They Believe in Creationism and Intelligent Design," *Guardian*, Education Pages, Aug. 15, 2006, 1.

6. Jon D. Miller, Eugenie C. Scott, and Shinji Okamoto, "Public Acceptance of Evolution," *Science* 313 (Aug. 11, 2006): 765–766.

7. Martin Beckford, "Put Creationism and Evolution on Equal Footing, Say Teachers," *Daily Telegraph*, Nov. 7, 2008, 1.

8. Its website is www.theosthinktank.co.uk/.

9. Nick Spencer and Denis Alexander, *God and Evolution in Britain Today* (London: Theos, 2009).

10. However, the methodology of the survey was also criticized, e.g., Sylvia Baker, "The Theos/ComRes Survey into Public Perception of Darwinism in the UK: A Recipe for Confusion," *Public Understanding of Science* 21 (2012): 286–293. See also the response: Nick Spencer and Denis Alexander, "Response to Dr Baker," *Public Understanding of Science* 21 (2012): 294–296.

11. Caroline Lawes, *Faith and Darwin: Harmony, Conflict or Confusion?* (London: Theos, 2009).

12. Ian Sample, "Four Out of Five Britons Repudiate Creationism: Belief Map Shows Support for Darwin's Theories; God and Evolution Can Be Compatible, Says Thinktank," *Guardian*, Mar. 2, 2009, 10. An interactive version of the "Belief Map of the UK" is available at www.guardian.co.uk/world/interactive/2009/mar/02/belief-map-uk-creationism.

13. British Centre for Science Education, "Northern Ireland Is the Creationist Capital of Europe," *BCSE website* (2010), www.bcseweb.org.uk/index.php/Main/UpdateOnNorthernIreland.

14. Ian Sample, "Four Out of Five Britons Repudiate Creationism," *Guardian*, Mar. 2, 2009, 10.

15. E.g., Richard Burgess, "Bringing Back the Gospel: Reverse Mission among Nigerian Pentecostals in Britain," *Journal of Religion in Europe* 4 (2011): 429–449; Katrin Maier and Simon Coleman, "Who Will Tend the Vine? Pentecostalism, Parenting and the Role of the State in 'London-Lagos,'" *Journal of Religion in Europe* 4 (2011): 450–470.

16. British Council, "Darwin Survey Reveals Divided Britain in Attitudes towards Evolution," British Council Press release, June 30, 2009.

17. Andrew Village, Sylvia Baker, Leslie Francis, and Jeff Astley, *The Bible, Creation and You Survey, 2011: Report on the Initial Findings* (York: St John University), www.yorksj.ac.uk/pdf/The%20Bible%20Creation%20and%20You%20survey%202011.pdf.

18. E.g., Peter Fulljames, "Science, Creation and Christianity: A Further Look," in *Research in Religious Education*, ed. Leslie J. Francis, William K. Kay, and William S. Campbell (Leomister: Gracewing, 1996), 257–266. Leslie J. Francis and John E. Greer, "Attitudes towards Creationism and Evolutionary Theory: The Debate among Secondary Pupils Attending Catholic and Protestant Schools in Northern Ireland," *Public Understanding of Science* 8 (1999): 93–103; Peter Fulljames and Leslie Francis, "Creationism among Young People

in Kenya and Britain," in *The Cultures of Creationism: Anti-evolutionism in English-Speaking Countries*, ed. Simon Coleman and Leslie Carlin (Aldershot: Ashgate, 2004), 165–173.

19. Ronald L. Numbers, *The Creationists: From Scientific Creationism to Intelligent Design*, expanded ed. (Cambridge, MA: Harvard University Press, 2006), 161–166.

20. Ibid., 166–175.

21. Ibid., 359. For an analysis of their discursive techniques and resources, see Simon Locke, *Constructing the Beginning: Discourses of Creation Science* (Mahwah, NJ: Lawrence Erlbaum, 1998).

22. Stephen Moss, "Defying Darwin," *Guardian*, G2 Supplement, Feb. 17, 2009, 6.

23. Numbers, *The Creationists* (2006), 358.

24. British Centre for Science Education, "The Biblical Creation Society," BCSE website (2008), www.bcseweb.org.uk/index.php/Main/BiblicalCreationSociety.

25. Stephen Bates, "Origins of Man: Far Away from Lofty Pulpits, a Small Band Will Gather to Welcome Their Champion; Star of Creationist Circuit Flies in Hoping to Stir the Faithful in Small Towns of Britain," *Guardian*, Apr. 18, 2006, 7; Moss, "Defying Darwin," 6.

26. Numbers, *The Creationists* (2006), 406.

27. Ronald L. Numbers, "Creationists and Their Critics in Australia: An Autonomous Culture or 'the USA with Kangaroos'?," in *The Cultures of Creationism: Anti-evolutionism in English-Speaking Countries*, ed. Simon Coleman and Leslie Carlin (Aldershot: Ashgate, 2004), 109–123.

28. James Randerson, "Revealed: The Vegetarian Eden That Was Home to Adam, Eve and T Rex; Dawkins's Worst Nightmare Takes His Literalist Biblical Message on a Tour of the UK," *Guardian*, Apr. 5, 2008, 15.

29. Clare Dean, "CTC to Host Creationists," *Times Education Supplement*, Jan. 25, 2002, 2.

30. Tania Branigan, "Top School's Creationists Preach Value of Biblical Story over Evolution: State-Funded Secondary Teachers Do Not Accept Findings of Darwin," *Guardian*, Mar. 9, 2002, 3.

31. Stephen Layfield, "Extracts from a Lecture by Head of Science Steven Layfield at Emmanuel College on September 21 2000," *Guardian*, Mar. 9, 2002, 3.

32. Michael Kallenbach, "School Creationism Is Exaggerated, Says PM," *Daily Telegraph*, Mar. 14, 2002, 12.

33. See, e.g., Michael Gross, "US-style Creationism Spreads to Europe," *Current Biology* 12 (2002): 265–266; Joachim Allgaier and Richard Holliman, "The Emergence of the Controversy around the Theory of Evolution and Creationism in UK Newspaper Reports," *Curriculum Journal* 17 (2006): 263–279.

34. Joachim Allgaier, "Scientific Experts and the Controversy about Teaching Creation/Evolution in the UK Press," *Science & Education* 19 (2010): 797–819.

35. Numbers, *The Creationists* (2006), 173–174.

36. James David Williams, "Creationist Teaching in School Science: A UK Perspective," *Evolution: Education and Outreach* 1 (2008): 87–95.

37. Department for Education and Employment (DfEE), *The National Curriculum—Science.* (London: The Stationary Office, 1999).

38. Joachim Allgaier, "Networking Expertise: Discursive Coalitions and Collaborative Networks of Experts in a Public Creationism Controversy in the UK," *Public Understanding of Science* 21 (2012): 299–313.

39. Qualifications and Curriculum Authority (QCA), *Key Stage 4, Programme of Study: Science* (2006), http://curriculum.qcda.gov.uk/key-stages-3-and-4/subjects/key-stage-4/science/programme-of-study/index.aspx?tab=2.

40. BBC news, "Sir Peter Vardy Settles Creationist Libel Action," *BBC news*, Jan. 14, 2010, www.bbc.co.uk/news/uk-england-12176333.

41. See, for instance, Jeevan Vasagar, "Creationist Groups Win Michael Gove's Approval to Open Free Schools," *Guardian*, July 17, 2012, www.guardian.co.uk/education/2012/jul/17/creationist-groups-approval-free-schools.

42. David Marley, "Creationism Materials Attacked by DfES," *Times Educational Supplement*, Nov. 17, 2006, www.tes.co.uk/article.aspx?storycode=2311153; Williams, "Creationist Teaching in School Science: A UK Perspective," 89–90.

43. David Marley, "Creationism Tag Puts Academies at Risk, Says Vardy," *Times Educational Supplement*, Nov. 24, 2006, www.tes.co.uk/article.aspx?storycode=2313863.

44. Graeme Paton, "How Genesis Crept Back into the Classroom: Science Students May Be Learning the Christian Concept of Creation under the Guise of a Debate on Darwinism," *Daily Telegraph*, November 28, 2006, 13.

45. Katherine Harmon, "Evolution Abroad: Creationism Evolves in Science Classrooms around the Globe," *Scientific American*, Mar. 3, 2011, www.scientificamerican.com/article.cfm?id=evolution-education-abroad.

46. Its website is www.c4id.org.uk/.

47. Riazad Butt, "UK Centre for Intelligent Design Claims It Will Focus on Science, Not Religion," *Guardian*, Science blog, Oct. 1, 2010, www.guardian.co.uk/science/blog/2010/oct/01/centre-intelligent-design-science-religion.

48. Duncan Campbell, "Academics Fight Rise of Creationism at Universities: More Students Believe Darwin Got It Wrong; Royal Society Challenges 'Insidious Problem,'" *Guardian*, Feb. 21, 2006, 11.

49. John Collins, "Creative Science in Our Universities," *Guardian*, Dec. 20, 2006, 33; Peter C. Kjærgaard, "Western Front," *New Humanist* 123:3 (2008): 39–41.

50. Numbers, *The Creationists* (2006), 421–427 (see also Riexinger, chapter 10 in this volume).

51. Ian Sample, "UCL Acts after Creationist Coup," *Guardian*, Feb. 23, 2008, 20.

52. "Harun Yahya Team in Glasgow," *Scotland's Chronicle* 27 (2011): 1–4.

53. Salman Hameed, "Face to Faith: The Abuse of an Imam for His Pro-evolution Views Plays to Anti-Muslim Prejudice," *Guardian*, Apr. 9, 2011, 37.

54. Williams, "Creationist Teaching in School Science: A UK Perspective."

55. Nick Collins, "Attenborough: Ban Creationism in Science Class," *Daily Telegraph*, Sept. 19, 2011, 6.

56. Robin Pharoah, Tamara Hale, and Becky Rowe, *Doubting Darwin: Creationism and Evolution Scepticism in Britain Today* (London: Theos, 2009).

57. Madeleine Bunting, "Why the Intelligent Design Lobby Thanks God for Richard Dawkins," *Guardian*, Mar. 27, 2006, 27.

58. Daniel Cressey, "Creationism Stir Fries Reiss," *Nature News*, Sept. 17, 2008, www.nature.com/news/2008/080917/full/news.2008.1116.html.

59. See their website www.darwinreconsidered.org/tour_dates_venues.html.

CHAPTER FOUR

The Low Countries

STEFAAN BLANCKE, ABRAHAM C. FLIPSE, AND JOHAN BRAECKMAN

In 2009 the world celebrated the Darwin year. In the Low Countries too, the two-hundredth anniversary of Charles Darwin's birthday and the one-hundred-fiftieth anniversary of the publication of *On the Origin of Species* were commemorated by countless events, exhibits, and publications. In Belgium, these celebrations did not inspire a public debate, let alone religious protest. In the Netherlands, however, creationists organized a leaflet campaign by which they intended to inform each and every household about their creationist alternative. This response did not come out of the blue. For decades, evolution has evoked heated debates both among Dutch Christians and in the public domain. In this chapter, we highlight the most relevant historical developments relating to creationism in both countries and cautiously provide an explanation for the remarkable difference in religious responses in the two small adjacent countries.

Religious Background

Today, both the Netherlands and Belgium are secularized countries, but they have distinct religious traditions. In Belgium, Roman Catholicism has been a dominant cultural factor since the establishment of the country in 1830. Although the influence of the church has waned considerably, Catholicism has left its cultural mark, particularly in Flanders, the Dutch-speaking part of the country. Nearly 60 percent of the 10.4 million Belgians regard themselves as Catholics, but only 5 to 10 percent of the population attends church regularly. There are small Muslim and Protestant minorities of about 4 percent each, and about 40 percent of the population does not believe in God.[1]

In contrast, since the time of the Reformation, the Netherlands has always been a religiously divided country. The majority of the population was Protestant (Calvinist), but there was a large Roman Catholic minority, concentrated in the southern provinces, and several other smaller minorities. The privileged position of the Reformed Church in the time of the Dutch Republic explains why today the Netherlands is still viewed as a Protestant country. In the course of the nineteenth century Dutch Protestantism split into several denominations.

Around 1900, of the 5 million inhabitants, 49 percent belonged to the Nederlandse Hervormde Kerk (Dutch Reformed Church), 35 percent to the Roman Catholic Church, and 8 percent to the smaller Reformed churches. Today 17 percent of the Dutch population (of about 16.5 million inhabitants) is Protestant, and 27 percent is Roman Catholic. In addition, 6 percent is Muslim. Almost half the population is not affiliated with any traditional religion.[2] It was especially among the members of the smaller Reformed denominations and the orthodox wing of the Dutch Reformed Church—most of whom did not belong to the social, ecclesiastic, or academic elite—that creationism found fertile ground. Although, compared to the United States, the creationist movement remained numerically small, creationism has become a visible and sometimes prominent phenomenon in the twentieth-century Dutch religious landscape.

The Dutch Calvinists' Struggle with Evolution

In the nineteenth century, leading theologians at the universities in the Netherlands tried to adapt Christian doctrines to modern science and to the historical-critical reading of the Bible. These "modernist" theologians managed to reconcile evolution with their faith. However, a considerable number of Dutch churchgoers—both in the Dutch Reformed Church and in various separatist churches—were less favorable to modern science and culture. The theologian, journalist, and statesman Abraham Kuyper became the charismatic leader of this marginalized group of (orthodox) Calvinists, who had no voice either within the churches or in Dutch society, which was dominated by a liberal elite. A process of emancipation started in which the Calvinists established their own private schools, a political party, newspapers, and many other organizations and institutions, including the Calvinist Free University in Amsterdam, founded in 1880. This resulted in a powerful orthodox subculture. Most of Kuyper's supporters were members of the seceded Gereformeerde Kerken in Nederland (Reformed Churches in the Netherlands). Under Kuyper's leadership, orthodox Calvinism experienced a revival, resulting in a worldview often denoted as "neo-Calvinism."[3]

The Roman Catholics, the socialists, and other groups followed the example of the Calvinists, which produced a pattern of social organization called *verzuiling* (pillarization): a "vertical division" of society into various religious and ideological groups. Between 1920 and 1960 these groups existed in relative isolation, but they also adapted to the general culture and tried to meet the challenges posed by modern society.[4]

The question of creation versus evolution was discussed vigorously within the Kuyperian, neo-Calvinist tradition. In contrast to pietistic, world-shunning Calvinist groups, the neo-Calvinists wanted to be both orthodox and modern, and they could not ignore the issue. However, they refused to adapt their faith as drastically to modern science as the liberal Protestants, which created a perpetual tension. The Dutch Roman Catholics also rejected evolution for a long time, but this was less determinative of their identity.[5]

The turn-of-the-century neo-Calvinists considered "the dogma of evolution" to be irreconcilable with the Christian belief in a providential God, because of its naturalistic, mechanistic, and ateleological character. Their criticism focused on the philosophical and social consequences that had been derived from evolutionary theory by Ernst Haeckel, Herbert Spencer, and others. However, they also addressed the discrepancy between the biblical creation story and the evolutionary account. Like most orthodox Protestants in the Anglo-Saxon world, Dutch neo-Calvinists were inclined to harmonize the findings of modern geology with the creation account in a "concordistic" way, for example, by using a day-age interpretation of Genesis 1.[6]

A First Wave of Creationism

The next generation of neo-Calvinists evaluated the relevance of late nineteenth-century theology and views of science in the light of new social and cultural developments. Some of the Calvinist scientists claimed that biological evolution was acceptable as a scientific theory, as long as it was not part of a mechanistic worldview.[7]

Most Calvinist theologians, however, followed a different path. They shifted the debate about the relation of faith to evolution and geology to the subject of the authority of scripture versus the authority of science. In the 1920s a controversy arose in the Reformed Churches about whether the story of the Fall (Genesis 2–3) should be taken literally or not. The Reverend J. G. Geelkerken had doubted the literal-historical reading of this story. In the end, the Synod of the Reformed Churches of 1926 decided to suspend Geelkerken. Several Dutch newspapers associated the Geelkerken case, which was often reduced to the question "Did the serpent really speak?" with the Scopes or monkey trial that had taken place in the United States the year before. The Geelkerken case, however, did not concern the teaching of evolution in public schools but revolved around a discussion within the churches about the interpretation of scripture. After all, the Dutch Calvinists had their own schools for secondary education, and evolutionary theory was simply neglected in their biology textbooks until the 1960s.[8]

The consequences of the theologians' attitude toward the sciences soon be-

Caricature comparing the Geelkerken case in the Netherlands ("Did the serpent really speak?") to the "monkey trial" in the United States in which a serpent and an ape are portrayed as "the interested parties" in "modern theological issues." *De Groene Amsterdammer,* Sept. 19, 1925. Atlas Van Stolk, Rotterdam.

came manifest. In 1930 the Free University professor of dogmatics Valentijn Hepp visited Princeton Theological Seminary to deliver the Stone Lectures on the topic of "Calvinism and the Philosophy of Nature." Hepp claimed that he did not accept the results of mainstream geology, and he approvingly referred to the work of George McCready Price. Nowadays the Canadian amateur geologist and Seventh-day Adventist Price is regarded as the founding father of twentieth-century young-earth creationism, but at that time support for Price's "flood geology" was rather limited, even in the United States. It is therefore remarkable that Hepp recommended Price's work, and he was not the only Dutch Calvinist who did so. In 1932 the Free University professor of Old Testament G. Ch. Aalders published an influential 552-page commentary on the stories of creation and the Fall in the book of Genesis. Criticizing modern geology and evolutionary theory, Aalders adopted several arguments from Price, and he stressed that catastrophes—especially the Deluge—provided a better explanation for the fossil record. By these and other publications, Reformed churchgoers became familiar with creationist arguments. And, as a result of pillarization, they could easily avoid a confrontation with mainstream science. Although most Calvinist scientists were critical of the views of Hepp and Aalders, a general discussion of the matter started only after World War II.[9]

Wider Acceptance of Evolutionary Theory

After 1945, Calvinist scientists gradually became more influential in the neo-Calvinist subculture, and a new generation of theologians was willing to engage in a renewed discussion about the theory of evolution. This initiated a debate about creation and evolution among a wider public. Especially Jan Lever, professor of zoology at the Free University, and his colleague Jan R. van de Fliert, professor of geology, argued that one could accept the biological theory of evolution and at the same time believe in a providential God. Their ideas caused quite a stir among many nonacademic Calvinists, but in the course of the 1960s their views gradually found acceptance. In the same period, several leading Calvinist theologians adopted increasingly liberal views and shifted their focus to other theological issues. In 1967 the verdict of the 1926 Synod concerning Genesis 2–3 was retracted, and around 1970 the debate seemed to have died out. The Free University shed its explicitly Calvinist character and gradually became less distinguishable from other Dutch universities. Analogously, the Reformed Churches transformed from a segregative, orthodox church into an open, pluralistic denomination. Seemingly, the Dutch Calvinists had finally come to accept the Darwinian theory of evolution.[10]

A Second Wave of Creationism

However, several small Reformed denominations remained orthodox and denounced the Reformed Churches and Calvinist organizations for capitulating to modernism and evolutionism. In these orthodox circles, the rise of creationism in the United States was noticed early on. On May 16 and 17, 1967, "concerned brothers" of the so-called Vrijgemaakt-Gereformeerde Kerken (Liberated Reformed Churches) organized a conference entitled "Creation-Evolution" to warn of the destructive impact of evolutionism in theology, ethics, and society and, more specifically, of the ideas of Lever and Van de Fliert. The lecturers invoked arguments against evolution that were explicitly drawn from recent publications by Morris and Whitcomb. A young geology student and minister's son, Nicolaas A. Rupke, gave a lecture entitled "Redating the Past." Rupke had learned about flood geology in the early 1960s and had contacted the aging Price in the United States. He joined the Creation Research Society, conducted creationist research work, and published several papers in the *CRS Quarterly*. In the autumn of 1968, Rupke left the Netherlands for the United States, where he later abandoned his creationist beliefs.[11]

In the Netherlands, however, the story of creationism continued. In 1969 a translation of Morris's *The Twilight of Evolution* was published, followed one year later by A. M. Rehwinkel's *The Flood in the Light of the Bible, Geology and Archaeology*. From 1974 onward, original Dutch creationist publications appeared, which often relied strongly on the work of American creationists. One of the most prolific Dutch authors was the biologist Willem J. Ouweneel, affiliated with the Plymouth Brethren, a flamboyant speaker and a rigorous polemicist. His books were widely read by both evangelicals and Calvinists.[12]

Although these books became quite popular, they were also criticized by fellow believers. Some orthodox Reformed theologians distanced themselves, on the one hand, from the theistic-evolutionist views of people like Lever and Van de Fliert but, on the other hand, also from the creationism of Morris, Whitcomb, Rehwinkel, and Ouweneel. One of the strict Liberated-Reformed theologians warned against the danger of "unwittingly drifting *from* Calvinism *into* fundamentalism." In his view, young-earth creationism was not compatible with Reformed theology. For many church members this middle course was all too subtle. One of the leading Calvinist creationists, J. A. (Koos) van Delden, a mathematician by training, wondered why the theologians did not wholeheartedly accept flood geology. The creationists, Van Delden argued, continued the work that had been initiated by the Reformed theologians of the early twentieth century. It was, in any case, better to follow Morris than Lever or Van de Fliert.[13]

Creationism Institutionalized

In the 1970s conservative Christians from several Reformed and evangelical churches joined forces in newly founded organizations. Particularly relevant were the activities of the evangelical broadcasting company, EO (Evangelische Omroep), founded in 1967, and the Stichting tot Bevordering van Bijbelgetrouwe Wetenschap (Foundation for the Advancement of Studies Faithful to the Bible), established in 1974. The latter founded the Evangelische Hogeschool (Evangelical College) and published the creationist journal *Bijbel en wetenschap* (Bible and Science). Van Delden and Ouweneel were among the founders of these organizations. In subsequent years a new "evangelical-Reformed" network materialized around the EO. Although the traditional "pillars" gradually crumbled and society became increasingly secularized, the neo-Calvinist tradition was in a way perpetuated by this new movement. Moreover, the opportunities that the pillarized structure of society still provided were exploited to the full. Prominent spokesmen of the evangelical-Reformed subculture fiercely criticized the developments in the Reformed Churches and in traditional Calvinist organizations in the 1960s and 1970s. To their discontent, the Calvinist leaders did not oppose the "revolt" of the sixties. As conservatives regarded the intrusion of evolutionism in Calvinist organizations as a mark of secularization, they responded by adhering to the strictest form of antievolutionism possible. The EO and affiliated organizations focused explicitly on the dissemination of young-earth creationism, and their statutory principles were more outspoken than orthodox Calvinist organizations had been before.[14]

In the 1970s, the Dutch creationists gained much attention, particularly with an EO television series entitled *Adam of Aap?* (Adam or Ape?), presented by Van Delden, and with a public debate between creationists and evolutionists, presided over by Ouweneel, that was attended by more than a thousand people. The English-born pharmacologist Arthur E. Wilder-Smith, who actively spread the creationist message in many European countries in this period, frequently appeared as an expert in EO television programs. The creationists did not succeed in converting the Dutch population to creationism, but they were extremely successful in making strict creationism generally accepted by members of several orthodox Reformed churches and the conservative wing of the Dutch Reformed Church. Eventually, the EO attracted more than half a million members. In a survey article on "Creationism in the Netherlands" (1978) in *Acts and Facts* of the Institute for Creation Research, Ouweneel proudly proclaimed: "In the last four years or so, creationism has developed so rapidly in the Netherlands that without doubt this country is assuming the lead in creationism at present in Europe." And in 1980 Van Delden wrote in *Bible and Science* that "the struggle with evolutionism lies behind us."[15]

Aware of their relatively strong institutional basis, the Dutch creationists turned their minds to Europe. The first European Creationist Congress was organized in August 1984, in Heverlee, Belgium, hosting creationists from various European countries, followed by two more conferences in 1986 and 1988. In subsequent years several congresses were organized in other countries. However, the Dutch creationists did not conquer Europe. On the contrary, even in the Netherlands the discussion died out.[16]

Intelligent Design in the Netherlands

In the meantime, the seeds were planted for the third creationist wave during the 1990s. In 1994 the physicist Arie van den Beukel published a book with the title *Met andere ogen* (With different eyes), in which he relied heavily on Michael Denton's *Evolution: A Theory in Crisis*, which is now considered a seminal work of the intelligent design movement. He argued that there was no hard evidence for evolution by natural selection and that therefore accepting evolutionary theory was nothing but an act of faith. Three years later, he wrote the introduction to the Dutch translation of Michael J. Behe's *Darwin's Black Box*, in which "irreducible complexity" in biochemical systems is presented as evidence for intelligent design. His writings attracted the attention of prominent and respected members within the evangelical-Reformed community who were dissatisfied with the young-earth creationist views common among their fellow believers. They welcomed intelligent design as an acceptable and scientifically justified means of reconciling their orthodox religious views with belief in an old earth.[17]

One of the first to endorse intelligent design as a valid alternative to young-earth creationism was Ouweneel. However, of greater importance were the conversions of Cees Dekker, a physicist who specialized in nanotechnology, and Andries Knevel, a former president of the EO. Dekker had never been convinced by young-earth creationism, but, being an evangelical, he was nevertheless influenced by it. He first publicly expressed his sympathy for intelligent design in his inaugural address as a professor at the Delft University of Technology in 2000. Referring to the works of Van den Beukel, Denton, Behe, and Phillip Johnson—cofounder of the Discovery Institute's Center for the Renewal of Science and Culture—he claimed that "there is remarkably little scientific support for such an important theory as Darwin's evolutionary mechanism" and that "evolution, defined as the explanation for the origin of life and the origin of biodiversity, is a dogma that, after careful examination, barely has any support." The same year, the mathematician Ronald Meester, in his inaugural address, stated that "on a popular level Darwin is still very much alive, but on an academic level, there are many, many doubts." He too referred to the works of Van den Beukel, Denton, and Behe. Dekker and Meester learned about each other's interest in intelligent design, and together with a group of fellow Christians, including the philosopher René van Woudenberg, they held monthly meetings to discuss topics relating to science and religion. Inspired by the discussions at these meetings, they compiled an edited volume on intelligent design that was published in 2005, *Schitterend ongeluk of sporen van ontwerp?* (Glorious accident or traces of design?). The volume, to which Dekker contributed three articles, Meester and Van Woudenberg two each, and Van den Beukel one, was strongly anti-Darwinian in content and tone. They referred repeatedly to the books of Behe, Johnson, and William Dembski, another leading figure in the American intelligent design movement.[18]

By then, intelligent design had fully entered the public arena. In March 2005, Maria van der Hoeven, a Catholic member of the Christian-Democratic party CDA and at the time minister of education, culture, and science, wrote on her weblog that she had had an interesting conversation with Dekker. She was particularly impressed by the way he reconciled science with religion and admitted that she felt unable to believe in "chance." Two months later, she stated that "it should be understood that evolutionary theory is incomplete and that we are still discovering new things" and that she hoped to start a dialogue between scientists and intelligent design proponents. Both scientists and politicians, however, heavily criticized her proposition, which led her to withdraw her plans. However, at the launch of the book edited by Dekker and his colleagues, on June 8, 2005, she was still hopeful that she would be able to organize a public debate. In her speech at this event, she expressed her wish to foster greater mutual respect between people with different philosophical backgrounds. The incident attracted international attention. A *Science* article asked ironically whether the Netherlands was becoming the Kansas of Europe.[19]

In October 2005 Dekker was invited to deliver a presentation at the "Darwin and Design" conference in Prague. According to Dembski, this conference "clearly demonstrated that the intelligent design controversy is not just an American phenomenon; it opened many doors to colleagues in Europe with whom the intelligent design community will be working extensively in the years to come." However, early in 2006, Dekker started to question the scientific merits of intelligent design openly. He was particularly disappointed about the fact that intelligent design did not result in any practical applications. He also claimed that he had been inappropriately associated with the movement. Soon, Dekker described himself as a theistic evolutionist. In August 2006 he wrote the foreword to the Dutch translation of Francis Collins's *The Language of God*; he stressed that he basically agreed with Collins's views that creation and evolution are reconcilable. In 2009 a Dutch book that advocated theistic evolution appeared with a laudatory foreword by Cees Dekker.[20]

A Dutch Wedge

Dekker was followed in his tracks, first to intelligent design and later to theistic evolution, by Andries Knevel. Knevel started working for the EO in 1978, and by 1993 he had become one of its three codirectors. During the 1990s, he had been drawn to intelligent design by the works of Van den Beukel, the reading of which he described as an awakening that made him conscious of other positions in regards to creation. Michael Behe's *Darwin's Black Box* caused the greatest shock. Knevel suddenly realized that he did not have to be a young-earth creationist to be a good Christian. He visited several American Christian scientists, including William Dembski and Walter Bradley, fellows of the Center for Science and Culture. Soon after, he abandoned his young-earth beliefs and accepted intelligent design.[21]

Knevel spoke favorably of intelligent design at the book launch of Dekker's *Schitterend Ongeluk*. Three days later he claimed that he regarded intelligent design as an acceptable means to reconcile science with a belief in the book of Genesis. However, other creationists did not feel the need to embrace intelligent design. Van Delden, who had stuck to his original young-earth beliefs, thought it foolhardy of Christians to regard intelligent design adherents as allies to their cause. Soon Knevel learned that intelligent design did not promote the reconciliation he had hoped for. Instead of reconciling Christian faith with science and "affectively attacking the Darwinian bastion," intelligent design engendered serious fractures within the Dutch evanglical-Reformed community.[22]

However, from the creationists' perspective, worse was to come. Dekker had started to question the scientific merits of intelligent design and called himself a theistic evolutionist. Knevel respected Dekker deeply; gradually he followed Dekker's shift, and by 2009 he had become a theistic evolutionist himself. On February 3, 2009, Knevel read out a prepared statement in an EO television program in which he announced that he was no longer a young-earth creationist or an adherent of intelligent design and that he regretted having misled his viewers. Many EO members, however, felt insulted by Knevel's confession because they thought he had presented his views as the result of improved judgment. Furthermore, they felt that "their" EO had wandered off the straight path. Bert Dorenbos, director of the EO between 1974 and 1987, described Knevel's statement as "an insult to God" and "an act of aggression." In response, Knevel apologized for the arrogant way in which he had presented his convictions and emphasized that his views were not those of the EO. The damage, however, had been done. In an open letter, Dorenbos detested the path the EO had taken under the guidance of people like Dekker, Knevel, and Ouweneel. Indeed, the EO was entirely divided on the issue of creation. In the United States, Phillip Johnson and his co-workers at the Center for Science and Culture had intended intelligent design to function as a wedge splitting the log of Darwinian naturalism and secular culture. Ironically, in the Netherlands intelligent design had worked as a wedge within the evangelical-Reformed community, by functioning as a halfway house and facilitating the transition of some of its prominent members to a theistic evolutionist position.[23]

The Darwin Year

The timing of Knevel's confession was not coincidental. The debate between creation and evolution had been put back on the agenda by the Darwin year commemorations. On January 6, 2009, Knevel had hosted a television show on the EO that featured a theistic evolutionist, Dekker, an atheistic Darwinian philosopher, and a young-earth creationist. By then, the young-earth creationists with an evangelical-Reformed but increasingly also with a pietistic-Calvinist or Pentecostal faith had regrouped and had initiated projects to counterbalance the impact of these festivities and to inform the public of an alternative to evolution. One project in particular garnered a lot of media attention. In November 2008, Christian newspapers reported that Kees van Helden, the president of the creationist group Bijbel en Onderwijs (Bible and Education) was rallying financial support to print an eight-page pamphlet with the title *Evolutie of Schepping. Wat geloof jij?* (Evolution or creation: What do you believe?). The leaflets were to be distributed in the mail to every household in the Netherlands around February 12, 2009, the two-hundredth anniversary of Darwin's birth. The project was backed by a committee of recommendation consisting mainly of reverends and pastors from various Protestant denominations and thirty creationist organizations from the Netherlands and Belgium. One of the supporters of the project was Johan Huibers, who in 2007 had finished building a replica of Noah's ark, which he used as a traveling exhibition to deliver the word of God. In 2012, having sold the first ark, he finished the construction of a much larger ark that had the dimensions mentioned in the Bible.[24]

However, by the time the leaflet was actually distributed, the move into the public sphere was already in part transforming into an internal debate. After Knevel had publicly disavowed creationism and intelligent design, Van Helden urged Knevel in a news report to restore his faith in biblical creation. Later that year, various creationist books were published to argue not only against the alleged shortcomings and immoral consequences of evolutionary theory but also

In 2009, at the start of the Darwin year, the Dutch creationists distributed this leaflet to almost every household in the Netherlands.

against the heresy of Knevel, Ouweneel, Dekker, and other "liberal" interpreters of the Bible. But the antievolutionary wave did not decay entirely. Van Helden started a civil initiative to collect signatures in support for "equal time" in education. The translation of a book by the Swiss creationist group ProGenesis, with the title *95 stellingen tegen evolutie* (95 theses against evolution), was promoted by posting the ninety-five theses by the entrance of the Free University—imitating Luther who allegedly posted his theses in 1517—because they blamed the formerly Calvinist university for having introduced evolution to the churches in the 1960s. In 2010 a group of creationists under the auspices of the young-earth organization Oude Wereld (Old World) founded *Weet magazine* (Know magazine), which was designed to resemble an ordinary popular scientific magazine. In 2012 it had about eight thousand subscriptions. The same organization also published several creationist books, including a translation of the German creationist textbook by Siegfried Scherer and Reinhard Junker, *Evolution. Ein kritisches Lehrbuch.* Although other publications and initiatives had attempted to stimulate the dialogue between science and religion, the Dutch Darwin year ended with a creationist movement that was stronger and more visible than it had been for decades.[25]

The Belgian Catholics and Evolution

In contrast with the creationist responses in the Netherlands, the Darwin year festivities did not inspire any negative religious reactions in Belgium. Instead, Catholic representatives and opinion makers considered the Darwin year an ideal opportunity to resume a rational dialogue between science and religion. Most argued that creation and evolution complement one another and that there exists no competition between science and religion. Cardinal Godfried Danneels, at the time the highest in rank in the Belgian hierarchy, described the relation between the two domains as the tracks of a railway that run in parallel and touch only in infinity. However, the editor of an influential Catholic weekly complained that "Darwinism" had become much more than a scientific theory and had turned into an ideology. Radical Darwinists, he claimed without providing any names, derive the most horrible moral directives from "natural selection" or "the survival of the fittest" on how to treat the ill and the weak or how to improve the human species. Although he distanced himself from creationism, he nevertheless resorted to arguments that are common in creationist discourse.[26]

The predominantly positive attitude toward evolution is in part explainable by the way in which, historically, evolution had been appropriated by Belgian Catholics. After Darwin's theory had been introduced in Belgium, there was a brief period when it was "vehemently belittled." However, already in 1875 the Société scientifique de Bruxelles was founded, followed by the publication of two journals, the *Annales de la société scientifique de Bruxelles* in 1875 and the widely read *Revue des questions scientifiques* in 1877. Jesuits were deeply involved in these initiatives, and they soon took the lead in defending Catholic evolutionism in Belgium. Around the turn of the century, this pro-evolutionary attitude took root at the Catholic University of Louvain, where a group of intellectuals supported the compatibility of faith and evolution, in both their lectures and their publications. Some of them even defended evolutionism through popular addresses, thus introducing evolution to the general public. In the 1930s most Belgian Catholic intellectuals accepted evolution, and later this pro-evolutionary stance became entrenched by the increasing popularity of the theistic evolutionary views of the Jesuit Teilhard de Chardin. By 1960 evolution had become incorporated in the curriculum of Catholic secondary comprehensive schools. The introduction in 1963 of a biology textbook that was used in the sixth grade explains at the outset that, "while the natural sciences are based upon experimental observations, theology depends on the fact of revelation. One should keep in mind that, if the natural sciences and theology each remain on their respective domain, there can be no contradiction." Other biology textbooks from that period contain similar passages.[27]

Creationism in Belgium

Strict creationism is almost exclusively found within non-Catholic denominations. Probably most common is Islamic creationism. In 2007 copies of the *Atlas of Creation* by the Turkish creationist Harun Yahya were delivered free of charge to schools, universities, and the editorial offices of several newspapers. Harun Yahya seemed to have gained at least some support in the Muslim communities. In February 2008 a talk show debate on national television featured Nordine Taouil, an imam from Antwerp. He stated that he believed that Allah had specially created the human species and that Adam and Eve really existed. He repeated the old creationist chestnut that evolutionary theory is but a hypothesis and maintained that scientists from the United States, Europe, and the Arabic world had convincing evidence that proved the theory wrong. Taouil explicitly referred to Harun Yahya. Islamic creationism, however, is not exclusively attributable to the influence of the Turkish creationist. In November 2009 it was reported that creationism was being taught at Lucerna College, a state-funded free school founded by Turkish immigrants. Witness reports of teachers and the materials used in religious education revealed that in religious classes, evolutionary theory was described as "an illogical belief that is not based on any scientific evidence" and that tests required pupils to render counterexamples to natural selection.[28]

Studies confirm that Belgian Muslim pupils have great difficulty accepting evolution. A study in Brussels, the Belgian capital, showed that one in five students rejected human evolution. Of this 20 percent, most were Muslims. A small study in Antwerp demonstrated that almost all young Muslim believed that Allah has created humans, whereas only one in ten Catholic and six in ten Jewish students endorsed creationist beliefs.[29] Mostly, creationism is taught locally at small, religiously inspired, but state-funded schools. In recent years, the media have reported the teaching of creationism in orthodox Jewish and Protestant schools. Evolutionary theory is also one-sidedly criticized and presented with a nonscientific alternative in some anthroposophical secondary schools.[30]

Active antievolutionism constitutes only a marginal phenomenon in Belgium. The best-known creationist organization is Creabel. It was founded in 1991 by Jos Philippaerts, who holds a PhD in chemistry, and some fellow believers with the assistance of David Rosevear, a British creationist of the Creation Science Movement. It soon had three hundred members. They published a creationist magazine and provided lectures in Baptist, Pentecostal, and evangelical churches. Catholic parishes were approached through Chris Hollevoet, a Catholic geologist. By the end of 2008, Creabel appeared on the list of organizations that supported the Dutch creationist leaflet campaign, and Philippaerts continues to lecture on creationism today, usually for friendly churches and organizations. On rare occasions he appears in the national media, where he is usually presented as a curiosity. In October 2011, Creabel celebrated its twentieth anniversary with a two-day symposium.[31]

Compared to the Netherlands, in Belgium creationism is no more than a marginal phenomenon. It is tempting to ascribe the difference between the two countries solely to their different religious backgrounds. Belgium is traditionally Roman Catholic, and, historically Catholics have taken less issue with evolution. The Netherlands can be regarded as a Protestant country, and creationism is predominantly a Protestant matter. This explanation certainly has merit, but it needs qualification. Although creationism is mostly associated with Protestantism, there are many versions of Protestantism that do not favor it. In Scandinavia, for instance, creationism has not gained a strong foothold in the national Lutheran churches or society. In the Netherlands itself, many Protestants cannot be regarded as creationists. Clearly, Protestantism does not directly imply creationism. The relation between Catholicism and evolution is not straightforward either. Historical research has shown that Catholics too had trouble accepting evolution, a process that has often been affected by local or national factors. Today, many Catholics in the United States (35 percent of Catholics in one survey) do not agree that "evolution is the best explanation for the origins of human life on Earth," and in Catholic Poland a creationist movement exists as well. Also, in recent years a conservative Catholic newspaper in the Netherlands has begun to tackle Darwinian evolution.[32]

This is not to deny that Protestant creationism is far more common than Catholic creationism, but an explanation in terms of Protestantism versus Catholicism needs to be supplemented by aspects of the concrete situations. It is possible that Protestantism offers only a fertile soil for creationism in combination with a particular relation between church and state. Only in a religious free market, as it exists in the United States—and, to a certain extent also in the Netherlands—will Protestantism give rise to creationism as a substantial social-religious phenomenon. In a context of religious and social pluralism, groups of believers may attempt to build their identity around strict antievolutionism. In the Netherlands this was reinforced by the pillarized structure of society in which religious and ideological organizations were accommodated by the state. However, as a consequence of this structure of society, the debates among the orthodox themselves were often more heated than those with the outside world, and the three "waves of creationism" that arose during the twentieth century can all be interpreted as reactions to modernizing tendencies within the orthodox Protestant subculture.

NOTES

1. http://theo.kuleuven.be/icrid/icrid_religies/icrid_religies_index.html#belgiekatholiek, accessed Sept. 25, 2012; Phil Zuckerman, "Atheism. Contemporary Numbers and Patterns," in *The Cambridge Companion to Atheism*, ed. Michael Martin (New York: Cambridge University Press, 2007), 51.

2. www.volkstellingen.nl, accessed Sept. 25, 2012; W. B. H. J. van de Donk et al., *Geloven in het publieke domein. Verkenningen van een dubbele transformatie* (Amsterdam: Amsterdam University Press, 2006).

3. Bart Leeuwenburgh and Janneke van der Heide, "Darwin on Dutch Soil: The Early Reception of His Ideas in the Netherlands," in *The Reception of Charles Darwin in Europe*, vol. 1, ed. Eve-Marie Engels and Thomas F. Glick (London: Continuum, 2008), 175–186; Arie L. Molendijk, "Neo-Calvinist Culture Protestantism: Abraham Kuyper's *Stone Lectures*," *Church History and Religious Culture* 88 (2008): 235–250; James D. Bratt, *Abraham Kuyper: Modern Calvinist, Christian Democrat* (Grand Rapids, MI: Eerdmans, 2013); Peter S. Heslam, *Creating a Christian Worldview: Abraham Kuyper's Lectures on Calvinism* (Grand Rapids, MI: Eerdmans, 1998). In this chapter "Dutch Calvinism" refers to the orthodox wing of the Reformed tradition in the Netherlands.

4. Peter van Rooden, "Long-Term Religious Developments in the Netherlands, 1750–2000," in *The Decline of Christendom in Western Europe, 1750–2000*, ed. Hugh McLeod and W. Ustorf (Cambridge: Cambridge University Press, 2002), 113–129, esp. 117–118.

5. Ab Flipse, "'De schepping zou er even wonderbaar om zijn.' Geschiedenis van het evolutiedebat in gereformeerde en rooms-katholieke kring," in *Botsen over het begin. Bavinck lezingen 2009*, ed. Koert van Bekkum and George Harinck (Barneveld: Nederlands Dagblad, 2010), 9–22; Ab Flipse, *Christelijke wetenschap. Nederlandse rooms-katholieken en gereformeerden over de natuurwetenschap, 1880–1940* (Hilversum: Verloren, 2014).

6. Abraham Kuyper, *Evolutie. Rede bij de overdracht van het rectoraat aan de Vrije Universiteit op 20 October 1899 gehouden* (Amsterdam: Höveker & Wormser, 1899); translation in *Abraham Kuyper: A Centennial Reader*, ed. James D. Bratt (Grand Rapids, MI: Eerdmans, 1998), 403–440; George Harinck, "Twin Sisters with a Changing Character: How Neo-Calvinists Dealt with the Modern Discrepancy between the Bible and Modern Science," in *Nature and Scripture in the Abrahamic Religions: 1700–Present*, vol. 2, ed. Jitse M. van der Meer and Scott Mandelbrote (Leiden: Brill, 2008), 317–370, on 325–328, 337; Abraham C. Flipse, "The Origins of Creationism in the Netherlands: The Evolution Debate among Twentieth-Century Dutch Neo-Calvinists," *Church History: Studies in Christianity and Culture* 81:1 (March 2012): 104–147, on 108–116.

7. Flipse, "Origins of Creationism," 119–120.

8. Maarten J. Aalders, *Heeft de slang gesproken? Het strijdbare leven van dr. J. G. Geelkerken (1879–1960)* (Amsterdam: Bert Bakker, 2013), 279–313. Newspapers on Geelkerken and Scopes: "De 'monkey trial' in Nederland," *Het Vaderland*, Sept. 8, 1925; "Fundamentalisme in Amerika en Nederland I. Meester Scopes en ds. Geelkerken," *Nieuwe Rotterdamsche Courant*, Oct. 20, 1925; "Fundamentalisme in Amerika en Nederland II. 'Bryan is not dead,'" *Nieuwe Rotterdamsche Courant*, Oct. 21, 1925. On biology textbooks: Maartje Brattinga, "'Zouden onze voorouders er zoo uitgezien hebben?' Hoe de evolutietheorie ontvangen werd in Nederlandse familiebladen en schoolboekjes, 1867–1974" (master's thesis, University of Amsterdam, 2006).

9. Valentine Hepp, *Calvinism and the Philosophy of Nature: The Stone Lectures Delivered at Princeton in 1930* (Grand Rapids, MI: Eerdmans, 1930), esp. 181, 185–186; cf. George McCready Price, *The New Geology: A Textbook for Colleges, Normal Schools, and Training Schools; and for the General Reader* (Mountain View, CA: Pacific Press, 1923), 5; G. Ch. Aalders, *De Goddelijke Openbaring in de eerste drie hoofdstukken van Genesis* (Kampen: Kok, 1932), 285–298.

10. Ab Flipse, *"Hier leert de natuur ons zelf den weg." Een geschiedenis van Natuurkunde en Sterrenkunde aan de VU* (Zoetermeer: Meinema, 2005), 159–169, 188–195; Flipse, "Origins of Creationism," 130–135; Hittjo Kruyswijk, *Baas in eigen Boek? Evolutietheorie en Schriftgezag bij de Gereformeerde Kerken in Nederland (1881–1981)* (Hilversum: Verloren, 2011), 215–236, 267–274.

11. Ab Flipse, "'Amerikaanse geleerden van formaat die deze dingen heel anders zien.' Nederlandse gereformeerden en het creationisme," in *Waar komen we vandaan? Anderhalve eeuw evolutiedebat in protestants-christelijk Nederland*, ed. Ab Flipse and George Harinck (Amsterdam: Historisch Documentatiecentrum voor het Nederlands Protestantisme, 2011), 31–39; Nicolaas A. Rupke, "Herdatering van het verleden. Inleidende Opmerkingen over een nieuwe Geochronologie," in *Creatie-Evolutie. Referatenbundel van de conferentie van Gereformeerden, met het thema "Creatie-Evolutie," gehouden op 16 en 17 mei 1967, in het Conferentieoord "De Pietersberg," te Oosterbeek* (Groningen: Veenstra & Visser, 1967); Ronald L. Numbers, *The Creationists: From Scientific Creationism to Intelligent Design*, expanded ed. (Cambridge, MA: Harvard University Press, 2006), 306–308, 367.

12. Henry M. Morris, *De evolutieleer. Een theorie op haar retour* (Groningen: De Vuurbaak, [1969]); A. M. Rehwinkel, *De zondvloed. In het licht van de Bijbel, de geologie en de archeologie* (Amsterdam: Buijten & Schipperheijn, 1970). The first four of the long series of Ouweneel's (creationist) books are *Wat is het nu? Schepping of evolutie?* (Winschoten:

Uit het Woord der Waarheid, 1974), *Operatie Supermens* (Amsterdam: Buijten & Schipperheijn, 1975), *De ark in de branding* (Amsterdam: Buijten & Schipperheijn, 1976), and *Vraag het de aarde eens* (Werkgroep Christelijk Boekenweekgeschenk, 1977).

13. J. Kamphuis, "Twee slechts gedeeltelijk benutte kansen," "Morris en Schilder," "Schilder en wij," *De Reformatie* 45 (1969–70): 23–24, 27–28, on 27. J. A. van Delden, "Bijbel en natuurwetenschap," *Nederlands Dagblad*, Sept. 19, 1970; J. A. van Delden, "De discussie rondom prof.dr. H. M. Morris. Bijbel en natuurwetenschap," *Nederlands Dagblad*, July 24, 1970, 5; J. A. van Delden, *Schepping en Wetenschap* (Amsterdam: Buijten & Schipperheijn, 1977). Another critical Reformed theologian: B. J. Oosterhoff, *Hoe lezen wij Genesis 2 en 3? Een hermeneutische studie* (Kampen: Kok, 1972), 36–58.

14. Van Rooden, "Long-Term Religious Developments," 122–123; Remco van Mulligen, "De Evangelische Omroep. Calvinistisch product of Amerikaanse kopie?," in *A Spiritual Invasion. Amerikaanse invloeden op het Nederlandse christendom*, ed. George Harinck and Hans Krabbendam (Barneveld: Vuurbaak, 2010), 123–144; J. A. van Delden, "Creationisme aan de Evangelische Hogeschool," *Bijbel en Wetenschap* 15 (1990): 14–16; J. A. van Delden, "Wat is het eigene van de Vrije Universiteit," *Bijbel en Wetenschap* 16 (1991): 222–223.

15. J. A. van Delden, *Adam of aap?* (Hilversum: Evangelische Omroep, 1977). For Wilder-Smith's influence in Germany, see Kutschera, chapter 6 in this volume. W. J. Ouweneel, "Creationism in the Netherlands," *Impact* supplement to *Acts & Facts* 7 (1978): i–iv; J. A. van Delden, "Van Harte," *Bijbel en wetenschap* 5 (Dec. 1980): 30.

16. J. A. van Delden, "Het eerste Europees Creationistisch Congres (nabeschouwing)," *Bijbel en Wetenschap* 9 (1984): 23–25; www.dutchcreationscience.com/artikelen-qin-de-maatschappijq/27-creationismebuitenland/68-europese-creationistische-congressen and www.creationconferences.co.uk, both accessed Oct. 24, 2012.

17. Arie van den Beukel, *Met andere ogen. Over wetenschap en zoeken naar zin* (Baarn: Ten Have, 1994); Michael Denton, *Evolution: A Theory in Crisis* (London: Burnett, 1985); Michael Behe, *Intelligent Design. De zwarte doos van Darwin* (Baarn: Ten Have, 1997); Stefaan Blancke, "Creationism in the Netherlands," *Zygon* 45: 4 (2010): 791–816.

18. Cees Dekker, *Het kleine is groots*, www.ceesdekker.net/files/oratiespeech.pdf; Ronald Meester, *100 % kans—de zin en onzin van de waarschijnlijkheidsrekening*, www.cs.vu.nl/~rmeester/oratievu.pdf, both accessed Oct. 10, 2012; Cees Dekker, "Uit wat ik ben en was," in *Geleerd en gelovig. 22 wetenschappers over hun leven, werk en God*, ed. Cees Dekker (Baarn: Ten Have, 2008), 328; Cees Dekker, Ronald Meester, and René van Woudenberg, eds., *Schitterend ongeluk of sporen van ontwerp. Over toeval en doelgerichtheid in de evolutie* (Baarn: Ten Have, 2005).

19. "Minister ontvangt boek over ID," www.kennislink.nl/publicaties/minister-ontvangt-boek-over-id; Michael Persson and Ben van Raaij, "Minister wil debat over evolutie en schepping," *Volkskrant.nl*, May 21, 2005, www.volkskrant.nl/vk/nl/2824/Politiek/article/detail/684761/2005/05/21/Minister-wil-debat-over-evolutie-en-schepping.dhtml; "Kritiek op evolutiedebat Van der Hoeven," *Volkrant.nl*, May 21, 2005, www.volkskrant.nl/vk/nl/2686/Binnenland/article/detail/684769/2005/05/21/Kritiek-op-evolutiedebat-Van-der-Hoeven.dhtml, all accessed Oct. 10, 2012; Martin Enserink, "Evolution Politics: Is Holland Becoming the Kansas of Europe?," *Science* 308:5727 (2005): 1394.

20. On the conference, see http://web.archive.org/web/20070203091217, www.darwinanddesign.org/; William Dembski, "'Not Just an American Phenomenon'—The Recent Prague ID Conference," Oct. 26, 2005, www.uncommondescent.com/intelligent-design/

not-just-an-american-phenomenon-the-recent-prague-id-conference; "Cees Dekker: ik ben geen ID-aanhanger," *Nederlands Dagblad*, Apr. 14, 2006, www.nd.nl/artikelen/2006/april/14/cees-dekker-ik-ben-geen-id-aanhanger, all accessed Oct. 10, 2012; Francis Collins, *De taal van God. Prominent geneticus verzoent geloof en wetenschap* (Kampen: Ten Have, 2006); René Fransen, *Gevormd uit sterrenstof. Schepping, ontwerp en evolutie* (Vaassen: Medema, 2009).

21. *Kerk in beweging*, www.eo.nl/programma/kerkinbeweging/2004-2005/page/-/mediaplayer/index.esp?aflid=6325049, accessed Oct. 11, 2012; Andries Knevel, *Avonduren. Dagboek van een bewogen jaar* (Baarn: Ten Have, 2007), 227–228. On intelligent design in Europe, see Forrest, chapter 12 in this volume.

22. *Kerk in beweging*; Knevel, *Avonduren*, 231.

23. Knevel, *Avonduren*; www.trouw.nl/tr/nl/4324/Nieuws/article/detail/1136077/2009/02/05/De-evolutie-van-de-Evangelische-Omroep.dhtml; for the English translation of the letter, see Blancke, "Creationism in the Netherlands"; for response by EO rank and file, see www.novatv.nl/page/detail/uitzendingen/6691; for reaction to Dorenbos, see http://dewerelddraaitdoor.vara.nl/media/43268; for the apologies of Knevel, www.eo.nl/programma/eonl/2008-2009/page/-/mediaplayer/index.esp?aflid=10238204; for the letter of Dorenbos, www.creatie.info/component/content/article/12–2004-nieuwsbrieven/445-open-brief-de-eo-op-drift-geraakt-door-drs-lp-dorenbos.html, all accessed Oct. 11, 2012; Blancke, "Creationism in the Netherlands."

24. Knevel's show, www.eo.nl/programma/hetelfdeuur/2008-2009/page/-/mediaplayer/index.esp?aflid=10135815; leaflet project, www.creatie.info; committee of recommendation, www.refdag.nl/media/2008/20081106_creatieAdressen_DEF.pdf; Johan Huibers' ark, www.arkvannoach.com/, all accessed Oct. 22, 2012.

25. Van Helden addresses Knevel, www.novatv.nl/page/detail/uitzendingen/6691#; H. A Hofman, *Het bittere conflict. Over schepping en evolutie in het jaar van Darwin en Calvijn* (Soesterberg: Aspekt, 2009); Zeger Wijnands, *God of Darwin. Kan wetenschap om de Bijbel heen?* (Elburg: Elpenburg, 2009); civil initiative, www.omroepflevoland.nl/nieuws/nieuwsbericht?NewsID=57800; ProGenesis, *95 stellingen tegen evolutie. Wetenschappelijke kritiek op het naturalistische wereldbeeld* (Doorn: Johannes Multimedia, 2009); *Weet magazine*, www.weet-magazine.nl/home; Oude Wereld, www.oude-wereld.nl, all accessed Oct. 22, 2012; Peter Borger, *Terug naar de oorsprong, of hoe de nieuwe biologie het tijdperk van Darwin beëindigt* (Doorn: De Oude Wereld 2009); Reinhard Junker and Siegfried Scherer, *Evolutie: het nieuwe studieboek* (Doorn: De Oude Wereld, 2010). For details on the Junker and Scherer book, see Kutschera, chapter 6 in this volume. Other publications: e.g., *Evolutie, cultuur en religie. Perspectieven vanuit biologie en theologie*, ed. Palmyre Oomen and Taede Smedes (Kampen: Klement, 2010).

26. Ernest Henau, "Dialoog," *Het Teken* 84 (May 2009), 289; Geert Van Coillie, "Schepping en evolutie verdiepen elkaar," *Tertio*, Nov. 25, 2009; "Darwin, het jaar van de waarheid," *Visie*, Sept. 25, 2009; "Ja, ik geloof in mirakels," *De Standaard*, Oct. 3–4, 2009; Jan De Volder, "Darwinisme tussen wetenschap en geloof," *Tertio*, Feb. 18, 2009.

27. Raf De Bont, *Darwins kleinkinderen. De evolutietheorie in België 1865–1945* (Nijmegen: Vantilt, 2008); Raf De Bont, "'Foggy and Contradictory': Evolutionary Theory in Belgium, 1859–1945," in Engels and Glick, *The Reception of Charles Darwin in Europe*, 1:188–198; Max Wildiers, "Woord vooraf," in Teilhard de Chardin, *Het verschijnsel mens* (Utrecht: Het Spektrum, 1963), 9–15; J. Doutreligne, *Van cel tot mens* (Antwerpen: Uitge-

versmij. N.V. Standaard-Boekhandel, 1963), 6; R. Claeys and W. Vandoninck, *Het mysterie van het leven* (Lier: Van In, 1968), 223.

28. www.youtube.com/watch?v=4MMpxlN-TdQ (part 1) and www.youtube.com/watch?v=LKhG94hIZx4&feature=related (part 2); "Wijdverspreid Turks boek geeft Darwin schuld van terrorisme," *De Morgen*, Mar. 1, 2007; "Scholen en universiteiten verrast met Atlas of Creation," *De Standaard*, Mar. 1, 2007: imam, www.youtube.com/watch?v=4MMpxlN-TdQ (part 1) and www.youtube.com/watch?v=LKhG94hIZx4&feature=related (part 2); "Integratie- of gettoschool?," *De Standaard*, Nov. 7–8, 2009.

29. Laurence Perbal, "Evaluation de l'opinion des étudiants de l'enseignement secondaire et supérieur de Bruxelles vis-à-vis des concepts d'évolution (humaine)" (master's thesis, Vrije Universiteit Brussel, 2005); An Bogaerts, "De ontkenning van de evolutietheorie door de Islam" (master's thesis, Lessius Hogeschool, 2005).

30. "Vraagtekens bij evolutieleer," *De Standaard*, Oct. 27, 2010; "Creationisme bestaat in joods onderwijs," *De Standaard*, Aug. 18–19, 2007; Vlaams Ministerie van Onderwijs en Vorming, *Verslag over de doorlichting van de Middelbare Steinerschool Vlaanderen te Gent* and *Verslag over de doorlichting van de Middelbare Steinerschool Vlaanderen te Antwerpen*, www.ond.vlaanderen.be, accessed Oct. 22, 2012.

31. Jos Philippaerts and Rudi Meekers, in an interview with Stefaan Blancke and Maarten Boudry, Dec. 4, 2008; www.creabel.org; www.klasse.be/leraren/archief/13679.

32. The Pew Forum on Religion and Public Life, *U.S. Religious Landscape Survey* (June 2008), 95, http://religions.pewforum.org/pdf/report2-religious-landscape-study-full.pdf#page=99; Henk Rijkers, "Cees Dekker moet eerlijk spreken over Intelligent Design," *Katholiek Nieuwsblad*, Aug. 20, 2012, www.katholieknieuwblad.nl, and many other articles; Stefaan Blancke, "Catholic Responses to Evolutionary Theory, 1859–2009," *Journal of Religious History* 37 (2013): 353–368; Stefaan Blancke, "Towards an Integrated Understanding of Creationism in Europe: Historical, Philosophical and Educational Perspectives" (doctoral thesis, Ghent University, 2011).

CHAPTER FIVE

Scandinavia

HANS HENRIK HJERMITSLEV AND PETER C. KJÆRGAARD

The Scandinavian countries, Denmark, Norway, and Sweden, are closely connected historically, culturally, and linguistically. They are also, according to polls, among the nations in which most people accept the theory of evolution. In 2005, for example, the European Commission's Eurobarometer Survey 63.1 concluded that 83 percent of the Danes, 82 percent of the Swedes, and 74 percent of the Norwegians agreed with the statement, "Human beings, as we know them today, developed from earlier species of animals," while a mere 13 percent of the Danes and Swedes and 18 percent of the Norwegians disagreed. Moreover, relatively few Scandinavian clergymen and politicians have suggested that creationism and intelligent design should be included in biology classes as scientific alternatives to evolutionary theory.[1]

This is no surprise, since sociologists and historians of religion have generally considered the Scandinavian countries to be among the most secularized in the world. In 2008 the American sociologist Phil Zuckerman went as far as entitling his book about the religious views of the Danes and Swedes *Society without God*. No doubt, Zuckerman is right when he argues that religion is much less visible in Denmark than in the United States. However, another US sociologist Andrew Buckser, who has studied secularization and religious life in Denmark, uses the Danish case to warn us against focusing too narrowly on supernatural belief and theological doctrines when studying religious practices. Thus, with Zuckerman's focus on people's views on certain theological doctrines such as original sin, the virgin birth, and the existence of hell, he understates the influence of religion in Scandinavia.[2]

Most Scandinavians are members of the national Evangelical-Lutheran churches that have dominated religious life in Scandinavia since the Protestant Reformation in the 1530s. Even though the support has been steadily declining since the 1980s, surveys in 2011 indicate that 80.4 percent of the 5.4 million Danes, 78.0 percent of the 4.9 million Norwegians, and 70.0 percent of the 9.4 million Swedes were still members of the national churches. While few church members (only 2% to 5%) attend service on a regular basis, a majority of Scandinavians have their children baptized in the national churches and attend service at least at Christmas. The national churches are primarily cultural institutions with strong symbolic meanings and are used primarily as markers of key moments in people's lives. Christianity is a central part of Scandinavian culture and tradition, but in their daily lives Scandinavians are not much concerned with religious doctrines, the reading of scripture, or the existence of God. In 2009 a Gallup poll revealed that only 19 percent of the Danes and 17 percent of the Swedes thought that religion was an important part of their daily lives, and a 2005 Eurobarometer survey stated that only 32 percent of the Norwegians, 31 percent of the Danes, and 23 percent of the Swedes believed in God.[3]

American-style activist and fundamentalist Protestantism is relatively marginalized in Scandinavia. Evangelicals, most prominently the low-church Inner Mission in Denmark and Norway and the Swedish Evangelical Mission, belong to the right wing of the national churches. An estimate of between 5 and 10 percent may accurately indicate the extent of those among the members of the national churches who sympathize with their evangelical wings. In contrast to the conservative theological views advocated by evangelicals, liberal views concerning issues such as female pastors, abortion, homosexuality, and indeed evolution dominate among mainstream clergymen and church members. Most pastors have received a theological degree from one of the faculties of theology at the state universities. In the twentieth century, academic theology in Scandinavia has been strongly influenced by, on the one hand, German biblical criticism and, on the other, liberal, existential, and dialectical theology. These modernist theological positions have rejected the infallibility of scripture as well as the tradition of natural theology and the ambition of finding explanations of natural phenomena in the Bible. Consequently, Lutherans with strong evangelical leanings often favor free churches outside the national churches. The Lutheran free churches emerged during a religious revival in the nineteenth century alongside other Protestant agencies such as Baptists, Methodists, Pentecostals, Seventh-day Adventists, and Jehovah's Witnesses. Moreover, all Scandinavian countries have significant minorities of Roman Catholics and Eastern Orthodox Christians, most of whom are first- or second-generation immigrants. In 2011, 93,000 Danish citizens were members of Christian churches outside the national church, while the equivalent numbers were 266,800 in Norway and about 700,000 in Sweden.[4]

Among non-Christian religions, Islam is by far the largest. Owing to immigration from Turkey, North Africa, and the Middle East since the 1960s, Muslims now constitute large minorities in the Scandinavian countries. In 2010 there were about 500,000 people with a Muslim background living in Sweden, 230,000 in Denmark, and 150,000 in Norway. However, the number of members of organized Muslim faith communities is much lower.[5]

Thus, Scandinavian antievolutionists are generally confined to the evangelical wing of the national churches, to free churches, and to Muslim groups; in addition, a few members of the spiritual Anthroposophical and Hare Krishna movements also advocate anti-Darwinian views. As such, creationism in Scandinavia is a relatively marginalized phenomenon compared to most other European countries and especially to the United States.

Early Protestant Responses to Darwinism

Scientists in Scandinavia began debating Charles Darwin's theory of evolution in the 1860s. Even though the countries were closely connected culturally and politically, the debates took place within national contexts. However, similar historical developments can be identified. While Scandinavian naturalists, with a few exceptions such as the Swedish professor of botany N. J. Andersson, remained skeptical of the theory of evolution in the 1860s, in the following two decades researchers at natural history museums and the five universities in Scandinavia at this time—Lund, Uppsala, Stockholm, Copenhagen, and Kristiania (Oslo)—began to integrate evolutionary theory in research and teaching. At the turn of the century, the theory of evolution was generally accepted among university-trained botanists and zoologists in Scandinavia, while the specific evolutionary mechanisms, for example Darwinian natural selection and Lamarckian direct adaptation, were still open to debate.[6]

The wider cultural debate about Darwinism, especially in Denmark and Norway, was influenced by the Danish literary critic Georg Brandes and his proclamation of the modern breakthrough in 1871. Brandes applied Darwinism in his atheist struggle against conservative values and religious beliefs. Among the members of the Brandes circle in Copenhagen, which also included Norwegian and Swedish writers, was the Danish translator of *On the Origin of Species* (1871–72) and the *Descent of Man* (1874–75), the freethinker, botanist, and author Jens Peter Jacobsen. In 1872 Jacobsen engaged in a controversy over Darwinism with Bishop D. G. Monrad, who ridiculed the so-called ape theory, which he argued was based on loose scientific speculations and needed substantial evidence. Among Christians, however, views varied. The national churches were divided between liberals who sought to reconcile evolution and Christianity and conservatives who remained skeptical of Darwinism. In Norway and Sweden, the first positive evaluations from Lutheran theologians of Darwin's theory of descent occurred in the 1880s after two decades of sporadic criticism.[7]

In Denmark, high-church and university theologians remained skeptical of evolution until the early twentieth century, while adherents of N. F. S. Grundtvig, who constituted a liberal faction within the national church, were engaged in polemics over evolution from the 1870s. Orthodox Grundtvigians remained critical of evolution well into the twentieth century. Meanwhile liberal neo-Grundtvigians legitimized the theory of evolution by referring to Grundtvig's so-called church view, which asserted the priority of the Apostolic Creed and the sacraments to scripture. Thus, Grundtvig's critique of traditional Lutheran scriptural theology made it possible for neo-Grundtvigians to accept theories, such as Darwinian evolution, which seemed to challenge a literal reading of scripture. Moreover, following the nineteenth-century Danish theologian Søren Kierkegaard and his protagonist, the nineteenth-century professor of philosophy Rasmus Nielsen, who combined the teachings of Grundtvig and Kierkegaard, liberal Grundtvigians made a radical distinction between knowledge and faith. This separation model of science and religion has provided liberal Danish Lutherans with the intellectual framework needed in order to embrace the theory of evolution.[8]

However, among the circa 100,000 Danish evangelicals connected to the Inner Mission, the case was different. Because of their literalist reading of the Bible, they were critical of the theory of evolution. From 1900 to 1920 the renowned professor of botany and conservative Christian Eugen Warming defended anti-Darwinian views in the evangelical press. Warming held strong Lamarckian views and rejected human evolution by advocating a theory of parallel lines of descent as an alternative to the Darwinian theory of branching evolution.

The German-born Jesuit amateur zoologist Amand Breitung also attacked Darwinism in this period. Like Warming, Breitung advocated Lamarckian views on evolution and parallel lines of descent, which he interpreted as being in accordance with the history of creation as recorded in Genesis 1. Both Warming and Breitung were prolific writers and reached large audiences with their critiques of Darwinism. As indicated by the cases of Warming and Breitung, the crucial and controversial point of the Danish debates on evolution and religion was the theory of common descent, popularly referred to as the "ape theory." This question defined the dividing line between those who embraced Darwinism and those who perceived themselves as in opposition to Darwinism. Thus, among critical commentators of Darwinism in the decades around 1900, no one defended strictly creationist views. However, many Scandinavian believers with evangelical leanings remained faithful to literalist readings of scripture, and among these factions of the national churches a large proportion was probably opposed to modernist ideas such as biblical criticism and evolutionary theory.[9]

In the 1920s, when the first wave of fundamentalist creationism flooded the United States, the relationship between evolution and religion was also debated in Scandinavia. At this time, evolutionary theory had been introduced as part of the natural history curriculum in upper secondary schools. In Denmark, the most vocal critic of Darwinism in this period was the celebrated fictional author Helge Rode, who engaged in heated polemics with defenders of evolution and free thought such as the literary critic Georg Brandes and the writer Johannes V. Jensen. During World War I, Rode had identified Darwinism and its doctrines of struggle for existence and survival of the fittest as the ideological background for amorality and militarism and the reason for the prevailing nihilism in culture and society. According to Rode, a "spiritual regeneration" was needed, and during the 1920s, when he was influenced by theosophy, spiritism, and Catholicism, he was finally convinced that a revitalization of the national Lutheran Church was the proper remedy. Rode supplemented his moral criticism of Darwinism with a discussion of its contemporary scientific status. Drawing on the authority of the renowned Danish geneticist Wilhelm Johannsen, whose studies of pure lines of beans had questioned whether small random variations could explain the evolutionary process, Rode argued that the theory of natural selection had been rejected by scientists and that the origin of species was still an unsolved puzzle. In the following collections of essays, which were widely distributed and much debated in the Copenhagen press, Rode expanded his argument. On the basis of his interpretation of Hugo de Vries's mutation theory, he now claimed that no evolution, but only degeneration, could be identified in the natural world and that there were no reasons to assume that humans were related to apes. The opposite could just as well be the case, according to Rode.[10]

In 1925 Rode's campaign against Darwinism was supported by the conservative Danish economist Knud Asbjørn Wieth-Knudsen, a well-known antifeminist and professor at the Norwegian Polytechnic College in Trondheim from 1922 to 1942. Wieth-Knudsen claimed that Darwinism was an outdated theory and that there existed an unbridgeable gap between evolutionary biology and the existence of human ethics. Rode and Wieth-Knudsen were not the only Danish commentators who criticized the theory of evolution. In fact, a range of prominent writers and critics participated in the debates in the 1920s. Many of them were skeptical of Darwinism on moral and philosophical grounds, while reservations based on literalist interpretations of scripture were rare. Unlike the creationists in the United States in the 1920s, the anti-Darwinists did not unite in order to activate the masses and campaign against the teaching of evolution. However, the widespread criticism of the theory of evolution worried some professional scientists, and in 1929 five lectures by university scientists delivered at the Society of Students in Copenhagen were published in order to clarify matters. According to the contemporary situation within the disciplines of geology, anatomy, embryology, paleoanthropology, and biology, practically all scientists, they emphasized, agreed that evolution had taken place. Contrasting the claims of Rode and others, the zoologist Ragner Spärck argued that the scientific discussions of the possible mechanisms driving the evolutionary process did not affect the validity of the general theory of descent originally advocated by Darwin.[11]

In the 1930s and 1940s, when political issues were pressing while scientists established the modern synthesis in biology by combining Darwinian selection theory, genetics, and mutation theory, the Darwinian debates faded in Denmark and Norway. However, evolution was also discussed among non-Christian faith communities. In the middle of the 1950s, for example, human evolution was criticized by Norwegian and Swedish supporters of Rudolf Steiner's esoteric theory of anthroposophy.[12]

The Emergence of Scientific Creationism in Scandinavia

In the 1970s, American flood geology, popularized by John C. Whitcomb Jr. and Henry M. Morris in their influential work *The Genesis Flood* (1961), was introduced to evangelicals in Scandinavia. In 1971 Knud Aage Back, a Christian schoolteacher, translated Hannington Enoch's work *Evolution or Creation* (1966) into Danish. Enoch, a Christian young-earth creationist affiliated with both the Creation Research Society and Evolution Protest Movement, was head of the department of zoology at Presidency College, University of Madras, and one of the few advocates of creationism in India at this time. Another Danish schoolteacher, Frode Thorngreen defended flood geology in his 1975 work *Syndfloden* (The Flood). In his argument, Thorngreen drew heavily on prominent US creationists such as Whitcomb and Morris, Byron C. Nelson, and Alfred M. Rehwinkel. Thorngreen's work was published by a small evangelical agency within the national church, *Ordet og Israel* (The Word and Israel), which defended biblical literalism, verbal inspiration, and the Second Coming of Christ.[13]

Meanwhile in Norway Odd Sæbö, a secondary-school teacher, became absorbed in US and British creationist literature. In 1975 Sæbö's work *Dogmet om evolusjonen,* whose title was borrowed from Louis T. More's *The Dogma of Evolution* (1925), was published by an orthodox Lutheran press. In line with the strategy of employees at the Institute of Creation Research in the United States, Sæbö questioned mainstream evolutionary biology and uniformitarian geology but refrained from explicitly referring to the Mosaic history of creation when arguing his case. In 1977 the Danish author Poul Hoffmann, a fundamentalist affiliate of the Inner Mission and a defender of biblical literalism, translated Sæbö's work into Danish in cooperation with his wife Kirsten Hoffmann. Poul Hoffmann, a law graduate from University of Copenhagen, was a prolific contributor to the evangelical press and acknowledged among fellow evangelicals for his biblical novels about Moses and Noah. In 1984 the Hoffmanns translated the prominent British creationist Edgar Harold Andrews's work *From Nothing to Nature* (1978). The translation was published by Credo, a publishing house controlled by the Danish nationwide student organization Kristeligt Forbund for Studerende, which has close ties to the evangelical Inner Mission within the national church. Andrews's work had been translated into Swedish as early as 1980 and also sold well in Norway, where a translation of a German work by another British creationist, the pharmacologist Arthur E. Wilder-Smith, arguably the most influential advocate of creationism in Europe, was published in 1982.[14]

In 1996 Inner Mission's publishing house Lohse sent out Poul Hoffmann's work *Dinosaurerne og syndfloden* (The dinosaurs and the Flood), in which he claimed that dinosaurs "lived a few thousand years ago and died during the Flood. One of its ancestors might be lying on a slope in Alaska starring into the red, setting sun. Cheer up, brother eagle. Falsehood and death will come to an end. Truth and paradise will come." Hoffman regarded Darwinism as a radical denial of Christianity, and he strongly criticized the evangelical bishop of the national church Karsten Nissen when, in 2009, he distanced himself from creationism and the teaching of intelligent design.[15]

Organizing Scandinavian Antievolutionism

In the late 1970s, Scandinavian creationists began to get organized. In Sweden Förening for Biblisk Skapelsestro (Biblical Creation Society) was established, and in Norway in 1979 the young mathematician Steinar Thorvaldsen organized a three-day creationist conference at the Norwegian University of Science and Technology in Trondheim with the participation of Wilder-Smith, who was a frequent visitor in Norway at the time. Also the biochemist Duane T. Gish, a prominent member of the Institute of Creation Research, visited Norway.[16]

In 1983 Bent Vogel and Peter Øhrstrøm, two science graduates from Aarhus

University in Denmark, and the arts graduate Ole Dichman launched the quarterly *Origo* as what was called "a scientific journal" and "an apologetic resource" for Protestants with evangelical leanings and an interest in natural science. The journal soon developed into a creationist society and expanded its activities to Norway. Steinar Thorvaldsen, now associate professor at the College of Education in Tromsø, joined the *Origo* staff in 1988, and in 1996 a Norwegian branch of the society was established with Peder A. Tyvand, professor of physics at the Agricultural University of Norway in Moss, as president. Now, editorial boards were organized in both Denmark and Norway, including academics with Lutheran, Pentecostal, and Adventist backgrounds.[17]

In 2010 *Origo* had around 950 subscribers. The Norwegian society runs the webpage origonorge.no, and since 2001 the Danish society has hosted the webpage skabelse.dk, which advertises lectures on science and evolution and a creationist textbook aimed at evangelical schools, of which there are thirty-six in Denmark. The textbook was meant to function as a supplement to the state-sanctioned biology textbooks. It advocates the theory of intelligent design and the concept of Basic Types of Life introduced by the American creationist Frank L. Marsh in the 1940s and popularized in Europe from the 1980s by the German professor of biology Siegfried Scherer, who argues that in the beginning there existed two hundred to three hundred kinds of animals (humans being one of them), which have developed into the animal world observed today through evolution by natural selection.[18]

The material on offer by *Origo* also includes a translation of *Icons of Evolution* by the Discovery Institute intelligent design advocate Jonathan Wells. Moreover, in the Darwin year of 2009, *Origo* published a children's book, a Darwin biography, and a critique of Darwinism. However, these publications were far from best sellers. Their circulation numbers were well under one thousand copies; in comparison, more than ten thousand copies of pro-evolutionary books on science and religion published in 2009 were distributed by the Danish national church and aimed at primary and secondary schools. As part of their anti-Darwin festivities in 2009, *Origo* affiliates registered the websites darwin2009.no and darwin2009.dk as a creationist alternative to university-sanctioned open educational resources such as evolution.dk.[19]

Unlike US and Turkish creationists, the journal *Origo* does not have strong financial backing. The editors, contributors, and lecturers are unpaid volunteers, and their websites do not include flashy audiovisual effects like many US and Turkish creationist pages. Though low in economic capital, the Danish and Norwegian creationists do score relatively high when it comes to cultural capital.

The editorial boards include scientists with master's or PhD degrees in biology, biochemistry, bioethics, geophysics, physics, engineering, and philosophy of science. Most contributors to the journal are old-earth creationists and intelligent design supporters and have more liberal views on scripture than mainstream US creationists generally do. In their arguments against Darwinian evolution, some Danish anti-Darwinists are careful to distance themselves from the heated rhetoric of the US and Turkish creationists, who accuse Charles Darwin and his theory of evolution of being the cause of modern evils such as terrorism, fascism, and communism. At a public conference in 2009, Peter Øhrstrøm, founding editor of *Origo*, a Pentecostal layman, and member of the Christian People's Party and the National Ethical Council in Denmark, even felt it necessary to warn against the demonizing of Darwin put forth by US and Turkish creationists. A professor of philosophy of science at Aalborg University, Øhrstrøm claimed that his critique was strictly scientific and philosophical and aimed at the methodology of scientific naturalism, which he regarded as the atheist and materialist ideological foundation of modern evolutionary biology. In order to advocate this antimaterialist agenda, *Origo* editors introduced the theory of intelligent design on Danish and Norwegian soil in 2000, when they devoted an issue of their journal to the theory. In 2007 Øhrstrøm succeeded in getting a book on intelligent design published by a Danish university press.[20]

Among *Origo* editors and contributors, however, views vary. While Øhrstrøm and another member of the editorial board, Kristian Bánkuti Østergaard, a biology graduate from Aarhus University, draw on arguments from intelligent design theorists such as Michael Behe, William Dembski, and Jonathan Wells, other contributors prefer mainstream American young-earth creationism. Unlike other countries, such as the Netherlands, the United States, and Turkey, these differences have not caused a fragmentation of the Danish-Norwegian creationist society, which allows different views on evolution and creation as long as the contributors share a Christian worldview and criticize the paradigmatic status of modern evolutionary biology.[21]

The hotbed of creationism in Denmark is the evangelical high school in the town of Ringkøbing in rural Western Jutland. Since the foundation of *Origo* in 1983, this small private high school has hosted several creationist conferences. In 1985, for example, Arthur E. Wilder-Smith delivered eight lectures in four days before an audience of up to one hundred people. Among the teachers in Ringkøbing are the former leader of the small Christian People's Party, Marianne Karlsmose, who in 2002 called for the teaching of creationism in Danish schools, and the webmaster of *Origo*'s Danish webpage and author of the previously mentioned creationist textbook, the Evangelical-Lutheran biologist Kristian B. Østergaard, who practices the "teach both sides" argument in his advanced biology classes. Another *Origo* editor, the Seventh-day Adventist and young-earth creationist Holger Daugaard, teaches biology at the Danish Adventist high school in the town of Vejle in Eastern Jutland. Daugaard has also launched a correspondence course on creation and evolution. In 1995 Daugaard and Bent Vogel, the founding editor of *Origo*, translated into Danish *What Is Creation Science?*, the 1987 volume by Institute for Creation Research biologist Gary E. Parker. The translation was distributed by Inner Mission's publishing house.[22]

The Norwegian and Danish societies collaborate with the Swedish creationists' society Genesis, a successor of the original Förening for Biblisk Skapelsestro, which has published the quarterly journal *Genesis* since 1988. The three Scandinavian societies organize joint conferences and often visit each other at their annual meetings. Typically, the conferences are hosted by Christian schools and free churches, and the lectures are supplemented by services and prayers. In 2011, for example, a Swedish three-day conference took place at the Pentecostal Church in Uddavalla with the leading Norwegian creationist professor Peder A. Tyvand among the speakers. In 2003 Swedish creationists welcomed people from the United States, Israel, the Netherlands, and Finland, among other countries, when Genesis arranged the eighth European Creationist Congress.[23]

Like US creationists, members of Scandinavian creationist societies are eager to popularize their views among teenagers, who are taught evolutionary theory at school. In 2005 collaboration between the journals *Origo* and *Genesis* resulted in a special issue aimed at teenagers and their parents, who were asked to hand the publication to their children's biology teachers. The issue was edited by the old Danish creationist warrior Knud Aage Back, while the Swedish young-earth creationist Mats Molén, a secondary-school teacher in biology and director of a creation museum established in 1996 in Umeå, had written the original text. Molén is among the most active creationist lecturers and writers in Sweden. He has written several creationist books, including *Livets uppkomst* (2000), which was translated into Danish by Poul Hoffmann's son Helge Hoffmann and published by Inner Mission's publishing house in 2010.[24]

The spokesman for Genesis is Anders Gärdeborn, who attracted some media attention in 2007, not so much because he published a creationist book but because he strongly criticized the decision by the Swedish conservative government to ban the teaching of creationism and intelligent design from biology classes in state-funded Christian schools. Its decision was made shortly after the passing of the Council of Europe Resolution 1580, which warned against "the dangers of creationism for education" in October 2007, and it was supported by the Christian Democrats, even though some of its evangelical members aired creationist views.[25]

Free Churches Fighting Darwin

While creationism is marginalized in the national church in Sweden, it has gained a foothold among members of charismatic free churches. Within the Swedish branch of the Word of Faith movement, Krister Renard, who is affiliated with Genesis and teaches physics at a high school run by the sect, has done much to popularize creationism and intelligent design. The Swedish National Agency for Education delivered a critical assessment in 2002 when its inspectors found out that schools run by Word of Faith introduced their students to creationist alternatives to evolutionary theory.[26]

Creationism is also advocated in free churches in Denmark. Pentecostal pastors are especially eager to promote creationism. Their arguments against evolution can be found on the internet, where their websites often link to the flashy pages of Ken Ham's prominent young-earth creationist society Answers in Genesis. Among some born-again Christian lay preachers with an evangelical and charismatic bent, creationism has also gained a foothold. For example, it generated some local attention when in 2009 the pastor of a small independent evangelical church in the village of Løkken in Northern Denmark put up a handwritten poster on the front of his church stating, "Darwin's theories have not been scientifically proven. Darwin's theories are religion to those who reject GOD." Also the sectarian Jehovah's Witnesses, which is among the largest Christian denominations outside the national churches in Scandinavia, have promoted their version of creationism in pamphlets and books offered free of charge.[27]

Intelligent Design

Intelligent design theory was taken up by Scandinavian creationists in 2000. However, in Denmark it received little attention until 2004, when Jakob Wolf, an associate professor of systematic theology at the University of Copenhagen, published a work that criticized the philosophical and scientific basis of modern evolutionary biology and suggested intelligent design as a valid alternative. Wolf's book was met with strong criticism from scientists, philosophers, and theologians, and even though initially some politicians and clergymen were pos-

Small-scale creationism. An independent church in Denmark aired its creationist sympathies in 2009, when the bicentennial of Charles Darwin's birth was widely celebrated in the media. Photo by Tomasz Sienicki, Licence: Creative Commons Attribution 3.0.

itive toward intelligent design, they soon left the advocacy of the theory to *Origo* affiliates and small religious denominations.[28]

Among non-Christian believers, no one has promoted anti-Darwinism as eagerly in Scandinavia as the Hare Krishna monk Leif Asmark Jensen, who founded the Danish Society for Intelligent Design and promoted the unorthodox ideas of fellow Vedic Michael Cremo, who argues that modern humans can trace their history back trillions of years to fully formed ancestors and that all professional archaeologists hide evidence in favor in this fact. In 2004 Asmark wrote a short introduction to intelligent design, and in 2006 he published his translation of Michael Cremo and Richard Thompson's best seller *Forbidden Archeology*. It attracted some media attention and a critique from university staff when in 2009 Asmark and Cremo lectured twice before small audiences at unofficial meetings at Aarhus University. Around 2010 Asmark began to collaborate with the *Origo* editor Knud Aage Back, who contributes to Asmark's website intelligentdesign.dk.[29]

Islamic Creationism

Unlike the Christian and Hindu anti-Darwinian campaigns in Scandinavia, the promotion of old-earth creationism by the Sunni Muslim Adnan Oktar, also known under his pen name Harun Yahya, is well funded and much debated in newspapers. As occurred in many other European countries between December 2006 and September 2007, the first and second parts of the English version of Oktar's *Atlas of Creation* were sent to politicians, scientists, high school teachers, and Lutheran pastors in Denmark. Between five hundred and one thousand Danes are estimated to have received a copy of *Atlas of Creation*. One of Oktar's anti-Darwinist works has been translated into Danish, and some of his creationist children's books can be found in both Danish and Norwegian. Moreover, in 2008 Oktar's well-designed websites were promoted by a Muslim society in Aarhus, and he was interviewed by the Swedish National Radio on October 14, 2008.[30]

Another strategy used by Oktar's disciples is manipulating web polls on evolution to make it look as though creationist views are gaining ground in response to the publication of *Atlas of Creation*. This has been done in Germany, France, and Denmark. In 2007 and 2009, polls on the webpage of the Danish tabloid *Ekstra Bladet* were manipulated by Turkish votes. As a result, the polls showed that 88 and 59 percent of the Danes rejected human evolution. In 2007 the change happened overnight, going from 78 percent accepting that "man descended from apes" to 88 percent rejecting it. When analyzing the votes, a journalist at *Ekstra Bladet* discovered that more than half of the four thousand votes were from computers in Turkey. In 2009, the journalist invited the readers to send their questions about evolution to Adnan Oktar and his public relations officer Seda Aral. Their detailed answers were published on the webpage a month later. Here Oktar claimed among many other things that "the forerunner of human beings millions of years ago was also human being" and that "living things have never changed. THERE EXISTS NOT A SINGLE TRANSITIONAL FORM [OF] FOSSIL to confirm the claims of Darwinists."[31]

The Oktar affiliate Dr. Cihad Gundogdu delivered a lecture at the University of Oslo in 2007 as part of a one-day conference arranged by a Muslim student organization. In May 2011 Gundogdu visited Copenhagen in Denmark and Malmö in Sweden for a similar event. The titles of the conferences were "The Collapse of the Theory of Evolution and the Fact of Creation." However, at his appearance in Denmark the few Muslims who attended the conference challenged Gundogdu's claim that he knew the exact date of Judgment Day, regarded as blasphemy

Avis | Web | TV | Mobil — Ekstra Bladet! Startside eAvisen RSS — Box EPN Jobzonen Meetic Side6 Turen går til

Ekstra Bladet

nationen!

Din mening og din viden batter noget! Tilmeld dig nationen! Læs mere her ... »

nationen! | 1224 - Tip nationen! | Skriv et læserbrev | Deltag i konkurrencer | Hvad er nationen!? | Min Side

Forside
Nyheder
Sport
Underholdning
112
nationen!
Døgnets nyheder
Web-tv
Forbrug
Erhverv
Biler
Ferie
Verden på vrangen
EBblog
Gallerier
Sex & samliv
Spørg Joan Ørting
Side 9 pigen
Massage & Escort
Poker
Download spil
Konkurrencer

Islamister spammer gymnasier

En seks kilo tung bog der fornægter darwinismen er landet på flere danske gymnasier. Blandt andet hos biologilærer Thorkild Laustsen

Frederik B. Andersen - 16:02 - 29. jun. 2007 — Send | Print

HUSK: Som medlem af nationen! kan du nedenfor kommentere emnet i denne artikel.

Mennesket stammer ikke fra aberne. Og darwinismen er skyld i både nazisme og kommunisme. Det er budskabet i en seks kilo tung bog fra islamister i Tyrkiet, som er blevet sendt til lærere på flere gymnasier i Danmark.

Blandt andet til biologilærer Thorkild Laustsen fra Horsens Gymnasium. Han så otte eksemplarer af bogen 'Atlas of Creation' ligge på lærerværelset.

Fornægter arternes udvikling
- Så fik jeg at vide, at en af bøgerne var sendt til mig. Resten var også til religions- og biologilærere. Det er jo en stor, flot bog - med masser af flotte billeder, siger Thorkild Laustsen, hvis søn Jeppe har taget et billede af bogen og lagt på sin blog.

n! NYT

LÆS OGSÅ ...
Banket på plads af islamister
LÆSERBLOG: Det er propaganda

AFSTEMNING
Tror du, mennesket nedstammer fra aberne?
Ja 78%
Nej 15%
Måske 6%

SØG PÅ EKSTRABLADET.DK — Find artikel — Søg

nationen!
NYHEDER — BIF i brand
NYHEDER — Tag mig hårdt
NYHEDER — SOSU'erne kan godt droppe drømmen om mere løn
UNDERSØGER — Her må du ryge
Adgang forbudt for fede
SPØRGER — 3F klipper støtten til Helle
SPØRGER — Kunne du fristes af en Google-telefon?
NYHEDER — Lykkelig vinder af nationen!-kit
SPØRGER — Hvorfor tager du ikke bussen?

Avis | Web | TV | Mobil — Ekstra Bladet! Startside eAvisen RSS — Box EPN Jobzonen Side6 Turen går til

Ekstra Bladet

nationen!

Din mening og din viden batter noget! Tilmeld dig nationen! Læs mere her ... »

nationen! | 1224 - Tip nationen! | Skriv et læserbrev | Deltag i konkurrencer | Hvad er nationen!? | Min Side

Forside
Valg '07
Nyheder
Sport
flash!
112
ekstrabladet.tv
nationen!
kup!
epn.dk
Biler
Ferie
Verden på vrangen
Gallerier
EBblog
Sex & samliv
Spørg Joan Ørting
Side 9 pigen
Massage & Escort
Poker
Download spil
Spil Online
Fodbold Manager
Konkurrencer

Islamister spammer gymnasier

En seks kilo tung bog der fornægter darwinismen er landet på flere danske gymnasier. Blandt andet hos biologilærer Thorkild Laustsen

Frederik B. Andersen - 16:02 - 29. jun. 2007 — Send | Print

HUSK: Som medlem af nationen! kan du nedenfor kommentere emnet i denne artikel.

Mennesket stammer ikke fra aberne. Og darwinismen er skyld i både nazisme og kommunisme. Det er budskabet i en seks kilo tung bog fra islamister i Tyrkiet, som er blevet sendt til lærere på flere gymnasier i Danmark.

Blandt andet til biologilærer Thorkild Laustsen fra Horsens Gymnasium. Han så otte eksemplarer af bogen 'Atlas of Creation' ligge på lærerværelset.

Fornægter arternes udvikling
- Så fik jeg at vide, at en af bøgerne var sendt til mig. Resten var også til religions- og biologilærere. Det er jo en stor, flot bog - med masser af flotte billeder, siger Thorkild Laustsen, hvis søn Jeppe har taget et billede af bogen og lagt på sin blog.

n! NYT

LÆS OGSÅ ...
Banket på plads af islamister
LÆSERBLOG: Det er propaganda

AFSTEMNING
Tror du, mennesket nedstammer fra aberne?

Ja 11%
Nej 88%
Måske 1%

SØG PÅ EKSTRABLADET.DK — Find artikel — Søg

nationen!
NYHEDER — Vi mangler internetadresser
NYHEDER — Dagens kommentator: Søren Espersen (DF)
NYHEDER — Æret være ånderne
Trådløs utroskab
Villy er uværdig
Programmet TV2 ikke turde sende
NYHEDER — Vi tjekker politikernes dagsform
Det kan blive for teknisk
UNDERSØGER — Her må du ryge

When Turkish votes manipulated web polls in 2007 it appeared as though Danes had turned creationists overnight. Screenshots by Peter C. Kjærgaard.

among many orthodox Sunni Muslims. Among the attendees, however, no one seemed to question the rejection of evolutionary theory.[32]

Muslim creationism is not only imported to Scandinavia from Turkey. In 2009 the Danish female Sunni convert Aminah Tønnsen, who is among the liberal interpreters of the Quran. argued that, while orthodox Muslims can easily accept an old earth, they should reject Darwin's theory of common descent and human evolution. Reports from high school teachers and students in Denmark and Sweden indicate that many Muslim students are critical of evolutionary biology, especially human evolution. However, quantitative studies are needed before we can draw conclusions on the impact of Islamic creationism on the Muslim minorities in Denmark, Norway, and Sweden.[33]

In the highly secularized Scandinavian countries, creationism is a relatively marginal phenomenon and has not gained a strong foothold in the national Lutheran churches, in the educational system, or in political life. However, since the 1970s flood geology, scientific creationism, and intelligent design theory have been imported from the United States, Britain, Germany, and India, and around 1980 two Scandinavian creationist societies were established with their own quarterly journals. These societies primarily attracted Christian natural scientists with evangelical and charismatic leanings. The active creationists in Scandinavia are restricted to the right wing of the national churches and to small sectarian Protestant free churches, such as orthodox Lutheran denominations, the Seventh-day Adventist Church, the Pentecostal Church, and the Word of Faith movement. In the 2000s, intelligent design theory was welcomed by both young- and old-earth creationists, who regarded the theory as a useful tool in their attempts to popularize antievolutionary theories in confessional schools. In 2007 Swedish politicians felt it necessary to ban the teaching of alternatives to evolutionary biology in state-funded religious schools, although this was more a response to a resolution from the Council of Europe than to an immediate threat to the Swedish curriculum. At this time, Muslim creationism received much attention owing to Adnan Oktar's campaign against evolutionary theory, which seems to have had some impact on the large Muslim minorities in the Denmark, Norway, and Sweden. In general, however, the influence of creationism, local as well as imported, is limited in Scandinavia and is likely to remain so without strong financial and political backing and greater activism among the relatively small number of religious fundamentalists in the region.

NOTES

1. European Commission, *Europeans, Science and Technology*, Special Eurobarometer 224, Wave 63.1 (Brussels: European Commission, 2005), accessed Oct. 31, 2011, http://ec.europa.eu/public_opinion/archives/eb_special_240_220_en.htm; Jon D. Miller, Eugenie C. Scott, and Shinji Okamoto, "Public Acceptance of Evolution," *Science* 313 (2006): 765–766; Niels Thimmer and Klaus Lange, "I Danmark holder Darwin," *MetroXpress*, Feb. 12, 2009, 10; Glenn Branch, "Creationism as a Global Phenomenon," in *Darwin and the Bible: The Cultural Confrontation*, ed. R. H. Robbins and M. N. Cohen (Boston: Penguin, 2009), 137–51; Ipsos Global @dvisory, *Supreme Being(s), the Afterlife and Evolution*, accessed Oct. 31, 2011, www.ipsos-na.com/news-polls/pressrelease.aspx?id=5217.

2. Phil Zuckerman, *Society without God: What the Least Religious Countries Can Tell Us about Contentment* (New York: New York University Press, 2008); Andrew Buckser, "Religion, Science, and Secularization Theory on a Danish Island," *Journal for the Scientific Study of Religion* 35:4 (1996): 432–441; Kjell Lejon and Marcus Agnafors, "Less Religion, Better Society? On Religion, Secularity, and Prosperity in Scandinavia," *Dialog: A Journal of Theology* 50:3 (2011): 297–307.

3. Center for Samtidsreligion, *Religion i Danmark 2011*, accessed Nov. 1, 2011, http://teo.au.dk/csr/rel-aarbog11/forside/; Danmarks Statistik, accessed Nov. 1, 2011, www.dst.dk/; Statistics Norway, *KOSTRA: Kirke 2010*, accessed Nov. 1, 2011, www.ssb.no/emner/07/02/10/kirke_kostra/; Statistics Norway, accessed Nov. 1, 2011, www.ssb.no/; Church of Sweden, accessed Nov. 1, 2011, www.svenskakyrkan.se/; Statistics Sweden, accessed Nov. 1, 2011, www.scb.se/; Buckser, "Religion, Science, and Secularization Theory on a Danish Island"; Zuckerman, *Society without God*; Bronislaw Szerszynski, "Understanding Creationism and Evolution in America and Europe," in *Science and Religion: New Historical Perspectives*, ed. Thomas Dixon, Geoffrey Cantor, and Stephen Pumfrey (Cambridge: Cambridge University Press, 2010), 153–174; Erik Sidenvall, "A Classic Case of De-Christianisation? Religious Change in Scandinavia, c. 1750–2000," in *Secularisation in the Christian World*, ed. Callum G. Brown and Michael Snape (Aldershot: Ashgate, 2010), 119–133; Gallup, *Religiosity Highest in World's Poorest Nations*, Gallup Global Reports, Aug. 31, 2010, www.gallup.com/poll/142727/religiosity-highest-world-poorest-nations.aspx; European Commission, *Social Values, Science and Technology*, Special Eurobarometer 225, Wave 63.1 (Brussels: European Commission, 2005), accessed Oct. 31, 2011, http://ec.europa.eu/public_opinion/archives/eb_special_240_220_en.htm.

4. Center for Samtidsreligion, *Religion i Danmark 2011*; U.S. Department of State, *International Religious Freedom Report 2009: Sweden*, accessed Nov. 1, 2011, www.state.gov/g/drl/rls/irf/2009/127339.htm; Statistics Norway, *Trus- og livssynssamfunn utanfor Den norske kyrkja, per 1. januar 2011*, accessed Feb. 17, 2012, www.ssb.no/trosamf/.

5. U.S. Department of State, *International Religious Freedom Report 2009: Sweden*; Center for Samtidsreligion, *Religion i Danmark 2011*; Lene Kühle, *Moskeer i Danmark*, accessed Jan. 16, 2010, www.teo.au.dk/cms/pluralismeprojektet/islamdk1; Statistics Norway, *Trus- og livssynssamfunn utanfor Den norske kyrkja, per 1. januar 2010*, accessed Nov. 1, 2011, www.ssb.no/trosamf/; Lars Østby, "Muslimsk flertall i Norge?," *Morgenbladet*, Aug. 26, 2011, www.morgenbladet.no/article/20110826/OIDEER/708269967.

6. Peter C. Kjærgaard, Niels Henrik Gregersen, and Hans Henrik Hjermitslev, "Darwinizing the Danes," in *The Reception of Charles Darwin in Europe*, vol. 2, ed. Thomas F.

Glick and Eve-Marie Engels (Cambridge: Continuum, 2008), 146–155; Thore Lie, "The Introduction, Interpretation and Dissemination of Darwinism in Norway during the Period 1860–90," in *The Reception of Charles Darwin in Europe*, vol. 2, ed. Thomas F. Glick and Eve-Marie Engels (Cambridge: Continuum, 2008), 156–174; Ulf Danielsson, "Darwinismens inträngande i Sverige," *Lychnos* (1963–64): 157–210; (1965–66): 261–334.

7. Danielsson, "Darwinismens inträngande i Sverige," 292–298; Thore Lie, "Fra Origin of Species til Arternes Oprindelse—darwinisme og utviklingslære i Norge (1861–1900)," in *Evolusjonsteorien: Status i norsk forskning og samfunnsdebatt*, ed. Nils Christian Stenseth and Thore Lie (Oslo: Gyldendal Norsk Forlag, 1984), 40–63.

8. Niels Henrik Gregersen and Peter C. Kjærgaard, "Darwin and the Divine Experiment: Religious Responses to Darwin in Denmark 1859–1909," *Studia Theologica* 63:2 (2009): 140–161; Hans Henrik Hjermitslev, "Protestant Responses to Darwinism in Denmark, 1859–1914," *Journal of the History of Ideas* 72:2 (2011): 279–303.

9. Kurt Ettrup Larsen, *En bevægelse i bevægelse: Indre Mission i Danmark 1861–2011* (Fredericia: Lohse, 2011), 419; Eugen Warming, *Udviklingslærens Standpunkt i Nutiden* (Copenhagen: Tænk Selv, 1910); Eugen Warming, *Nedstamningslæren* (Copenhagen: Udvalget for Folkeoplysningens Fremme / G.E.C. Gads Forlag, 1915; 2nd ed. 1927); Hans Henrik Hjermitslev, "Danes Commemorating Darwin: Apes and Evolution at the 1909 Anniversary," *Annals of Science* 67:4 (2010): 485–525, esp. 510–512; Amand Breitung, *Abeteoriens Bankerot og vor populære Darwinisme* (Copenhagen: Høst & Søn, 1899); Amand Breitung, *Udviklingslæren og Kristentroen—Et "Baade—Og" imod Vilh. Rasmussens og andres "Enten—Eller" i "Vor Ungdom," "Frem" og "Verdensudviklingen"* (Copenhagen: T.T.R. Thomassen s Bogtrykkeri, 1905); Hjermitslev, "Protestant Responses to Darwinism in Denmark, 1859–1914," 294–295.

10. Helge Rode, *Krig og Aand* (Copenhagen: Gyldendal, 1917); Helge Rode, *Regenerationen i vort Aandsliv* (Copenhagen: Gyldendal, 1923) (quote); Helge Rode, *Pladsen med de grønne Træer: Den religiøse Strømning i Nutidens Aandsliv* (Copenhagen: Gyldendal, 1924; 6th ed. 1926); Helge Rode, *Det store Ja* (Copenhagen: Gyldendal, 1926); Helge Rode, *Den sjette Dag: Smaa Indlæg om Udvikling og Skabelse* (Copenhagen: Gyldendal, 1927); Jesper Vaczy Kragh, *Kampen om Livsanskuelse* (Odense: Syddansk Universitetsforlag, 2005), 186.

11. Knud Asbjørn Wieth-Knudsen, *Darwinismen i Likvidation: Antidarwinistiske Causerier* (Copenhagen: Jul. Gjellerups Forlag, 1925); Vaczy Kragh, *Kampen om Livsanskuelse*; Ronald L. Numbers, *The Creationists: From Scientific Creationism to Intelligent Design*, expanded ed. (Cambridge, MA: Harvard University Press, 2006); Ragner Spärck et al., *Udviklingslæren* (Copenhagen: Jespersen og Pio, 1929), 138.

12. Per Bergan, "Norsk evolusjonsopplysning og evolusjonsdebatt i vårt århundre," in *Evolusjonsteorien: Status i norsk forskning og samfunnsdebatt*, ed. Nils Christian Stenseth and Thore Lie (Oslo: Gyldendal Norsk Forlag, 1984), 64–70.

13. J. C. Whitcomb Jr. and H. M. Morris, *The Genesis Flood: The Biblical Record and Its Scientific Implications* (Philadelphia: Presbyterian and Reformed Publishing, 1961); Hannington Enoch, *Udvikling eller skabelse*, trans. Knud Aage Back (Copenhagen: Forlaget Perspektiv, 1971); Numbers, *The Creationists*, 419; Frode Thorngreen, *Syndfloden* (Holbæk: Ordet og Israels Forlag, 1975; 3rd ed. 1984).

14. Odd Sæbö, *Dogmet om evolusjonen* (Oslo: Luther Forlag, 1975); Odd Sæbö, *Udviklingslæren—Videnskab eller tro?*, trans. Kirsten and Poul Hoffmann (Holbæk: Dansk Luthersk Forlag, 1977); www.poulhoffmann.dk, accessed Nov. 7, 2011; E. H. Andrews, *Ur intet,*

trans. Kerstin Gislén and Erik Gislén (Angered: Det står skrevet, 1980); E. H. Andrews, *Livet blev til*, trans. Kirsten Hoffmann and Poul Hoffmann (Copenhagen: Credo, 1984); A. E. Wilder-Smith, *Naturvitenskap uten evolusjon. Eksperimentelle og teologiske innvendinger mot evolusjonsteorien*, trans. Willy Fjeldskaar (Oslo: Luther Forlag, 1982); Jens Gabriel Hauge, "Evolusjon og kristen skapertro," in *Evolusjonsteorien: Status i norsk forskning og samfunnsdebatt*, ed. Nils Christian Stenseth and Thore Lie (Oslo: Gyldendal Norsk Forlag, 1984), 286–295; Numbers, *The Creationists*, 358–368.

15. Poul Hoffmann, *Dinosaurerne og syndfloden* (Fredericia: Lohses Forlag, 1996), 77 (quote); Hans Henrik Hjermitslev, "Dansk kreationisme," *Religion* 1 (2010): 16–27, 19–20.

16. Hauge, "Evolusjon og kristen skapertro," 292; Numbers, *The Creationists*, 367–368.

17. Hjermitslev, "Dansk kreationisme"; "Norsk medarbejder ved ORIGO," *Origo* 6:2 (1988): 2; "Visionen om ORIGO som en apologetisk ressource," *Origo* 100 (2006): 4–5 (quotes); Peder A. Tyvand, "Skapelsesdager viktigere enn skabelsen?," *Origo* 85 (2003): 25–26; www.skabelse.dk and www.origonorge.no, both accessed Nov. 6, 2011.

18. Numbers, *The Creationists*, 148–155, 409; See also Kutschera, chapter 6 in this volume.

19. Hjermitslev, "Dansk kreationisme"; www.origonorge.no; www.skabelse.dk; www.darwin2009.no; www.darwin2009.dk, all accessed Nov. 6, 2011; Kristian Bánkuti Østergaard, *Evolution—hvad din biologibog ikke fortæller, Origo* 105 (2007); Kristian Bánkuti Østergaard, *Lærervejledning til temahæftet "Evolution—hvad din biologibog ikke fortæller,"* accessed Jan. 16, 2010, http://darwin2009.dk/data/LaerervejledningTilEvolutionshaefte4.0.pdf; Jonathan Wells, *Evolutionens ikoner* (Herning: Origo, 2006); Knud Aa. Back, *Humlebien kan ikke flyve: design i naturen? Det 'et spørgsmål om fysik* (Herning: Origo, 2009); Jostein Andreassen, *Darwinbogen* (Herning: Origo, 2009); Peder A. Tyvand, *Darwin 200 år—en festbrems* (Herning: Origo, 2009); *Dig og mig og Darwin: Tro og Viden—Om livets fakta og forunderlighed* (Løgumkloster: Landsnetværket af Folkekirkelige Skoletjenester og Skole-Kirke Samarbejder i Danmark, 2009); *Stykker af himlen: Tro og Viden—Om livets fakta og forunderlighed* (Løgumkloster: Landsnetværket af Folkekirkelige Skoletjenester og Skole-Kirke Samarbejder i Danmark, 2009); *Verden set med to øjne: Tro og Viden—Om livets fakta og forunderlighed* (Løgumkloster: Landsnetværket af Folkekirkelige Skoletjenester og Skole-Kirke Samarbejder i Danmark, 2009); Peter C. Kjærgaard, "The Darwin Enterprise: From Scientific Icon to Global Product," *History of Science* 48:1 (2010): 105–122; Casper Andersen, Jakob Bek-Thomsen, Mathias Clasen, Stine Slot Grumsen, and Peter C. Kjærgaard, "The Unexpected Learning Experience of Making a Digital Archive," *Science & Education* (2011), online Nov. 5, 2011, www.springerlink.com/content/k235108270125175/.

20. Maziar H. Etemadi and Peter Øhrstrøm, *Intelligent Design: An Intellectual Challenge?* (Aalborg: Aalborg University Press, 2007).

21. www.skabelse.dk, accessed Jan. 14, 2010; *Origo* 1–121 (1983–2011); Jørgen Lerche Nielsen, "Religion, evolution og abeprocesser," in *Naturens Historiefortællere*, vol. 2, ed. N. Bonde, J. Hoffmeyer, and H. Stangerup (Copenhagen: Gad, 1987), 469–496; Torben Riis, "At læse i naturens bog: en samtale med professor dr. Scient, Peter Øhrstrøm om darwinisme, skabelsestro og intelligent design," *Katolsk Orientering*, Jan. 11, 2006; Henri Nissen, "I tro fatter vi . . . ," accessed Jan. 14, 2010, domino-online.dk/?p=1324; Holger Daugaard, *Skabelse eller udvikling*, vols. 1–10 (Nærum: Korrespondanceskolen); Stefaan Blancke, "Creationism in the Netherlands," *Zygon* 45:4 (2010): 791–816; Numbers, *The Creationists*; Martin Riexinger, "Islamic Opposition to the Darwinian Theory of Evolu-

tion," in *Handbook of Religion and the Authority of Science*, ed. James R. Lewis and Olav Hammer (Leiden: Brill, 2010), 483–510.

22. Numbers, *The Creationists*, 411; Lerche Nielsen, "Religion, evolution og abeprocesser"; Hjermitslev, "Dansk kreationisme"; Gary E. Parker, *Skabelse og videnskab*, trans. Bent Vogel and Holger Daugaard (Fredericia: Lohses Forlag / Credos Forlag, 1995).

23. www.origonorge.no; www.genesis.nu; http://old.genesis.nu/8thecc/mike_story .html, all accessed Nov. 6, 2011.

24. *Origo* 96 (2005); Mats Molén, *Vårt Ursprung: om universums, jordens och livets uppkomst samt historia* (Haninge: Salt och Ljus, 1988; 4th ed. 2000); Mats Molén, *Livets Uppkomst—evolution och vetenskap: om jordens samt livets uppkomst och historia* (Haninge: XP media, 2000); Mats Molén, *Evolutionslåset: hur västvärlden kan öppnas för evangelium* (Haninge: XP media, 2004); Mats Molén, *När människan blev ett djur: om evolutionsteorins konsekvenser för individ och samhälle* (Haninge: XP media, 2005); Mats Molén, *Og Gud skabte Darwin—Evolution og videnskab*, trans. Helge Hoffmann (Fredericia: Lohse/Credo, 2010).

25. Anders Gärdeborn, *Intelligent skapelsetro: en naturvetare läser Första Moseboken: vetenskap och religion: biblisk skapelsetro, evolution, intelligent design* (Haninge: XP Media, 2007); Hjermitslev, "Dansk kreationisme," 21; "Debatten i Sverige," *Origo* 107 (2007): 20–23; Per Kornhall, *Skapelsekonspirationen: Fundamentalisternas angrepp på utvecklingsläran* (Stockholm: Leopard Forlag, 2008), 145–152.

26. Kornhall, *Skapelsekonspirationen*, 153–175.

27. Hjermitslev, "Dansk kreationisme," 23; www.skabelsen.dk (Lars Poulsen: Apostolsk Kirke in Vejle, young-earth creationism); www.oprindelse.dk and www.darwin sludder.dk (Ole Michaelsen: young-earth creationism); skabelsesberetningen.dk (Ole Madsen: gap theory old-earth creationism); www.apologetik.dk (David Jakobsen); www .amen.nu (Simon Griis); Simon Griis, *Nyt lys over skabelsen*, compendium and 12 cd's, www.gudogvidenskaben.dk; www.oplevjesus.dk (Torben Søndergaard); all pages accessed Jan. 15, 2010; *Livet—Hvordan er det kommet her? Ved en udvikling eller en skabelse?* (Holbæk: Vagttårnet, 1985); *Findes der en skaber?*, special issue of *Vågn op*, Sept. 2006.

28. Jakob Wolf, *Rosens råb: Intelligent deisgn i naturen. Opgør med darwinismen* (Copenhagen: Anis, 2004); Astrid Nonbo Andersen, "Debatten om Intelligent design," *Aktuel Naturvidenskab* 5 (2006): 28–29; Torben Hammersholt Christensen and Søren Harnow Klausen, eds., *Darwin eller intelligent design* (Copenhagen: Anis, 2007).

29. Numbers, *The Creationists*, 420; Hjermitslev, "Dansk kreationisme," 23–24; Leif Asmark Jensen, *Intelligent design: et nyt syn på udviklingen* (Mørkøv: Skou & Asmark, 2005); Michael Cremo, *Forbudt arkæologi* (Mørkøv: Skou & Asmark, 2006); Stefan Frello, *Intelligent Design*, www.intelligentdesign.skysite.dk; www.intelligentdesign.dk; both accessed Nov. 6, 2011.

30. Hjermitslev, "Dansk kreationisme"; Peter C. Kjærgaard, "Western Front," *New Humanist* 123:3 (2008): 39–41; Harun Yahya, *Atlas of Creation*, vols. 1–2 (Istanbul: Global Publishing, 2007); Halil Arda, "Sex, Flies and Videotape: The Secret Lives of Harun Yahya," *New Humanist* 124:5 (2009), accessed Feb. 17, 2012, http://newhumanist.org.uk/2131/sex-flies-and-videotape-the-secret-lives-of-harun-yahya; Martin Riexinger, "Propagating Islamic Creationism on the Internet," *Masaryk University Journal of Law and Technology* 2:2 (2008): 99–112; Peter C. Kjærgaard, "Vårflue afslører manipulation," www.videnskab .dk, June 8, 2008; *Hvad er Islam?* (Brabrand: Ligheds- og Broderskabsforeningen, 2008); Harun Yahya, *Til forstandige mennesker: Himlenes og jordens tegn*, trans. Omar Louborg

(Aarhus: Islamisk Videns- og Informationscenter, 2003); www.harunyahya.com, accessed Nov. 6, 2012.

31. Thomas Harder, "Mysteriet om IP-adresserne," www.ekstrabladet.dk, Sept. 13, 2007; Kate P. Clark, "Gymnasielærer vil bruge omstridt bog i undervisningen," *Jydske Vestkysten*, Sept. 1, 2007, 4; Thomas Harder, "Tillykke Darwin, din gamle fusker," www.ekstrabladet.dk, Feb. 12, 2009; Thomas Harder, "Svar fra Anti-darwinisten Adnan Oktar," www.ekstrabladet.dk, Mar. 6, 2009; www.harunyahya.com, accessed Jan. 16, 2010 (quote); Kjærgaard, "The Darwin Enterprise," 15.

32. www.harunyahya.com, accessed Nov. 6, 2011; Søren K. Villemoes, "Hvordan bliver en giraf en giraf," *Weekendavisen*, June 10, 2011, 4.

33. Amirah Tønnsen, "Islam og livets oprindelse," accessed Nov. 6, 2011, www.livetsoprindelse.dk/Faglige%20artikler.aspx; Kornhall, *Skapelsekonspirationen*, 276–277.

CHAPTER SIX

Germany

ULRICH KUTSCHERA

Surveys of public opinions indicate that most Germans accept evolution. In the frequently cited poll by Miller, Scott, and Okamoto, more than 70 percent of the Germans responded that evolution may be true, whereas slightly more than 20 percent believed that it is probably or definitely false. A recent Ipsos Mori poll found that 12 percent subscribed to a creationist view on human evolution, whereas 65 percent accepted an evolutionary account. Nonetheless, Germany houses one of the best-organized, vocal, and active groups of creationists in Europe.

On December 19, 2011, the widely distributed German newspaper *Süddeutsche Zeitung* reported that intelligent design creationism was being taught in biology classes at a state-sponsored Christian school, the Freie Christliche Bekenntnisschule in Gummersbach. Only two days later, the Christian media-magazine *Pro* responded to this "biased accusation." The anonymous authors of the *Pro* article admitted that in more than one hundred *Bekenntnisschulen* (faith schools) that currently exist in Germany, "gaps in the theory of evolution are discussed and the possibility of God's creation is taught." In addition, the authors justified the teaching of intelligent design creationism as an alternative to evolutionary theory with reference to a section of the German School law, wherein it is stated that Christian associations can present their religious doctrines in a way corresponding to their literal interpretation of the Bible. It should be noted that, three decades ago, there were only four evangelical private schools in West Germany. As such, the small German antievolution agenda of the early 1980s developed into an influential movement.[1]

In this chapter, after sketching the religious landscape in Germany, I outline the origin and evolution of the modern creationist movement in this country, with a focus on the activities and publications of the Studiengemeinschaft Wort und Wissen (Study Community Word and Knowledge), the most influential creationist group in German-speaking countries of Europe that include Germany, Austria, and Switzerland. I devote special attention to the "theory of Basic Types," which has been introduced and explained at length by two members of the organization in a textbook, which is especially designed to influence the teaching of evolutionary theory in German classrooms. Also considered are the prevalence of creationism in German schools, creationist activities in the "Darwin year 2009," and the impact of religious education on the distribution of creationist beliefs in Germany.

The Religious Landscape

Although there are substantial differences between former East and West Germany, on average among those Germans who express a religious preference, one-third belongs to the Protestant Church, another third is Roman Catholic, and about 4 percent of the population is Islamic. The country, however, is highly secular. Many Germans, particularly in former East Germany, describe themselves as nonbelievers, and in many states in the East antireligious people form a majority. The exact numbers vary from survey to survey, but approximately 35 percent of Western Germans and more than 70 percent of Germans in the East can be considered atheists or agnostics.[2]

In Germany, creationism is largely situated within the evangelical movement. Note, however, that in the German language, *evangelisch* means Protestant, whereas *evangelikal* stands for a more or less literal interpretation of the Bible and should be translated as evangelical. Hence, the Evangelische Kirche in Deutschland (EKD) means the "Protestant Church in Germany," which is an association comprising twenty-two Lutheran, Reformed (Calvinist), and Unified Protestant regional church bodies in Germany (*Landeskirchen* or regional churches). This is the church to which one in three Germans belongs. The members of this church should not be confused with the *Evangelikalen*, who represent not more than 1.5 million people, that is, 2 to 3 percent of the German population. However, many evangelicals do belong to a mainstream Protestant church, although most are members of the so-called Freikirchen (free churches). Most Free Christian *Bekenntnisschulen* are evangelical (i.e., strictly Bible-based) but nevertheless are connected to the EKD federation.[3]

Finally, from my personal experience in discussions with German creationists, the distinction between Christians who are evangelisch or evangelical is not always clear. In other words, a continuum exists between Protestantism, a religious view that is compatible with the concept of God-directed (theistic) evolution, and fundamentalism, a Bible-based ideology that cannot be reconciled with the fact of descent with modification and the common ancestry of all forms of life. However, despite some vocal creationist movements, creationism in Germany remains a marginal phenomenon, mainly concentrated in just a few German states (*Bundesländer*).

The Rise of Modern Creationism in Germany

In 1959, when scientists around the world celebrated the centenary of the publication of Charles Darwin's *On the Origin of Species*, a book entitled *The Erroneous Way of Darwinism* was published that marked the beginning of the first antievolution agenda in Germany. The author, the zoologist Robert Nachtwey, argued that Darwin's theory leads to immorality and that it explains nothing except for the occurrence of degenerations. Instead, he proposed an alternative theory that he called "creative Lamarckism" and claimed that "life phenomena and developmental processes" can be explained only by a "mysterious, creative, and teleological principle that is already present in each and every cell." Moreover, Nachtwey claimed that evolutionary theory had provided the basis for the ideology of the Nazis, an argument that is still being used today by creationists in Germany and elsewhere. During the 1960s and 1970s, the British chemist Arthur E. Wilder-Smith, who lived in Switzerland, introduced American-style young-earth creationism into Germany by writing numerous creationist books and articles in German and by translating the works of Henry Morris and other American creationists. Thus, Wilder-Smith may have laid the foundation for a German creationist movement. Between 1990 and 2003, there were several creationist organizations and groups active in Germany, including the Jehovah's Witnesses and the Studiengemeinschaft Wort und Wissen, the first German creationist organization. The latter was founded in 1979 by Horst W. Beck, a theologian and engineer who first had accepted evolution but converted to creationism in 1978 after having attended the lectures of Willem Ouweneel, a Dutch creationist. The organization initiated the creationist monthly journal *Factum*, which is still published today in Switzerland. In addition, it sponsored conferences and maintained contact with American creationists.

Today, the organization is based in Bayersbronn in southern Germany. It is supported by private sponsors and has about two hundred members. In 2000 the society spent about 532,000 Deutsche Mark (approximately €266,000 or $344,000) for the employment of staff members and the support of projects for the distribution of creationism in Europe. According to the doctrines of the society, the firm belief in Jesus Christ and the biblical account (the "Word") provide crucial knowledge about the real world, notably the origin and evolution of organisms. Nonetheless, the organization has repeatedly distanced itself

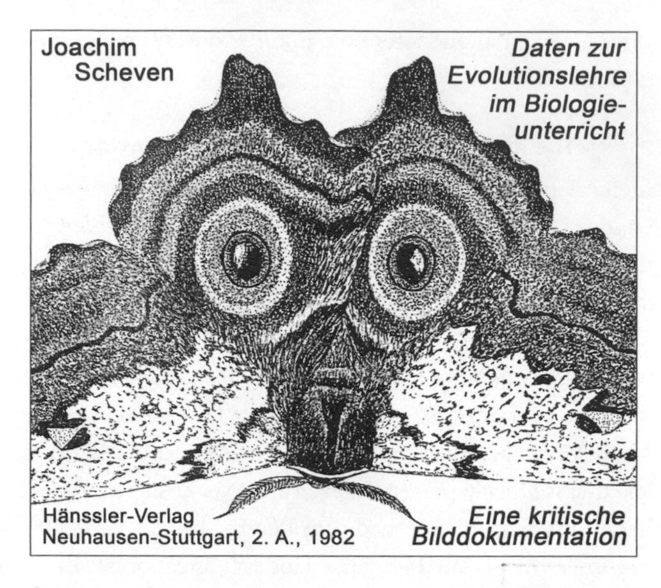

The origin of young-earth-creationism in Germany. Drawing of an East African moth (*Athletes semialba*). The "owl-face" may signal to birds that their prey may be dangerous. According to the author, J. Scheven, this "perfect design" must have been created by the biblical God. Adapted from Joachim Scheven, *Daten zur Evolutionslehre im Biologie-unterricht* (Neuhausen-Stuttgart: Hänssler-Verlag, 1982).

from creationism. In order to promote its beliefs, the organization issues various publications including a newsletter and a journal with the title *Studium Integrale Journal.* Its main goal is to promote the teaching of creationism in public schools in the three German-speaking countries of Europe: Germany, Austria, and Switzerland.[4]

In 1982, one of the other founding members of the society, Joachim Scheven, a German schoolteacher, wrote the booklet *Data on the Teaching of Evolution in Biology Classes,* which was published by the Study Community Word and Knowledge and is probably the first German creationist textbook. Scheven started off his career as an "evolutionist," obtaining his PhD in paleontology at the University of Munich. However, later he adopted a version of creationism that was generally considered to be typical of the United States. On the book's cover, Scheven explained that he was a biblical literalist, believing that the earth was only a few thousand years old and that all species on the planet, including mankind, had been created as described in the book of Genesis.

Scheven's book contains numerous illustrations, including a drawing by the author that depicts one of the miraculous acts of the biblical God, resulting in "perfectly designed" creatures. Also, Scheven argued that there was not a shred of evidence for the theory of evolution and that therefore the entire philosophy of evolutionary thinking is without factual basis.[5]

Scheven's little textbook covered many aspects of evolutionary biology from a biblical perspective. Six years later, in 1988, he published a monograph entitled *Megasuccessions and Climax in the Tertiary: Catastrophes between the Deluge and the Ice Age.* In this illustrated volume, Scheven argued that the 47-million-year-old fossils of the Grube Messel in Darmstadt, Germany, should be assigned to the time after the Great Flood or the Deluge, which, according to the Bible, occurred only a few thousand years ago.[6]

In the 1990s Scheven founded the creationist journal *Leben. Deutsches Schöpfungsmagazin* (Life: A German creation magazine), and from 1984 to 2001 he managed a creationist museum Lebendige Vorwelt (Living World before the Present) in Hagen-Hohenlimburg that displayed living fossils to show that evolution had not occurred, a tactic also used in Harun Yahya's *Atlas of Creation.* Today, both the journal and the museum, which were praised by the Study Community Word and Knowledge, no longer exist. Another book, the first in a series of three private publications by Scheven, accompanied this failed project. In this monograph, the author argued that insects, preserved in amber, prove that the earth is young and that evolution had not occurred. In the opinion of Scheven, these amber fossils are proof for creation as described in the Bible.[7]

In 2007 Scheven published a large volume on geology and paleontology from a young-earth creationist perspective, which he distributed to public schools free of charge. However, owing to its crude anti-science content and direct references to the Bible (including e.g., a three-thousand-year-old earth, the Deluge, and one ice age), this "very young-earth creationism" did not have much impact in Germany or elsewhere, despite the support of his daughter Esther Scheven (an employee at the Deutsche Nationalbibliothek, Frankfurt) and some private sponsors from evangelical communities.[8]

Finally, in 2010, Scheven published yet another book, called *A Biology Teacher Discovers Creation.* In this illustrated monograph, the author describes his experience as a naturalist in Africa and presents his readers with excellent drawings of tropical butterflies, which he regards as proof of "God's creations." These high-quality pictures are similar to those he had published in his earlier textbook three decades ago.[9]

When comparing the five books Scheven published over the past three decades, it becomes clear that, in his first work, the author described his US-style (very) young-earth creationism much more clearly than in his later publications, in which he focused more on selected topics from geology and biology and mixed well-established scientific facts with his biblical belief. However, the impact of these books remained limited to the evangelical communities of Germany, whose members are convinced that "Darwin was wrong" anyway. The lack of a clear concept (i.e., an "antievolutionary theory") and, more importantly, the absence of a powerful lobby backing these activities may account for the marginal success of Scheven's antievolutionary agenda. However, because his booklet of 1982 can be considered to be the precursor of a book published four years later that had much more impact, Scheven did ultimately exert at least some influence, albeit indirectly.

The Basic Types of Life

In October 1986, the first edition of a textbook that summarized and considerably extended the conclusions of Scheven's booklet of 1982 was issued by the Study Community Word and Knowledge. *Origin and History of the Organisms: Data and Interpretations for Biology Classes* was published by Ulrich Weyel, an evangelical Christian associated with the Evangelische Allianz in Gießen, a town in the state of Hessen. The book's authors, Reinhard Junker and Siegfried Scherer, were supported by a team of ten coauthors, who were also members of the Study Community Word and Knowledge. However, none of them was a professional evolutionary biologist. In the acknowledgments, the authors thanked Horst W. Beck for organizing the lecture series that had given rise to the book's publication. They also added a brief motto for their richly illustrated, low-priced college-level book: "There exists an alternative to the (unproven) assumption of macroevolution that is motivated by the revelations of the Bible—the theory of creation."[10]

In this illustrated monograph, Junker, a former biology teacher, theologian, and employee of the society, and Scherer, a microbiologist at the Technical University of Munich, introduced their theory of the "Grundtypen des Lebens" (Basic Types of Life), which comprised a mix of young-earth creationism and intelligent design. They also referred to an article by Scherer and Thomas Hilsberg that had been published earlier in a German scientific periodical, the *Journal für Ornithologie*, in which they proposed that the order of the Anatidae (ducks, geese, and swans) could be interpreted as a created "Basic Type." In several articles, the first author, Junker, explicitly pointed out that the Basic Types of Life had been created "by the word of the God in the Bible about 10,000 years ago." However, in the textbook, this short time interval is nowhere mentioned. Nonetheless, it sets the time frame of their model according to which the created

Covers of the first (1986) and sixth editions (2006) of the critical textbook authored and edited by R. Junker and S. Scherer. Note the "Basic Types of Life" on the cover of the first edition, to which is juxtaposed a scheme of the theory of evolution via natural mechanisms (the phylogenetic tree indicating common descent of vertebrates).

Basic Types diversified via rapid microevolution (i.e., high-speed speciation processes) and thus gave rise to the enormous biodiversity of today. This also holds for the approximately 7 billion people that have all descended from the originally created couple, Adam and Eve, whom the authors regard as the Basic Types of humankind. Unsurprisingly, as this model is based on the belief that the Creator or the Designer acted by means of independent miracles, it is depicted in mini-trees without common descent.[11]

The concept of the Basic Types of Life, however, was not original. Junker and Scherer referred to American young-earth creationist Frank L. Marsh as the true originator and spiritual father of the Basic Types concept, derived from the Genesis "kinds," which he called *Baramins*. Marsh was a member of the Los Angeles–based Society for the Study of Creation, the Deluge, and Related Science, which was founded in 1938. As a teacher at an Adventist school in the Chicago area, he had earned a MS in zoology in 1935. In 1940 he completed his PhD in botany, becoming the first Adventist of the United States with an advanced degree in the biological sciences. As a biblical literalist, Marsh regarded

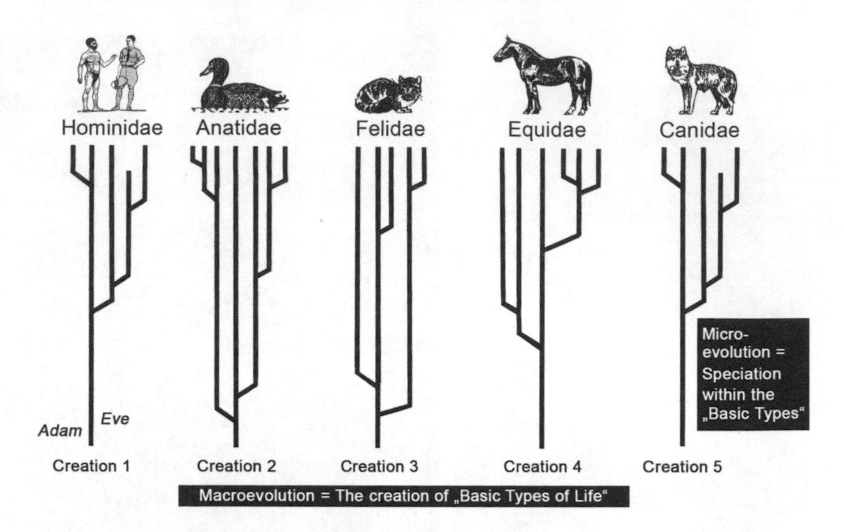

Scheme illustrating the "Basic Types of Life," according to the "Adam and Eve" model, as envisioned by the German creationists R. Junker and S. Scherer. The God of the Bible created animals (and plants) after their own kind, without common descent. These "ground types" were equipped with an inbuilt capacity for variation: via rapid microevolutionary processes, new species evolved at high rates within less than ten thousand years. Adapted from Ulrich Kutschera, *Evolutionsbiologie*. 3. Auflage (Stuttgart: Verlag Eugen Ulmer, 2008).

the statement "after their kind" written down in the book of Genesis as a biological law that separated the created kinds from each other. Marsh had developed his concept of basic kinds in a book entitled *Evolution, Creation and Science* that was consequently sent out by the publisher to two leading evolutionary biologists, the German American zoologist Ernst Mayr and the Russian American geneticist Theodosius Dobzhansky. Both scientists later became known as two of the major architects of the evolutionary synthesis. Mayr, who was an atheist, declined to comment on this theistic antievolution pamphlet, but Dobzhansky, who was a nondogmatic Russian Orthodox Christian, wrote a devastating review. Thirty-two years later, in 1976, Marsh published his final opus. Under the title *Variation and Fixity in Nature: The Meaning of Diversity and Discontinuity in the World of Living Things, and Their Bearing on Creation and Evolution,* Marsh summarized his ideas about the Basic Types of Life and concluded that "the Bible knows nothing about organic evolution." Moreover, he argued that his theory should be "respected as an alternative view-point to evolutionism." The German model of the Basic Types by Junker and Scherer is much indebted to this American concept.[12]

The theory of Basic Types also resonates within intelligent design. In his seminal book *Darwin on Trial*, published in 1991, Phillip E. Johnson, the father of the American intelligent design movement, stated that the "[Creationists'] doctrine has always been that God created basic kinds, or types, which subsequently diversified. The most famous example of creationist microevolution involves the descendants of Adam and Eve, who have diversified from a common ancestral pair to create all the diverse races of the human species." Since 2003, this American best seller has also been available in a German translation, which is distributed and promoted by the Study Community Word and Knowledge. One of its prominent members, pharmacologist and young-earth creationist Peter Imming, wrote the preface, in which he praised the work as highly valuable and full of insights, and he warmly recommended it to Christians of all denominations. Official documents published by the Study Community Word and Knowledge show that the book *Darwin im Kreuzverhör* is regarded as a welcome supplement to its own model of the Basic Types of Life.[13]

In 1992 Junker and Scherer actively sought international collaboration, when Scherer presented their theory at the Third International Conference on Creationism in Pittsburg under the headline "Basic Types of Life, Genesis Kinds Not Species." In a report about this meeting, Scherer summarized the views of himself and his colleagues as follows:

> Siegfried Scherer and colleagues employ the term "Basic Type" (also called "Ground Types"), coined by Frank Marsh some decades ago, to denote a Genesis Kind, and outlined their findings at the 3rd International Conference on Creationism in Pittsburg . . . The "Basic Types" include, for example, the Anatidae (ducks, geese and swans), Equidae (horses and zebras), Cercopithecidae (old-world monkeys), Canidae (dogs, wolves and foxes), Felidae (cats, tigers, leopards), Maloideae (apples and relatives), Aspleniaceae (spleenwort ferns) and Triticeae (wheat, barley, oats and rye) . . . Scherer and colleagues have recently brought ideas together into a publication, *Typen des Lebens*, which contains an overview, then a chapter focusing on each "Basic Type." This is an important book. It may provide the most significant contribution yet offered in this decade to a creationist research programme.[14]

Scherer referred to a monograph edited by himself in which the theory of the Basic Types of Life was described in much more detail, and which also contained an English summary. The volume had a positive review in the international scientific journal *Flora*, in which the reviewer concluded that "the Basic Types should be tested" and recommends that the ideas described in this book be followed up. Later, in 2002, the same reviewer would write a positive review of the fifth edition of the Junker and Scherer textbook, again in the same journal. By that time, the German creationists had some success in publishing in the international scientific literature. Scherer and a colleague managed to publish a News and Comment article in the highly esteemed journal *Trends in Ecology and Evolution*. They did not explicitly refer to the concept of Basic Types, but they wrote that "the hypothetical descent of mankind from 'mitochondrial Eve' has been much debated . . . Nobody was actually there . . . If molecular evolution is really neutral at these sites, such a high mutation rate would indicate that Eve lived about 6,500 years ago." In an article in the respected German scientific journal *Naturwissenschaftliche Rundschau*, the author invoked the Junker and Scherer model of speciation to explain the apparent young age of particular species of cichlids, noting that "the biologists Junker and Scherer regard the explosive speciation events as a process that is caused by polyvalent basic types with a built-in capacity for variation." When informed of this surreptitious promotion for a creationist model, the editor pledged not to accept such a manuscript again. The same thing occurred after the creationist Wolf-Ekkehard Lönnig and a colleague had managed to publish a review article in a high-impact book series, in which they discussed the possibility of a "partly predetermined generation of biodiversity and new species." They claimed that the origin of higher systematic categories depends on the "genesis of irreducibly complex structures" and referred to the publications by the American intelligent design proponents Michael Behe and William Dembski. To conclude, they argued that we should "continue to welcome the plethora of different and diverging ideas and hypotheses on the origin of life . . . wherever they may lead."[15]

Evolution: A Critical Textbook

The first edition of the Junker and Scherer schoolbook of 1986 had sold well and was followed by two modified editions, published in 1988 and 1992 under the same title. In 1998, however, the authors changed their strategy and published an extended version under a new, more general title with a new cover: *Evolution. Ein kritisches Lehrbuch* (Evolution: A critical textbook). This fourth edition was promoted by a popular video (also available on DVD) produced by Fritz Poppenberg, *Is the Bible Right? There Is No Evidence for the Theory of Evolution*. It was mainly targeted at pupils and students and presented the standard creationist arguments against evolutionary theory, again including the allegation that there exists a straight connection between the ideas of Darwin and Adolf Hitler. The film features Scherer and the intelligent design creationist Wolf-Ekkehard Lönnig, a retired biologist from the Max Planck Institute for Plant Breeding

Research in Cologne, Germany, and member of the Jehovah's Witnesses, and has been translated into several languages. Two sequels followed in 2003 and 2005.[16]

In the sixth edition of their textbook, Junker and Scherer describe their Basic Types as a "scientific alternative to the occurrence of novel phenotypic variants, species, and ultimately new body plans via natural processes" (i.e., macroevolution). The "Creator" (God in the Bible) is also called the "Designer," so that one can appropriately label their current model as "young-earth intelligent design creationism." Indeed, whereas the first edition of the textbook offered standard young-earth creationist creeds, later editions had become increasingly influenced by intelligent design. Tellingly, in the meantime, Scherer had become a fellow of the American Discovery Institute, the Seattle-based think tank that promotes intelligent design. In their latest editions, Junker and Scherer explicitly started copying concepts of American intelligent design adherents, such as Michael Behe's irreducible complexity, illustrating it with a detailed description and the image of the flagellum of the bacterium *Escherichia coli*, the poster child of the American intelligent design movement. Also, the language of the textbook had become increasingly scientific-like. They used the term "genetisch polyvalente Stammformen" (genetically polyvalent ground types) to refer to the "Schöpfungseinheiten" (i.e., the created units of life). One original idea, however, is that complex structures of extant organisms (e.g., flowers, eyes) may be regarded as "design signals" emitted by the supernatural Creator or Designer. This idea of "spiritual signals" sent by the biblical God to his devout believers represents one of the few novel concepts introduced by the German creationists.[17]

Evolution: A Critical Textbook can be considered a success for the Study Community Word and Knowledge. Several editions have been translated into Serbian, Finnish, Portuguese, Dutch, Italian, and Russian. Also, it was awarded a "German school book prize" that was sponsored by Christian evangelicals, and the textbook has indeed been introduced into the German school system. One of the textbook's major accomplishments is that it seems to offer an opportunity for European religious conservatives to question undesirable aspects of evolutionary theory without committing themselves to creationism, which they associate with fundamentalism. In 2002 the minister-president of the German state Thüringen, the Christian Democrat Dieter Althaus, praised the Junker and Scherer textbook as "a good example of value-based education." Even prominent Catholics referred to the book as a scientific work doubting particular tenets of evolutionary theory. In a published lecture entitled "Die Christenheit" (The

Christian religion) delivered on November 27, 1999, at the Sorbonne in Paris, Joseph Cardinal Ratzinger, who in 2005 became Pope Benedict XVI, referred to the textbook by Junker and Scherer. In this speech, which is still available on the internet, Ratzinger quoted from the preface of the fourth edition of 1998 and summarized some of the creationist standard arguments "against macroevolution." Hence, the leader of the Catholic Church from 2005 to 2013 believed in the accuracy of some of the comments and conclusions in the Junker and Scherer book and thus might have helped to distribute German intelligent design creationism among his flock of Christian believers.[18]

Creationism in German Schools

Copies of the textbook had been distributed free of charge to public schools, where they have been deposited in the libraries. The 2006 edition was even accompanied by a webpage that provided supplementary data and information for pupils and students. These efforts have not gone unnoticed. In September 2006, the television network Arte broadcast a documentary revealing that creationism was being taught in two secondary schools in the town of Gießen in the state of Hessen, one an evangelical private (but partly state-funded) school, the other a public school. The creationist textbook used in the classroom was *Evolution: A Critical Textbook*. The documentary was picked up by the media, forcing the state minister of culture, who was also responsible for the state's education, the theologian and Christian Democrat Karin Wolff, to comment on the events in an interview with the German news agency. Although she emphasized that she did not support creationism, which she considered to be a fundamentalist position, she declared that private schools "can broaden education by introducing particular issues and forms of education," and she thought it "meaningful to discuss questions that transcend and unite the different courses in class" in order to avoid confronting students with different views in biology and religious classes. Her suggestion to include the biblical creation story in the biology classes provoked severe protest from both politicians and biologists, who requested a bolder statement from her condemning the teaching of creationism. The only consequence of this incident was that the Junker and Scherer textbook, which was not permitted in schools of Hessen, was no longer used in biology classes.[19]

However, creationism is still being taught at more than a hundred evangelical schools (*Bekenntnisschulen*), which are supported by the German government (e.g., in Berlin, Braunschweig, Detmold, Düsseldorf, and Gummersbach). In biology classes, evangelical teachers use the Junker and Scherer textbook and its descendant, *Creatio. Biblical Creationism* (2005), authored by Alexander vom Stein, a Christian fundamentalist associated with the Study Community Word and Knowledge, to teach their pupils the model of the Basic Types of Life.[20]

In addition, there is a state-sponsored evangelical college, the Freie Theologische Hochschule in Gießen, where Junker, together with two other theologians, teaches apologetics, which is also open to the general public. At the Gustav-Siewerth-Akademie, a private college in Weilheim-Bierbronnen, members of the Study Community Word and Knowledge teach their views. Moreover, Horst W. Beck and Erich Blechschmidt have published monographs in the book series of the Gustav-Siewerth-Akademie. Finally, the Campus für Christus in Gießen is of some importance. This association of evangelical Christians distributes, via the Professoren-Forum webpage, articles and books of Horst W. Beck, Michael Behe, and other intelligent design creationists.[21]

In the Freie Evangelische Schule Berlin (Free Evangelical School of Berlin), the official curriculum contains two statements, which can be translated from the original German version as follows:

1. Man is not a product of matter that developed by chance events, but a special creation of the biblical God, as described in the Bible. God created man in his image, and Jesus Christ will save man and forgive him all of his sins.
2. In biology classes, the thesis of evolution, from unicellular microbes or nonliving matter, up to humans, is unacceptable, because this concept is not verifiable by the scientific method. The idea of evolution is, for scientific and theological reasons, to be rejected. Pupils should learn the model character of evolution but also of scientific creationism, and discuss the pros and cons of both views. Moreover, they should learn that evolutionary development from lower to higher forms of life in the course of time is incompatible with the Bible, because the first human pair was created by God and everything was good. Only after the occurrence of the biblical sin did the good creation deteriorate.

In Germany, more than thirty-three thousand individuals, about 1 percent of all German pupils, are educated according to these principles.[22]

The Intelligent Creator in the Shadow of Darwin 2009

In the year 2009, which marked the two-hundredth anniversary of Charles Darwin's birth and the one-hundred-fiftieth anniversary of the publication of his monograph *On the Origin of Species* (1859), numerous valuable books and articles on evolution appeared in print. In addition, numerous writings and pamphlets authored by proponents of all branches of creationism were published.

With respect to the textbook and doctrine of the Basic Types, I briefly summarize the contents of a selection of these antievolutionary publications.

In his book *Traces of God in his Creation? A Critical Analysis of Design-Arguments in Biology*, Junker refers to the "Basic Types of Life" and concludes that "creation" may be viewed as the work of a "Designer." However, he states that this cannot be detected via the methods of the natural sciences. The author defines "design" as "goal-orientation and purposefulness of an acting being." In a small volume entitled *What Now, Mr. Darwin?*, Vom Stein describes the "Basic Types" model, with reference to the God of the Bible, but adds nothing original to this topic. Finally, in *Did God Create via Evolution?*, Werner Gitt, an engineer, an active member of the Study Community Word and Knowledge, and an associate of the magazine *Factum*, summarizes his arguments in favor of a literal interpretation of the Bible and against evolution. Like his colleagues of the Study Community Word and Knowledge, Gitt regards evolution as an atheistic ideology that cannot be proved by experiments.[23]

In contrast to these rather moderate writings, the ProGenesis society, based in Aeugst am Albis, Switzerland, distributed on the Reformation Day, October 31, 2009, a book entitled *95 Theses against Evolution*. In this antievolutionary pamphlet, the standard arguments of the intelligent design creationists are summarized, with references to the Junker and Scherer book. As an alternative to the "not verifiable" evolutionary development of the organisms, the seven authors (Gitt and another engineer, a lawyer, a physician, a chemist, and two theologians) present the created "Basic Types of Life." Moreover, they describe "new proof for the existence of God," which is, however, only a recycled version of the older creationist arguments. On October 31, 2011, an updated version of this pamphlet was distributed via the internet free of charge throughout Germany and Switzerland.[24]

Some Christians have argued that the belief in biblical creation and the acceptance of evolutionary theory are perfectly reconcilable. In *And God Created Darwin's World*, the Christian biologist Hansjörg Hemminger rejects the arguments of Junker and Scherer. Instead, he argues that both the creation of man in a diffuse, Bible-based sense and evolution by natural selection can be accepted and united—they represent opposite sides of the same coin. Also, Hemminger argues that creationists and "God-less, naturalistic scientists" are to the same extent "dogmatic believers," because these "evolutionary atheists" reject the author's moderate views of an "undetectable God behind extant life." The opinion of Hemminger and others, that "Man was created in the image of God and thereafter evolved via natural processes," is widespread in Germany. It may be interpreted as a form of "theistic (i.e., God-directed) evolutionary thought" or, more precisely, as "soft creationism."[25]

Finally, W.-E. Lönnig, who is well known in Germany for this intelligent design sympathies, argued in *The Affair of Max Planck that Never Existed* that design concepts must be viewed as alternatives to the atheistic research programs of the German "Darwinists." He clearly identifies the Designer as the biblical God (Jehovah, in his view).[26]

Creationism and Religious Education

In 2005 a comprehensive poll provided a detailed picture of the public's attitude toward creationism and evolution in Germany. The total number of respondents was 1,520 (age fourteen to ninety-four years). The unpublished data of September 2005 were reanalyzed in June 2007. Three statements were read to the respondents, who had to select the one that most closely reflected his or her own conviction:

1. God created all forms of life directly, as described in the Bible (creationism).
2. Life on earth was created by a supernatural being (or God) and thereafter developed over a long period of time. This process was guided by a higher intelligence (or God) (intelligent design).
3. Life on earth evolved without the interference of God (or a higher being) by natural processes (naturalistic evolution).

The results showed that the majority of the German population (60.9%) accepts naturalistic evolution, whereas 37.7 percent consider themselves to be creationists or adherents of intelligent design. However, most striking were the differences between the western and eastern parts of Germany: in the former East Germany, which was communist and thus atheistic, and where science education had been part of the ideological struggle with the West, creationism and intelligent design are unpopular (16.5%). In the states of former West Germany, however, where Christian churches have been allowed to teach their religious beliefs in public schools, creationists and adherents of intelligent design constitute a much larger percentage of the population (42.4%).

The religious teachings at home and in schools may account for the creationist worldview of many adult Germans. Juveniles are exposed to the biblical creationist story in kindergarten and elementary school. Science education, however, starts much later, and evolutionary biology, with a strong focus on Darwin's life and work, is taught only briefly in high school classes. The modern view of evolutionary biology as an integrative branch of the natural sciences is largely ignored. Once creationism has been picked up by young people, it becomes very difficult to override such beliefs by science education. This conclusion is supported by a poll in 2006 that has shown that about 8 percent of biology students who want to become schoolteachers are adherents of young-earth creationism (n = 721, western Germany; data courtesy of Dittmar Graf, University of Dortmund).

In the United States, the real cause of creationism is not Christianity itself but fundamentalism (i.e., the commitment to strict literal inerrant interpretation of a "holy text" such as the Bible). However, Christian fundamentalists, such as the members of the Study Community Word and Knowledge, currently represent only approximately 3 percent of the German population. Hence, fundamentalism may only in part be responsible for the occurrence and spread of antievolutionism in Germany. Religious education, along with a lack of basic knowledge in evolutionary biology, might be responsible for the fact that approximately 38 percent of the German population adheres to creationism or intelligent design, but more studies are required to further support this conclusion.[27]

Although many Germans accept evolution, and modern Germany is a highly secular state with a substantial group of nonbelievers, Germany is also the home country of a minority of highly vocal and active creationists. American-style young-earth creationism was promoted during the 1960s and 1970s by Arthur E. Wilder-Smith, who wrote prolifically and procured many German translations of American creationist books. The modern creationist movement, however, took off only after a group of German evangelicals gathered forces and established the Study Community Word and Knowledge by the end of the 1970s. Soon thereafter, in 1982, Joachim Scheven, one of the founding members of this organization, published a booklet which can be considered as the first German creationist textbook designed for public schools. His ideas laid the basis for another publication (1986) intended as a comprehensive creationist textbook, written by Siegfried Scherer and Reinhard Junker, who are both prominent members of the Study Community Word and Knowledge. In this book, they developed an alleged case against the documented fact of macroevolution on the basis of the concept of the Basic Types of Life.

The textbook of 1986 went through another two editions, and with the fourth version of 1998, they altered the layout, included increasingly more intelligent design terminology, and changed the title into *Evolution: A Critical Textbook*. Copies of this copiously illustrated volume were consequently sent free of charge to many private and public schools, in some of which teachers used the book to balance the teaching of "atheistic" evolutionary theory. In September 2006, a television network broadcast a documentary in which it was revealed that creationism was being taught in a least two secondary schools in the town of Gießen in the federal German state of Hessen. The response by the minister of culture, Karin Wolff (a theologian), caused a small media row as she claimed that she did not think it was necessary for her to intervene because she thought that students would be less confused if they could discuss questions that united different views on the origin of life. Today, creationism is still being taught at more than a hundred evangelical schools.

The textbook of Junker and Scherer can be considered to be quite a success. It was awarded a German evangelical "schoolbook prize" in 2002, and over the past few editions it has been translated into several European languages. This medium is thus still widely used, and demanded, by Christian readers. However, its major accomplishment is probably that it is able to convince conservative Christians that it offers them an acceptable alternative to naturalistic evolutionary theory that is not associated with American fundamentalism.[28]

With the teaching of creationism in German evangelical schools, the availability of a much-demanded and translated textbook, and the appearance of some creationist articles in the international scientific literature, the agenda of the German anti-macroevolutionists appears to have been firmly established. Although creationists in Germany represent a much smaller percentage of the population than their fellow believers in the United States, a minority of religious enthusiasts have been able to exert its influence on the teaching of evolution in biology classes of public schools.[29]

NOTES

1. Jon D. Miller, Eugenie C. Scott, and Shinji Okamoto, "Public Acceptance of Evolution (Supporting Online Material)," *Science* 313: (2006): 765–766, accessed Oct. 3, 2012, www.sciencemag.org/content/suppl/2006/08/08/313.5788.765.DC1/Miller.SOM.pdf; Ipsos Global @dvisory, *Supreme Being(s), the Afterlife and Evolution*, accessed Oct. 3, 2012, www.ipsos-na.com/news-polls/pressrelease.aspx?id=5217; Ann-Kristin Schäfer, "Wenn die Bibel zum Gesetz wird," *Süddeutsche Zeitung*, Dec. 19, 2011, www.sueddeutsche.de/karriere/naturwissenschaften-an-evangelikalen-schulen-wenn-die-bibel-zum-gesetz-wird-1.1238547; Anonymous, "Wenn christliche Schulen Christliches lehren," *Pro-medienmagazin*, Dec. 21, 2011.

2. Forschungsgruppe Weltanschauungen in Deutschland, *Religionszugehörigkeiten, Bundesländer Bevölkerung ab 18 Jahren 2004*, accessed Oct. 3, 2012, http://fowid.de/fileadmin/datenarchiv/Religionszugehoerigkeit_Bundesl%E4nder,%202002,2004.pdf; Phil

Zuckerman, "Atheism. Contemporary Numbers and Patterns," in *The Cambridge Companion to Atheism*, ed. Michael Martin (New York: Cambridge University Press, 2007), 50.

3. Oda Lambrecht and Christian Baars, *Mission Gottesreich. Fundamentalistische Christen in Deutschland*, 2nd ed. (Berlin: Ch. Links Verlag, 2009).

4. Robert Nachtwey, *Der Irrweg des Darwinismus* (Berlin: Morus-Verlag, 1959); Thomas Schirrmacher, "The German Creationist Movement," accessed Oct. 3, 2012, www.icr.org/article/german-creationist-movement; Ulrich Kutschera, *Streitpunkt Evolution. Darwinismus und Intelligent Design*, 2nd ed. (Berlin: Lit, 2007), 132–133; Ulrich Kutschera, "Von Darwin zu Einstein: Der Evolutions- und Photonenglaube," in *Kreationismus in Deutschland. Fakten und Analysen* (Berlin: Lit, 2007), 21–29; Dittmar Graf and Christoph Lammers, "Evolution und Kreationismus in Europa," in *Evolutionstheorie—Akzeptanz und Vermittlung im europäischen Vergleich*, ed. Dittmar Graf (Heidelberg: Springer 2011), 9–28; Letter from the Word and Knowledge Society, author's private collection, Mar. 20, 2001; www.wort-und-wissen.de, accessed Oct. 2, 2012.

5. Joachim Scheven, *Daten zur Evolutionslehre im Biologieunterricht* (Neuhausen-Stuttgart: Hänssler-Verlag, 1982).

6. Joachim Scheven, *Megasukzessionen und Klimax im Tertiär: Katastrophen zwischen Sintflut und Eiszeit* (Neuhausen-Stuttgart: Hänssler-Verlag, 1988).

7. Joachim Scheven, *Bernstein-Einschlüsse. Eine untergegangene Welt bezeugt die Schöpfung. Erinnerungen an die Welt vor der Sintflut* (Hofheim a. T.: Kuratorium Lebendige Vorwelt, 2004).

8. Joachim Scheven, *Sehen lernen wo der Erdboden aufgedeckt ist. Vor uns die Sintflut. Stationen biblischer Erdgeschichte. Eine Kritik der aktualistischen Geologie* (Hofheim a. T.: Kuratorium Lebendige Vorwelt, 2007).

9. Joachim Scheven, *Wer Ihrer achtet. Ein Biologielehrer entdeckt die Schöpfung. Afrikanische Erinnerungen* (Hofheim a.T.: Kuratorium Lebendige Vorwelt, 2010).

10. Reinhard Junker and Siegfried Scherer, *Entstehung und Geschichte der Lebewesen. Daten und Deutungen für den schulischen Bereich* (Gießen: Weyel Lehrmittelverlag, 1986).

11. Siegfried Scherer and Thomas Hilsberg, "Hybridisierung und Verwandtschaftsgrade innerhalb der Anatidae: Eine systematische und evolutionstheoretische Betrachtung," *Journal für Ornithologie* 123 (1982): 357–380; Kutschera, *Streitpunkt Evolution. Darwinismus und Intelligentes Design*.

12. Ronald L. Numbers, *The Creationists: From Scientific Creationism to Intelligent Design*, expanded ed. (Cambridge, MA: Harvard University Press, 2006), 148–153; Ulrich Kutschera and Karl J. Niklas, "The Modern Theory of Biological Evolution: An Expanded Synthesis," *Naturwissenschaften* 91 (2004): 255–276; Ulrich Kutschera, "Dogma, Not Faith, Is the Barrier to Scientific Enquiry," *Nature* 443 (2006): 26; Frank L. Marsh, *Evolution, Creation and Science* (Washington, DC: Review and Herald Publishing Association, 1944); Frank L. Marsh, *Variation and Fixity in Nature: The Meaning of Diversity and Discontinuity in the World of Living Things, and Their Bearing on Creation and Evolution* (Mountain View: Pacific Press Publishing Association, 1976).

13. Phillip E. Johnson, *Darwin on Trial* (Washington, DC: Regnery Gateway, 1993); German edition: *Darwin im Kreuzverhör* (Bielefeld: Christliche Literatur-Verbreitung, 2003).

14. Siegfried Scherer, "3rd International Conference on Creationism—Report. Basic Types of Life," accessed Oct. 2, 2012, www.biblicalcreation.org.uk/scientific_issues/bcs052.html.

15. Siegfried Scherer, ed., *Typen des Lebens* (Berlin: Pascal Verlag, 1993); Focko Weberling, Book Review, *Flora* 191 (1996): 229–230; Focko Weberling, Book Review, *Flora* 197 (2002): 490–491; Ulrich Kutschera, "The Basic Types of Life: Critical Evaluation of a Hybrid Model," *Reports of the National Center for Science Education* 26:4 (2006): 31–36; Ulrich Kutschera, "Darwinism and Intelligent Design: The New Anti-evolutionism Spreads in Europe," *Reports of the National Center for Science Education* 32:5–6 (2003): 17–18; Laurence Loewe and Siegfried Scherer, "Mitochondrial Eve: The Plot Thickens," *Trends in Ecology and Evolution* 12:11 (1997): 422–423; Wolf-Ekkehard Lönnig and Heinz Saedler, "Chromosome Rearrangements and Transposable Elements," *Annual Review of Genetics* 36 (2002): 389–410.

16. Kutschera, *Streitpunkt Evolution. Darwinismus und Intelligentes Design.*

17. Reinhard Junker and Siegfried Scherer, *Evolution. Ein kritisches Lehrbuch*, 6th ed. (Gießen: Weyel Lehrmittel-Verlag, 2006).

18. Matthias von Bartsch, Simone Kaiser, and Steffen Winter, "Die Grenzgängerin," *Der Spiegel*, July 29, 2007, 44–45, http://wissen.spiegel.de/wissen/image/show.html?did=52263645&aref=image036/2007/07/14/ROSP200702900440045.PDF&thumb=false; Cardinal Joseph Ratzinger, "Die Christenheit, die Entmythologisierung und der Sieg der Warheit über die Religionen, " accessed Oct. 2, 2012, http://ivv7srv15.uni-muenster.de/mnkg/pfnuer/Ratzinger-Wahrheit.html.

19. www.schulbuchpreis.de/preistraeger.html; Frank Papenbroock and Peter Moors, *Von Göttern und Designern. Ein Glaubeskrieg erreicht Europa* (blm Filmproduktion, 2006), www.youtube.com/watch?v=oq81Prmzpso; Britta Mersch, "Vor uns die Sintflut," *Spiegel Online*, Sept. 19, 2006, www.spiegel.de/schulspiegel/wissen/kreationismus-in-deutschland-vor-uns-die-sintflut-a-437733.html; "Kulturministerin fällt auf Kreationisten herein," *Spiegel Online*, Oct. 31, 2006, www.spiegel.de/schulspiegel/hessische-schulen-kultusministerin-faellt-auf-kreationisten-herein-a-445487.html; "Kreationisten im Aufwind," *Zeit Online*, Oct. 31, 2006, www.zeit.de/online/2006/44/Kreationisten; "Kreationisten im hessischen Biologie-Unterricht," *Die Welt*, Nov. 1, 2006, www.welt.de/wissenschaft/article91539/Kreationisten-im-hessischen-Biologie-Unterricht.html; von Bartsch, Kaiser, and Winter, "Die Grenzgängerin."

20. Alexander vom Stein, *Creatio. Biblische Schöpfungslehre* (Lychen: Daniel-Verlag, 2005).

21. www.professorenforum.de, accessed Oct. 15, 2012.

22. *Humanistischer Pressedienst*. Kreationistischer Hokus Pokus, accessed Oct. 16, 2012, http://hpd.de/print/12775.

23. Reinhard Junker, *Spuren Gottes in der Schöpfung? Eine kritische Analyse von Design-Argumenten in der Biologie* (Holzgerlingen: SCM Hänssler, 2009); Alexander vom Stein, *Was nun, Mr. Darwin?* (Lychen: Daniel-Verlag, 2009); Werner Gitt, *Schuf Gott durch Evolution?*, 8th ed. (Bielefeld: CLV Christliche Literatur-Verbreitung, 2009).

24. ProGenesis, *95 Thesen gegen die Evolution. Wissenschaftliche Kritik am naturalistischen Weltbild* (Bielefeld: Christliche Literatur-Verbreitung, 2009).

25. Hansjörg Hemminger, *Und Gott schuf Darwins Welt. Schöpfung und Evolution, Kreationismus und Intelligentes Design* (Gießen: Brunnen-Verlag, 2009).

26. Wolf-Ekkehard Lönnig, *"Die Affäre Max Planck," die es nie gegeben hat* (Münster: Verlagshaus Monsenstein und Vannerdat OHG, 2011); Ulrich Kutschera, ed., *Kreationismus in Deutschland. Fakten und Analysen* (Münster: Lit-Verlag, 2007).

27. Ulrich Kutschera, "Creationism in Germany and Its Possible Cause," *Evolution: Education and Outreach* 1:1 (2008): 84–86.

28. Ulrich Kutschera, *Evolutionsbiologie*. 3. Auflage (Stuttgart: Verlag Eugen Ulmer, 2008).

29. Ulrich Kutschera, *Design-Fehler in der Natur. Alfred Russel Wallace und die Gott-lose Evolution* (Berlin: Lit-Verlag, 2013); Julia Kern, "Kreationismus: Adam, Eva und der Stegosaurus. Menschen-FAZ, 22.6. 2013," accessed Sept. 9, 2013, www.faz.net/aktuell/gesellschaft/menschen/kre-a-ti-o-nis-mus-.

CHAPTER SEVEN

Poland

BARTOSZ BORCZYK

Around 2005, two independent polls contained data about the acceptance of evolution in Poland. The first one was part of a worldwide survey, the results of which were summarized in *Science*, a leading journal. The second one was restricted to Poland only. Both surveys showed that approximately 30 percent of Poles did not accept evolution, and of those who did, most regarded it as a process guided by God. Analysis of the data revealed that antievolution sentiments are more prevalent among Poles who are older and less educated.[1]

My discussion of the history of creationism in Poland covers three periods: first, between 1859, the year in which Charles Darwin published his *On the Origin of Species*, and 1939, the start of World War II; second, the period between 1946, just after World War II had ended, and 1989, the fall of the Iron Curtain; and third, between 1990 and 2012. The period between 1939 and 1945 is not included because during World War II creationism was simply not an issue in Poland. Each of these three periods corresponds with historical events that had a great impact on the Polish sociological context, which, in turn, influenced the perception of the evolution-creation controversies in Poland.

The Historical Background

Between 1795 and 1918, Poland was erased from the European map by consecutive conquests by the Russian, Prussian, and Austrian empires. This era was marked by several uprisings. In 1806 Polish soldiers and politicians joined Napoleon's campaign against Prussia and Russia, which led, in 1807, to the establishment of the Duchy of Warsaw, consisting mainly of Polish land ceded by the Kingdom of Prussia. However, after Napoleon's defeat in 1815, the Russians occupied the Duchy and consequently divided it into three parts: the Congress Kingdom, with the Russian tsar as its constitutional king; the Grand Duchy of Posen, which became politically integrated with Prussia; and the Free City of Krakow. In 1831 the Russian Empire crushed an armed rebellion in the Congress Kingdom, costing the lives of about 1 percent of the Polish population. The next insurrection took place in 1846 in the Free City of Krakow, which was consequently annexed by Austria. Two years later, in 1848, an uprising in the Grand Duchy of Posen was also crushed, and in 1863 a revolt in the Congress Kingdom of Poland resulted in the deportation of many Polish insurgents, including naturalists and intelligentsia who were engaged in the rebellion, to Siberia.

Poland regained its independence after World War I. The reconstruction of the nation, however, ended abruptly in 1939 when Nazi Germany started World War II by invading Poland. After 1945, Poland, again, became a sovereign state, at least officially. In reality, however, it joined the fate of all countries lying to the east of the "Iron Curtain," coming under the political influence of the Soviet Union. The situation changed after 1989, a time of great political change across Central and Eastern Europe. The landmarks of this era are the Polish union leader Lech Wałęsa and his Solidarność (Solidarity), the fall of the Berlin Wall in Germany, and the "Velvet Revolution" in former Czechoslovakia. The political transformation from communism to democracy brought along many changes, including the end of censorship, the possibility of unrestricted traveling around the world, and a free market. The concomitant enthusiasm for everything Western, in particular American, however, was short-lived.

In order to understand the evolution-creation controversy in Poland, one needs to appreciate some of the aspects of everyday life between 1945 and 1990, a period during which the creationist movement developed into its modern form. In 2012 an average monthly gross salary in Poland was 3,225 Polish Zloty (PLN), which equals €770 or $1,024. However, salaries were much lower between 1945 and 1990, amounting to no more than $10 or $20 per month; for reasons of propaganda, the official data contained higher numbers and a better exchange rate. In a retrospective paper on his research career, Leszek Berger, a prominent Polish biologist, wrote that the official exchange rate was 18.92 PLN for $1, although in reality it was between 180 and 200 PLN. With press censorship, it was nearly impossible, or at least very risky, to publish any information that opposed the official propaganda or questioned the position of the state. In biology, Lysenkoism was the official ideology and could not be criticized well into the 1950s. For a long time, creationist publications were published only in very small numbers, but by the end of the 1980s, when the pressure of the regime started to wane, the output increased considerably. At some point during the Cold War, even private correspondence was censored, and some periodicals were more closely watched than others, but I have no information about articles favorable to creationism being censored. However, it would not be surprising because censorship was part and parcel of the strategy the government deployed to isolate Poland from Western influences. Also, in the early 1980s, everyday life in Poland was deeply affected by an economic crisis. This resulted in food regulations for a number of years and the introduction of martial law by an army junta, which lasted more than a year and a half (December 13, 1981, to July 22, 1983).[2]

The Religious Landscape

Exact numbers of religious groups in Poland are hard to come by. Fragmentary information can be found in the *Concise Statistical Yearbook of Poland*, which, however, covers only Christian denominations and two Buddhist movements. Islam and Judaism are not included. However, despite the lack of data, it is safe to state that, with 89 percent of the Polish population, the Catholic Church is by far the largest denomination; almost all of the Catholics, 99 percent, belong to the Roman Catholic Church, while the remaining 1 percent belongs to the Greek Catholic, the Armenian Catholic, or the neo-Unitarian Church. The second-largest religious group, with only 1 percent of the population, is the Eastern Orthodox Church. Protestant churches, Jehovah's Witnesses, and other smaller denominations jointly amount to another 2 percent. The remaining 8 percent comprises atheists, agnostics, and adherents to other religions, but this information is not included in the recent *Statistical Yearbook of Poland*. As to the other large monotheistic religions, Judaism and Islam, the number of Muslims in Poland is estimated to be between 10,000 and 20,000 and the number of Jews varies between 10,000 and 100,000, depending on the sources. These numbers have not changed much over the past two decades. However, the number of Catholics, especially of Roman Catholics, needs to be qualified. In the official statistics, the number of 33,695,233 souls includes all people who have been baptized. However, some people simply follow tradition or submit to family pressure, so not all Catholics are necessarily true believers. It is assumed that churchgoers total approximately 42 percent, and this number has been dropping over the past few years. Smaller churches are not affected by this problem because they have more accurate statistics regarding their adherents.[3]

Before World War II, Polish society was more religiously diverse. Between 1921 and 1931, Roman Catholics (including Armenian Catholics) were in the majority with slightly more than 60 percent of the population, followed by Greek Catholics (nearly 12 percent), Eastern Orthodox (around 11 percent), Jews (about 10 percent), and Protestants (about 3 percent), with other groups not exceeding 1 percent. At that time, Poland was home to several nationalities, each of which had its own religion. Most of the Polish were Roman Catholic, whereas Russians were usually Eastern Orthodox and Ukrainians Greek Catholics. As a consequence, the southeastern part of Poland was mainly populated by Greek Catholics, the eastern part by Eastern Orthodox, and central and western Poland by Roman Catholics. Protestants lived mainly in the northern part, for instance in the city of Gdańsk, in former western Poland, and in the southwest.[4]

In the 1940s and 1950s, the communist regime discriminated against the Catholic Church, but after that period, both found a way to coexist peacefully. The election of the Polish priest Karol Wojtyła (John Paul II) to the papacy further strengthened the position of the church, and ever since the fall of the Iron Curtain, the church has played a very significant role in Polish public life and politics. Today, political parties in Poland need to secure their good relations with the church in order to be successful. Recently, however, the power of the church seems to be waning. Nonetheless, owing to pressure from the Polish Episcopate after the political transformations in 1990, the Catholic Church managed to enter the public educational system. Today, Catholic doctrine is still being taught in state schools in "religion" classes, a name suggesting that Catholicism is the one and only religion in Poland. Although the course is not compulsory—a student can opt for an alternative course of ethics—there is no real freedom of choice, especially in smaller towns. This issue has been raised recently at the European Court of Human Rights in Strasbourg.[5]

The Reaction to the Publication of *On the Origin of Species*

The first writings on the theory of evolution that were intended for a lay audience were published soon after the publication of *On the Origin of Species*. The first Polish scientist to comment on the theory, Leon Kąkolewski, wrote a short and simplified but quite correct description of Darwin's theory of evolution. He based his account on the famous incident between Samuel Wilberforce and Thomas Huxley over the apish ancestry of man at a meeting in June 1860 in the Museum of Natural History in Oxford. The following years several more detailed popular reviews of Darwin's theory were published.[6]

The response to Darwin's theory was different in each of the three parts of occupied Poland. In the Congress Kingdom (under Russian occupation), the theory became readily and widely accepted, especially in Warsaw; Galicia (the southeastern part of Poland under Austrian occupation) accepted the theory about twenty years after the publication of the *Origin*; in the Grand Duchy of Posen (the western part of Poland, under Prussian occupation), however, conservative forces soon ridiculed Darwin's idea, which blocked any serious discussion until the 1920s. The early acceptance of the theory of evolution in Warsaw was due to a great popularizing activity of professor and naturalist Benedykt Dybowski and his co-workers. As early as in 1862, Dybowski, then professor of zoology at the Warsaw Main School, presented natural selection and other Darwinian concepts as valid aspects of evolutionary theory. He was among the first academics to incorporate Darwin's theory of evolution into his courses. After Dybowski's exile to Siberia, his former students continued to popularize evolutionary theory.[7]

Early opposition to Darwin's theory was anchored in scientific, religious, and social concerns. Scientific objections were raised against particular aspects of the new and revolutionary theory as part of academic discussions. However, these criticisms were never directed at the theory of evolution as a whole and soon disappeared. Religious opposition mainly came from the Catholic Church hierarchy and conservative publicists who regarded the theory of evolution as a serious challenge to the creation story in the book of Genesis.

Social and religious concerns usually went hand in hand. Darwin's theory was often associated with progressive social movements, positivism, and early socialism and communism. Generally, it was considered to constitute a threat to the old, traditional way of life and the social order. As a result, the conservative part of society, which consisted primarily of landowners, did not favor the theory. Darwin's theory was also considered to be immoral, resulting in a decline of society. In 1869 a solicitor defending a murderer asked the court for a low penalty arguing that his client was influenced by the immoral and materialistic works of Darwin and Auguste Comte, numbing his human sensibilities. Others attributed the occurrence of mental disease to the reading of the *Origin*.[8]

Most authors tackling the idea of evolution were Catholic priests. Feliks Wartenberg, for instance, published a series of articles in *Tygodnik Katolicki*, a Roman Catholic weekly, later edited as a single brochure, in which he depicted the theory of evolution as immoral, antisocial, and unfounded. Józef Szujski, a publicist, and Michał Nowodworski, a Catholic priest, took a similar approach. They maintained that science confirmed the biblical account of the origin of life and the creation of man. The criticism got stronger when in 1871 Darwin published *The Descent of Man*.[9]

One anecdote beautifully illustrates the evolution versus religion conflict in Poland. Benedykt Dybowski, professor of zoology at the Lviv University, had been invited to deliver the opening lecture for the academic year 1885–86. Dybowski seized on the occasion to praise the theories of Darwin and Ernst Haeckel, thereby offending three archbishops, some theologians, and the political and intellectual elite of Lviv. Dybowski, who was a passionate lecturer, also strongly criticized the concept of teleology, saying "down with teleology" (*precz z teleologią*). However, he was probably misunderstood as saying "down with theology" (*precz z teologią*), upon which the officials just mentioned demonstratively left the lecture hall. Nevertheless, Dybowski carried on with his lecture, compared Darwin to Isaac Newton, Nicolaus Copernicus, and Johannes Kepler, and criticized intolerance, emptiness, and fanaticism, while addressing the remaining officials. Following his lecture, several articles, pro and contra Dybowski, were published in local newspapers and weeklies. The "Dybowski's affair" even reached Vienna. The Catholic Church hierarchy complained about Dybowski in a letter to the Ministry of Confessions and Education, which consequently requested the original typescript of the lecture. The minister considered Dybowski's lecture to be "becoming neither to the position of professor nor to the character of a university ceremony" and advised the rector of Lviv University to put a ban on the teaching of evolutionary theory by Dybowski. Dybowski, however, refused to comply and said he would rather resign as a professor and return to the prison camp on Kamchatka Peninsula (a reference to his years of exile in Siberia) than abandon the teaching of evolution. He was a well-known and respected naturalist, so, in order to avoid a scandal, the officials stopped harassing him. Moreover, because the minister was Austro-Hungarian and thus considered to be an enemy agent, Dybowski's refusal to give in had strengthened his reputation as a Polish patriot. As such, the incident turned out to promote rather than hinder the popularization of the theory of evolution in Poland.[10]

Creationism between 1945 and 1989

Between 1945 and 1989, there did not exist a creationist movement worth mentioning in Poland. No popular books or press articles on creationism or anti-evolutionism were published. Because of the harsh economic and political circumstances, the subject did not arouse any public debate. However, the subject did prompt some discussion in the Polish Catholic Church and in the faculties of philosophy at the universities. In the 1950s and 1960s, this debate was still marginal but gradually became more vigorous, as the regime's pressure diminished. For the most part, the debate took place within the Catholic Church, with arguments for and against evolution being of a religious nature. The articles and books on the subject were written in specialized philosophical and theological language, which made them nearly inaccessible to laypeople. As such, these publications were not intended as apologetic tools—contrary to typical creationists' publications. The main issue addressed in these Catholic publications was the creation of man, which was usually not understood in a literal biblical sense but as an ontological leap by which man had become endowed with a soul.[11]

A slow rise in creationist activity was observed in the 1980s. Members of small religious groups such as Jehovah's Witnesses, Seventh-day Adventists, and Pentecostals published dozens of articles in religious magazines that reached only a limited readership. Most of these articles were translated from foreign publications, and they usually propagated *biblical* creationism. Only a few promoted scientific creationism.[12]

Creationism after 1990

During the 1990s, Polish creationism expanded rapidly. The first to popularize creationism was Maciej Giertych, a professor at the Institute of Dendrology of the Polish Academy of Sciences, who, in multiple articles, reviewed the creationist book *The Crumbling Theory of Evolution* by J. W. G. Johnson, for an ultra-Catholic magazine (*Rycerz Niepokalanej*). He also supported the publication of a Polish translation of the book. In 1993 he sent a videotape entitled *Evolution: Fact or Belief?* to many Polish public schools, free of charge, thereby instigating a public debate, mainly in the Catholic press. One Catholic newspaper reported, falsely, that the Ministry of Education had recommended the videotape for use in the classroom. In response, the Ministry of Education prepared a special booklet in which it addressed the disinformation in the videotape and urged scientists to respond to the ideas promoted by Giertych. However, this intervention unintentionally made creationism more visible to the larger public. As such, Giertych's campaign possibly inspired other creationists to take action.[13]

In 1993 Mieczysław Pajewski and fellow creationists formed an "initiative group" of future founders of an officially registered organization. Pajewski was the group's most enthusiastic member, its "spiritual mover," as he called himself. He ran a bulletin titled *Na początku . . .* (In the beginning), which was distributed to all group members and to people who possibly sympathized with the group's activities. In 1995 the group was officially registered under the name of "The Polish Anti-Macroevolution Society" (its Polish name is Polskie Towarzystwo Kreacjonistyczne, literally the "Polish Society for Creationism"), with Pajewski as its first president. Soon, the organization became the most important creationist organization in Poland. In its official statement, the organization describes itself as "an interdenominational organization with a view to defending, developing, and propagating the ideas of creationism (scientific and biblical creationism) by means of scientific research, popularization, and teaching." Also, despite the many and often-contradictory accounts of creation and the development of life, "the Society does not adopt an official stance on some lesser controversies among creationists (e.g., disputes arising between young- and old-earth creationists as regards the age of the earth and universe or the extent of the Flood) and tolerates the creative multitude of inconsistent ideas among its own members." Membership of the society is open to everyone, regardless of one's faith.[14]

Interestingly, the Polish Anti-Macroevolution Society interprets the word *creationism* in such a manner that it allows it to include all varieties of antievolutionism and, in the same breath, exclude theistic evolution, which considers evolution to be God's tool for creating life:

> Creationism as it is used by the Polish Anti-Macroevolution Association not only refers to a general thesis of cosmological creationism that the Universe appeared (was created) unexpectedly but also means that the first life on earth and, successively, all bigger types of live organisms (especially human) appeared unexpectedly (this is called biological creationism). Biological creationism may be (but is not necessarily) supplemented by physical creationism (astronomical creationism) by saying that particular kinds of heavenly bodies and/or the earth appeared (were created) unexpectedly. Creationism, therefore, means antievolutionism because it excludes the idea of evolution (at least the biological one) in both its atheistic and theistic interpretation, postulating a number of so-called acts of Special Creation.[15]

In one manifesto, explaining the need for the Polish Anti-Macroevolution Society, Pajewski maintains that the creationist view of life accounts better for the facts than does evolutionary theory and, therefore, should be considered as a serious alternative for evolution. Pajewski also mentions some nonscientific reasons:

> Toward the end of the twentieth century, societies are in a crisis, a fact that can be established with the naked eye: increasing crime wave, drug addiction, the weakening of family ties, the decline in status of authorities such as church, state, and family, the increasing anarchy in social life, the spread of moral relativism joined with a consumptionistic attitude, which leads to a higher number of abortions, and so on. We consider this to be in fact a moral crisis that is caused by a gradual loss of faith in God—the Creator, who not only created the world and man but also gave him rules for a good life: a proper attitude toward God, our fellow humans, and the world. Propagating a creationist understanding of the world could stop that dangerous trend and reverse it. And if that is not possible, it could at least help some people to regain their sense of moral order.

These statements can be considered the credo of the society, since they are printed on the inside cover of each issue of *Genesis Problems* (*Problemy Genezy*;

formerly *Na początku . . .*). In 2006 similar views were advanced by Eugeniusz Moczydłowski, who by then had become president of the society. Although many of these reasons for supporting creationism are of a sociological nature (e.g., preventing crime, drug use), they are closely linked to religious concerns. After all, because the problems of society are caused by a common loss of faith, the opposition to evolutionism is more religiously than sociologically inspired.[16]

Pajewski believed that the main tasks of the society consisted of editing a bulletin and a journal; writing, translating, and publishing books; delivering talks; and doing scientific research. At some point, he even made plans to found the society's own research institute. Furthermore, he thought that the organization had to strive for the implementation of creationism into the official curriculum, along with evolutionism. However, Moczydłowski did not share his opinion on this matter and stated in an interview that "the Polish Anti-Macroevolution Society claims that neither creationism nor intelligent design should be taught in biology classes. The proper place for such concepts is in religion and philosophy classes."[17]

Another creationist organization, the Biblical Creationism Society, was founded in 2004 in Poznań by members of various Protestant churches. Although the organization claims to be interdenominational, the membership is open only to evangelical Protestants, who are required to accept the confession of faith. Among its official aims listed in the society's statement are "preaching to all people about God whom the Bible declares Holy Creator of heaven and earth, and popularizing creationism as a scientific alternative for evolutionary explanations of the origin of man and the world." Its website (www.stworzenie.org) gives more detailed information about its views on creationism. It advocates young-earth creationism, which is based on a literal interpretation of the Bible. The members of the society link the greatest tyrannies of the twentieth century to the acceptance of an evolutionary view and the abandonment of God. Thus, its activity is at least partly motivated as an act to save the World.

The Biblical Creationism Society sponsors meetings, which are called creationist conferences and are held in various venues, sometimes in university buildings. Among the invited speakers are British, German, and American creationists such as Paul Garner, John Peet, James B. Jordan, Werner Gitt, Sylvia Baker, Geoff Barnard, and Andy McIntosh. Events are scheduled in various Polish cities, including Gdańsk, Poznań, Zielona Góra, and Szczecin.[18]

The rise of the Biblical Creationism Society correlates with a sudden rise in incidents relating to intelligent design that have attracted Polish creationists' attention in the early 2000s and the revival of evangelical churches in Poland.

In 2005 Maciej Giertych, with the support from both the Biblical Creationism Society and the Polish Anti-Macroevolution Society, launched a campaign directed against the teaching of evolution. This project also received support from the Ministry of Education, which was then headed by Roman Giertych, the son of Maciej. His deputy minister was Mirosław Orzechowski, who publicly subscribed to a creationist worldview. In an interview for a popular newspaper, he claimed that evolution is a lie and a concept of an old, faithless man (Charles Darwin) who probably lost his inner fire because he was a vegetarian. In reaction to these statements, several open letters were published in which prominent Polish scientists, the Council of the University of Warsaw, and even important members of the Polish Catholic Church voiced their disagreement. Furthermore, more than a thousand Polish scientists signed an open letter addressed to the prime minister demanding the dismissal of Mirosław Orzechowski.[19]

Creationism has also been propagated by an influential group of philosophers, led by Kazimierz Jodkowski, a professor from the University of Zielona Góra. Jodkowski has always refrained from officially declaring himself a creationist or evolutionist. He justifies his stance by referring to his "research position," which requires him to be neutral with respect to both sides of the conflict. Nonetheless, many creationists readily interpreted the neutrality of a respected philosopher of science as supporting their case and listed his books and papers as pro-creationist publications; evolutionists too often consider Jodkowski to be a creationist. However, sometimes Jodkowski is called a "radical evolutionist." Whatever his position, Jodkowski and his co-workers are seriously contributing to the recent rise of creationism in Poland, mainly by lending credibility to creationists. They have written papers for *Genesis Problems*, the journal of the Anti-Macroevolution Society, in which they often take a positive stance toward intelligent design and other forms of creationism. In an earlier paper and book, Jodkowski claimed that scientific creationism is truly scientific and argued for methodological pluralism, allowing supernatural explanations to be part of science. His former PhD student Dariusz Sagan, who is now assistant professor at the University of Zielona Góra, has published several papers advocating intelligent design, finding its explanations more accurate than evolutionary ones. Also, Sagan and Robert Piotrowski have translated numerous books and papers dedicated to the subject, such as Phillip Johnson's *Darwin on Trial*, Michael Behe's *Darwin's Black Box*, and James Porter Moreland and John Mark Reynolds's *Three Views on Creation and Evolution*. Furthermore, Jodkowski and Sagan have written and edited a series of philosophical books dedicated to the evolution-creationism controversy. The series was published by Megas Press, owned by Eugeniusz Moczydłowski.[20]

These philosophers have drawn the attention of their colleagues. However, the resulting philosophical discussions are often based on a flawed understanding and limited knowledge of the biological sciences on both sides. For instance, Zbigniew Pietrzak, a philosopher from the University of Wroclaw who does not take sides, claims that the morphology of anacondas (*Eunectes* sp.) has not changed for more than 200 million years. However, the oldest fossils of snakes have been found in deposits of the lower Cretaceous (*Lapparentophis*, 160 to 140 million years ago). The Boidae lineage, to which anacondas belong, did not evolve until 70 to 68 million years ago. Similar errors are made concerning many other species and even entire lineages, such as crocodilians, coelacanths, and turtles.[21]

Smaller creationist groups in Poland, which are usually associated with small Protestant churches, often work together and sometimes cooperate with larger organizations such as the Biblical Creationism Society. Jehovah's Witnesses are also well known for actively propagating creationism.

The Polish Catholic Church and Creationism

Recently, the Catholic Church in Poland has become more engaged in the public debate concerning evolution versus creation. Creationism previously had been mentioned only in philosophical and theological publications, and the subject rarely entered the public domain. Publications that did speak to the wider public usually stated that the theory of evolution and the Catholic view on the creation of life are compatible. Some authors even argued that the Catholic Church never had any problem with the theory of evolution and that some saints such as St. Augustine could be considered precursors of evolutionary thought. However, the introduction of creationism in Polish public space has invoked Catholics to discuss this issue more openly. At first, the church hierarchy strongly criticized religious attacks on evolution, but today, it does not publicly oppose the creationist views put forth by some of its officials, and it refrains from making a clear statement about its position on creationism. Some have explained the church's hesitation as a marketing tool by which the church allows Catholics to believe whatever they want concerning the origin of life. However, it might also be considered a reaction to the rise of fundamentalist churches in Poland after 1990, which tend to interpret scripture literally. In a recent book, Michał Chaberek, a Dominican monk, reviewed the position of church officials, as well as recent popular books on this topic. He showed himself to be critical of theistic evolution and concurred with the ideas of the intelligent design movement. However, this book is an exception, as many other Catholic publications present the Catholic Church as "evolution friendly."[22]

The fact that the Catholic Church does not take a stance on evolution is not without consequences. A recent report on religion classes in Polish schools shows that some teachers still teach the literal interpretation of Genesis. One textbook that is used in the second class of lower secondary school instructs students "to explain, why man could not have evolved from an ape and why the ape could not have descended, through the process of degeneration, from man." This textbook has been approved by both the Ministry of Education and the Catholic Church. In 2008 the Catholic University of Lublin released a post-conference book, entitled *Ewolucjonizm czy Kreacjonizm* (Evolutionism or creationism), in which most of the papers were critical of evolution. Some of the authors, including professors of the Catholic University of Lublin, were creationists, advocating the coexistence of dinosaurs and humans on the basis of the Paluxy River footprints. The book was financed by the Ministry of Science and Higher Education.[23]

International Creationist Organizations

In recent years, the Polish creationist market has been invaded by large, international (usually US-based) creationist organizations. For instance, Kent Hovind, commonly known as Dr. Dino, runs a website in Polish that offers numerous DVDs with Polish subtitles. *Answers in Genesis* has translated some of the articles on its website into Polish and has made the Polish translation of Ken Ham's best seller *The Lie: Evolution* freely available. Creation Ministries International too translated almost fifty articles on its website into Polish. Also, on YouTube one can watch numerous videos and other sources propagating creationism with Polish subtitles or dubbing, free of charge. How these groups will fare in Poland is hard to predict. On the one hand, some Polish creationists are also members of one of the international organizations, such as *Answers in Genesis* or the *Institute for Creation Research,* and use these channels to promote their views. On the other hand, it is difficult to put an estimate on the value of the Polish creationist market. The Polish society is not very prosperous (although the financial situation is gradually improving), and people are mostly Catholic (whereas most of the creationist societies are based on Protestant-tradition Christianity) so opportunities to make money may be rather scarce. At the moment, only Kent Hovind sells products, mainly DVDs, that are especially designed for Polish customers.

Polish creationist organizations also put products for sale on their websites, but their supply is limited to a few books and the DVD *Unlocking the Mystery of Life* with Polish subtitles.[24]

Modern Polish creationism is almost solely imported from the United States and Western Europe. Several reasons account for this. First, the communist regime prevented creationism from entering the public debate and, as a result, disrupted the continuity between earlier Polish creationists and their modern successors. Second, despite dissident voices, the Catholic Church, which has been very influential in Poland, has abandoned a literal interpretation of the book of Genesis in favor of a more compatible position concerning the relation between evolution and the creation of man.[25]

So far, only a few Polish creationists have played a role at an international level. Maciej Giertych is probably the best known. His campaigns were usually nothing more than public relations stunts and make no original contribution to the field of creationism. Nonetheless, he has been interviewed in productions such as *Evolution: Fact or Belief* and *Expelled: No Intelligence Allowed.* A much more serious contribution is a PhD thesis with the title "Phenotype-Genotype Dichotomy: An Essay in Theoretical Biology," which was written in 1975 by Piotr Lenartowicz, a Jesuit monk. Lenartowicz claims—as do other Polish creationists—that, because of his dissertation, he should be considered a precursor of the modern intelligent design theory. However, his dissertation is not cited very often by any of the mainstream design proponents. Together with Jolanta Koszteyn, who works at the Institute of Oceanography of the Polish Academy of Science, Lenartowicz has also published creationist papers in English, which is unique among Polish creationists.[26]

Popular books published by Polish creationists are usually based on US publications. For instance, Pajewski's *Stworzenie czy Ewolucja* (Creation or evolution) draws heavily on Henry M. Morris and Gary E. Parker's book *What Is Creation Science?* Another book, *Zagadka Biblijnego Potopu* (The mystery of the Flood), although not stated explicitly, is in many aspects based on John Whitcomb and Henry Morris's *The Genesis Flood.* More original is the book *Czy istniały dinozaury?* (Did dinosaurs exist?), which is written from a biblical, not a scientific, creationist perspective.[27]

The Polish Anti-Macroevolution Society's journal *Problemy Genezy* publishes numerous translations of creationist papers along with original articles written by Polish creationists. Originally, the journal was published only in electronic form and was mainly edited and written by Mieczysław Pajewski. However, the journal underwent a true metamorphosis, from being a slightly satirical

A cover of the creationist book *Czy istniały dinozaury?* (Did dinosaurs exist?), written by a Polish author and addressed to the general public. This publication advocates biblical creationism. Courtesy of Imelda Chlodna.

magazine to a more professional journal today. *Problemy Genezy* now publishes mainly philosophical articles, short reviews of current foreign creationist articles, translations, and reprints of old papers, authored by early creationists or naturalists, such as St. George Jackson Mivart and Władysław Biegański, who did not entirely agree with Darwin's ideas. Occasionally, special issues are edited by professional philosophers, such as Robert Piotrowski, who works at the University of Zielona Góra, the same institution as Jodkowski and Sagan.[28]

Creationism in the Public Domain

Creationism is not as visible in Poland as in the United States, although it regularly enters the public domain. In response to Polish translations of antireligious books such as Richard Dawkins's *The God Delusion*, Daniel Dennett's *Breaking the Spell*, and Christopher Hitchens's *God Is Not Great*, conservative publishers have released translations of books written by creationists or intel-

PROBLEMY
GENEZY

miesięcznik założony w roku 1993

Tom XVI Numer 11-12 (237-238)
listopad-grudzień 2008

Wł. Biegański NEO-TELEOLOGIA
Wł. Biegański METODYKA TELEOLOGII
M. Pajewski KREACJONIZM. XVIII.
M. Ostrowski AKTUALNOŚCI

Warszawa

A cover of *Problemy Genezy* (Problems of Genesis), bimonthly journal of the Polish Anti-Macroevolution Society. This journal, formerly titled *Na początku . . .* (At the beginning . . .), was started in 1993.

ligent design proponents, including *Darwin's Black Box, Darwin on Trial, Was Darwin Wrong?, The Lie: Evolution,* and *Intelligent Design 101.* In 2009, in reaction to the festivities surrounding the Darwin year, *Najwyższy Czas!* (High time!), the biweekly journal of the political party called Unia Polityki Realnej (Real Politics Union), ran a column with the title *Creationism Corner.* The ultraconservative magazines *Cel* and *Polonia Christiana* published special issues that were dedicated to creationism. Religious journals, such as *Idź Pod Prąd* (Against the current), a journal of the Church of New Testament from Lublin, or *Duch Czasów* (The spirit of the times), a journal of the Church of Seventh-day Adventists, publish pro-creationism articles regularly. The issue of creationism is only occasionally discussed on radio or TV shows. Except for the incident in 2006 involving Maciej Giertych and Orzechowski, creationism has not entered mainstream politics. However, some creationists are members of smaller political parties, and they would surely put the subject on the agenda if they manage to become elected to the Polish Parliament, although this is not very likely.[29]

The first ninety years after the publication of *On the Origin of Species*, antievolutionism in Poland was mostly, but not exclusively, religiously inspired. During the period between 1945 and 1990, it was almost entirely absent from the public debates in Poland, which halted the development of a creationist movement. However, after the fall of the Berlin Wall, when censorship and other forms of regime pressure had been removed, the creationist movement reemerged in Poland. Contrary to the first antievolutionary wave in the nineteenth century, this modern form of creationism was not local but mostly imported from Western Europe and the United States. Today, modern Polish creationism, which is almost an exclusively Christian affair, can be considered to be about twenty years old, when counting from the foundation of the Polish Anti-Macroevolution Society. However, the phenomenon still remains marginal when compared to creationist movements in the United States. Moreover, Polish creationists have not as yet contributed any original concept to the public debates about evolution and creation abroad.

ACKNOWLEDGMENT

I thank Agnieszka Lachowska-Tsang for linguistic improvement to the text.

NOTES

1. Teodozja Długołęcka, "Miejsce Darwinizmu w działalności Przeglądu Tygodniowego (1866–1890)," in *Materiały do Dziejów Myśli Ewolucyjnej w Polsce*, ed. K. Petrusewicz and A. Straszewicz (Warszawa: PWN, 1963), 11–111; Irena Lipińska-Zubkiewicz, "Problematyka ewolucyjna na łamach Ateneum (1876–1901)," in ibid., 113–181; Wanda Grębecka, "Materiały do recepcji darwinizmu na łamach prasy rolniczej," in ibid., 183–287; Henryk Dominas, "Stosunek publicystów czasopism socjalistycznych (Równości, Przedświtu, Walki Klas, Światła) do socjaldarwinizmu i darwinizmu," in ibid., 289–350; Leszek Kuźnicki, "Percepcja Darwinizmu na ziemiach Polskich w latach 1860–1881," *Kosmos* 58 (2009): 279–285; Kazimierz Jodkowski, *Metodologiczne Aspekty Kontrowersji Ewolucjonizm-Kreacjonizm* (Łódź: Wydawnictwo Naukowe Uniwersytetu Marii Curie-Skłodowskiej, 1998), 114–119; Kazimierz Jodkowski, *Spór Ewolucjonizmu z Kreacjonizmem. Podstawowe Pojęcia i Poglądy* (Warszawa: Wydawnictwo Megas, 2007); Bartosz Borczyk,

"Creationism and Teaching of Evolution in Poland," *Evolution: Education and Outreach* 3 (2010): 614–620. The polls on the acceptance of evolution are in Jon D. Miller, Eugenie C. Scott, and Shinji Okamoto, "Public Acceptance of Evolution," *Science* 2006 (313): 765–766; TNS OBOB. "Ewolucja po polsku," 2006.

2. Ronald Numbers, *The Creationists: From Scientific Creationism to Intelligent Design*, expended ed. (Cambridge, MA: Harvard University Press, 2006), 412–416; Central Statistical Office, *Concise Statistical Yearbook of Poland, 2011* www.stat.gov.pl (2011), 172–173. For a discussion of the economic problem, see Berger Leszek, "Problemy z taksonomią żab zielonych," *Przegląd Zoologiczny* 52–54 (2008–10): 113–122; The exchange rates are given as for February 23, 2012. The information about salaries prior to 1990 comes from personal communications with Maria Ogielska and Jan Kusznierz.

3. Central Statistical Office, *Concise Statistical Yearbook*, 114–135; Anonymous 2009, "Spada liczba praktykujących katolików," *Tygodnik Katolicki Niedziela On-line*, www.niedziela.pl/wiad.php?p=200901&idw=309; www.muzulmanie.com/portal/.

4. Elżbieta Olczak, ed., *Atlas Historii Polski* (Warszawa: Świat Książki, 2003).

5. Borczyk, "Creationism and Teaching of Evolution in Poland"; Joanna Podgórska, "Raport: wierz boś nie zwierz. Lekcje religii po polsku," *Polityka* 2680 (2008): 44–49.

6. Wanda Grębecka, "Pierwsza informacja o teorii Darwina w prasie Królestwa Polskiego," *Kosmos* A9 (1960): 601–602; Tadeusz Matecki, "Teorya Darwina," *Ziemianin. Tygodnik rolniczo-przemysłowy* 48 (1864): 1–5.

7. Gabriel Brzęk, "Benedykt Dybowski i Józef Nusbaum-Hilarowicz jako pionierzy darwinowskiego ewolucjonizmu w Polsce," *Przegląd Zoologiczny* 26 (1982): 19–35; Gabriel Brzęk, *Benedykt Dybowski. Życie i Dzieło* (Warszawa and Wrocław: Polskie Towarzystwo Ludoznawcze, Wydawnictwo "Biblioteka Zesłańca," Stowarzyszenie "Wspólnota Polska," 1994), 219–296; Henryk sen. Hoyer, "Krytyczny pogląd na Darwinizm," *Ateneum* 2 (1876): 169–194.

8. Brzęk, "Benedykt Dybowski i Józef Nusbaum-Hilarowicz jako pionierzy darwinowskiego ewolucjonizmu w Polsce," 21–22.

9. Tadeusz Wartenberg, "O teoryi Darwina," *Tygodnik Katolicki* 7 (1866): 429–433, 437–441, 475–477, 485–487, 497–501, 505–507, 513–518; Stefan Pawlicki, *Studia nad darwinizmem* (Kraków: Wydawnictow Dzieł Katolickich Wł. Milkowskiego, 1875); Zbigniew Kępa, "Recepcja Darwinizmu na ziemiach polskich w latach od 1859 do 1884," *Zagadnienia Filozoficzne w Nauce* 18 (1996): 29–51; Kuźnicki, "Percepcja Darwinizmu na ziemiach Polskich w latach 1860–1881."

10. Benedykt Dybowski, "Rzut oka na historyczny rozwój zoologii," *Kosmos* (1885): 1–23. Also see Ludwik Wciórka, "Spór między ewolucjonizmem a kreacjonizmem jako problem filozoficzny," *Poznańskie Studia Teologiczne* 1 (1972): 279–296; Stanisław Kowalczyk, "Chrześcijańska a marksistowska myśl o stworzeniu," *Communio* 2 (1982): 86–92.

11. Józef M. Dołęga, *Kreacjonizm i Ewolucjonizm. Ewolucyjny Model Kreacjonizmu a problem Hominizacji* (Warszawa: Akademia Teologii Katolickiej, 1988).

12. Jodkowski, *Metodologiczne Aspekty Kontrowersji Ewolucjonizm-Kreacjonizm*, 114–117.

13. Adam Łomnicki, ed., *Opinia o filmie wideo "Ewolucja: Rzeczywistość czy Domniemanie?"* (Kraków: Universitas Press, 1994).

14. Polish Anti-Macroevolution Society, Statut Polskiego towarzystwa kreacjonistycznego, www.creationism.org.pl/info/status, §7 & §8; Borczyk, "Creationism and the Teaching of Evolution in Poland."

15. Polish Anti-Macroevolution Society, "Aneks do Statutu," www.creationism.org.pl/info/aneks.

16. Mieczysław Pajewski, "Dlaczego potrzebne jest Polskie Towarzystwo Kreacjonistyczne?," *Na Początku* . . . 1 (1993): 1–2; Jacek Słaby, "Pytania o prawdę. Z dr. Eugeniuszem Moczydłowskim, prezesem Polskiego Towarzystwa Kreacjonistycznego, rozmawia Jacek Słaby," *Cel* 7 (2006): 26–28.

17. Mieczysław Pajewski, "Jak wyobrażam sobie członkostwo w Polskim Towarzystwie Kreacjonistycznym?," *Na Początku* . . . 3 (1993): 17–20; Adam Wielomski, "Eugeniusz Moczydłowski: Kreacjonizm czy ewolucjonizm? Wywiad z dr Eugeniuszem Moczydłowskim, prezesem Polskiego Towarzystwa Kreacjonistycznego w sprawie teorii Darwina," Archiwum Konserwatyzm.pl.

18. Biblical Creationism Society, www.stworzenie.org.

19. Almut Graebsch, "Polish Scientists Fight Creationism," *Nature* 443 (2006): 890–891; A. Pezda, "Wiceminister edukacji: poradzimy sobie bez tolerancji. Wywiad z Mirosławem Orzechowskim," *Gazeta Wyborcza* 10 (2006): 14; Borczyk, "Creationism and the Teaching of Evolution in Poland."

20. For different views on creationist vs. evolutionist positions of Jodkowski and his co-workers, see, e.g., Karol Sabath, "Kreacjonizm a sprawa polska," *Świat Nauki* (Sept. 2002): 73; Krzysztof Rajewicz, "Przyrodnicze a Teologiczne Ujęcie Genezy i Rozwoju Człowieka" (PhD dissertation, Papieski Wydział Teologiczny we Wrocławiu, 2008); Bańbura Jerzy, "Niebezpieczeństwa Kreacjonizmu," *Kosmos* 58 (2009): 595–602. Some titles from Megas Press are Kazimierz Jodkowski, ed., *Teoria Inteligentnego Projektu—Nowe Rozumienie Naukowości?* (Warszawa: Wydawnictwo Megas, 2007); Dariusz Sagan, *Spór o nieredukowalną złożoność układów biochemicznych* (Warszawa: Wydawnictwo Megas, 2008); Michael Behe, *Czarna skrzynka Darwina* (Warszawa: Wydawnictwo Megas, 2008).

21. Zbigniew Pietrzak, "O nieporozumieniach dotyczących teorii ewolucji i epistemologicznych tego konsekwencjach," in *Ewolucja, Filozofia Lectiones & Acroases Philosophicae III, Religia,* ed. Damian Leszczyński (Wrocław: Polskie Forum Filozoficzne, 2010), 203–230.

22. Andrzej Siemieniewski, *Stwórca i Ewolucja Stworzenia* (Warszawa: Fronda, 2011); Andrzej Siemieniewski, *Ścieżką Nauki do Boga. Nauki Przyrodnicze i duchowość w Starożytności i Średniowieczu* (Warszawa: Fronda, 2009); Józef Życiński, *Bóg i Ewolucja. Podstawowe Pytania Ewolucjonizmu Chrześcijańskiego* (Lublin: Towarzystwo Naukowe Katolickiego Uniwersytetu Lubelskiego, 2002).

23. Siemieniewski, *Stwórca i Ewolucja Stworzenia*; Siemieniewski *Ścieżką Nauki do Boga. Nauki Przyrodnicze i duchowość w Starożytności i Średniowieczu*; Życiński, *Bóg i Ewolucja. Podstawowe Pytania Ewolucjonizmu Chrześcijańskiego*; Marek Słomka, *Ewolucjonizm Chrześcijański o Pochodzeniu Człowieka* (Lublin: Gaudium 2004); Radosław Tyrała, *Dwa Bieguny Ewolucjonizmu. Nauka i Religia w Poznawczym Wyścigu Zbrojeń* (Kraków: Nomos, 2007); Borczyk, "Creationism and the Teaching of Evolution in Poland"; Michal Chaberek, *Kościół a Ewolucja* (Warszawa: Fronda, 2012); Podgórska, "Raport: wierz boś nie zwierz. Lekcje religii po polsku"; Piotr Jaroszyński, ed., *Ewolucjonizm czy Kreacjonizm* (Lublin: Fundacja Lubelska Szkoła Filozofii Chrześcijańskiej, 2008).

24. The website addresses of international creationist organizations that offer creationist material in Polish: www.drdino.pl, www.answersingenesis.org/worldwide/translations and www.creation.com/polish/.

25. Jodkowski, *Metodologiczne Aspekty Kontrowersji Ewolucjonizm-Kreacjonizm*; Borczyk, "Creationism and the Teaching of Evolution in Poland"; Radosław, *Dwa Bieguny Ewolucjonizmu. Nauka i Religia w Poznawczym Wyścigu Zbrojeń.*

26. Piotr Lenartowicz, "Phenotype-Genotype Dichotomy: An Essay in Theoretical Biology" (PhD diss., Gregorian University, 1975).

27. Mieczysław Pajewski, *Stworzenie czy ewolucja.* (Bielsko-Biała: Wydawnictwo Duch Czasów, 1992); Jonathan Wells, *Ikony Ewolucji* (Gorzów Wielkopolski: W Wyłomie, 2007); Leszek S. Pasiud, *Zagadka Biblijnego Potopu* (Wrocław: Wydawnictwo A Propos, 2008); Aleksander Barański, *Czy Istniały Dinozaury? Czyli Kreacjonizm Biblijny dla Początkujących* (Aleksander Barański Publishing House, 2010).

28. E.g., Robert Piotrowski, ed., "Numer specjalny. Brytyjska filozofia przyrody. I. Mivart Romanes Stokes," *Problemy Genezy* 16 (2008): 1–84.

29. E.g., Maciej Giertych, *O Ewolucji w Szkołach Europejskich* (Krzeszowice: Dom Wydawniczy Ostoja, 2008).

CHAPTER EIGHT

Greece

EFTHYMIOS NICOLAIDIS

After the East-West schism of Christianity, officially proclaimed in 1054, the Eastern Orthodox Church was led by the patriarch of Constantinople, considered as the *primus inter pares* of the Eastern patriarchs. The history of this church can thus be traced in the cultural context of Byzantium, until the gradual separation of the Russian Church, which became officially independent in the sixteenth century. From then on, historians of the Orthodox Church have to consider the two traditions of the Russian Church and of the Patriarchate of Constantinople. After the Ottoman conquest of Constantinople in 1453, the patriarch of Constantinople became the head of all Orthodox Christians of the empire. This included almost all Orthodox people except Russians. According to Ottoman rules, the patriarch of Constantinople also controlled the education of the Orthodox subjects, and he had the power to dismiss professors whose teaching was found to be subversive for Orthodoxy.

After the creation of the Orthodox national states following the national revolutions in southeastern Europe during the nineteenth century, the role of the Patriarchate of Constantinople became far less important. Although keeping the title of ecumenical, it eventually lost control over the gradually more independent national Orthodox churches. The Greek Church was the first to become independent in 1833. The Bulgarian Church followed in 1870, the Serbian in 1879, the Romanian in 1885, and the Albanian in 1922. Since that period, the Orthodox Church has been divided in numerous parts, representing mainly national entities.

The fragmentation of the Greek Orthodox Church led to different reactions toward questions about science and religion, such as the issue of evolution and creation. Because these reactions depended on local policies and cultures, they should be studied separately. In southeastern Europe, except for the various reactions of the national churches, there were also different responses within the frame of a specific national church. Indeed, although there is a head in each Orthodox Church—for example, the archbishop of "Athens and all Greece," the patriarch of Bulgaria, of Serbia, and so forth—there have always been different trends, opinions, and lively debates.

During the Cold War, all Orthodox countries except Greece belonged to the Soviet bloc. Contacts between the Greek Church and the other Orthodox churches became rare, which deepened the differences in approach on numerous matters, including science. This chapter presents the various trends of the Greek Orthodox Church with regard to creationism, based on the Greek national church as an example.

Most of Greece's 10 million citizens adhere to the Greek Orthodox dogma, except for a small Muslim minority located in northeast Greece and a small Catholic minority, located mainly in the Aegean islands. This latter minority originates from the time of the Latin conquest of the last Byzantine centuries. Statistics show that Orthodox believers make up more than 80 percent of the Greek population. Very few, a mere 3 percent, consider themselves atheists. The Greek Orthodox Church is considered the official dogma of the Greek state. The members of the parliament and the government swear on the Bible in the presence of the archbishop of Greece, and only a few left-wing parliamentarians refuse to do so. A course on religion is taught in elementary and secondary schools. This is mainly a catechist course on the Orthodox dogma. The Greek Orthodox Church has often interfered in politics. A recent example is the debate on the introduction of new identity cards in 2000. The new cards did not mention the personal religious belief of the holder as had previously been the case. In protest of this change, the archbishop of Greece collected more than 2 million signatures to demand that the new cards be revoked by the government. The archbishop did not win the case, and the cards remained secular. But the affair shows the power and significant role of the Orthodox Church in Greek society. The archbishop is a powerful figure within the church. This has, however, not precluded differences and heated discussions within the church. The history of creationism in Greece is a case in point.[1]

The Historical Background: Darwin, Haeckel, and the Greek Orthodox Church

Creationism was challenged in the Greek Orthodox Church by the followers of Darwin. The theory of how species changed was presented for the first time in Greece by Alexander Theotokis, who was a student of Georges Cuvier's successor at the Museum of Natural History in Paris, Henri Ducrotay de Blainville. His early evolutionary views were published in his book *General Zoological*

Tables, or Forerunner to Greek Zoology in 1848. Ten years after Charles Darwin's *On the Origin of Species,* we find the first mention of Darwin's ideas in Greece. This concerned paleontological findings. In the years from 1855 to 1860, the French paleontologist Albert Gaudry had conducted excavations in fossil deposits dating from the Miocene at Pikermi in Attica near Athens. In 1869, in the *Attic Calendar,* an almanac containing texts of general interest, an anonymous author wrote: "Apart from the great number of animals discovered, the collection found in Pikermi gives force and scientific significance to the theory of Darwin on zoogeny." Darwin was aware of Gaudry's research and mentions the "discoveries of Attica" in both the fifth edition of the *Origin of Species* from 1869 and *The Descent of Man* published in 1871. A few years later, in 1873, the chemist and assistant professor at the University of Athens, Leandros Dosios, presented the theory of evolution to an educated Athenian public in the context of a lecture titled "The Struggle for Existence."[2]

Three years later, the first serious attack against the theory of evolution came from an assistant professor of the faculty of theology at the University of Athens, Spyridon Sougras, who wrote *The Most Recent Phase of Materialism: Darwinism and Its Lack of Foundation*: "The question of evolution or changes in animals and vegetables has attracted from the start the attention of all scientists, because of its imbecilic and horrible results, meaning the common origin of man and the monkey. It also attracts religious interest, for theology ought also to be concerned with this question, since by its simple suppositions alone, it risks inducing error among those who are totally ignorant of this theory."[3] Sougras strongly feared that Darwinian theory would push the Greeks toward socialism and communism, and in the sequel to his book he attacked Karl Marx and the German Jewish socialist Ferdinand Lassalle.

Creationism and antievolutionism were also proclaimed by the official church but without a formal condemnation of the theory of evolution. The Greek Church, which became the privileged ally of the right-wing party, left it to laypeople who were in opposition to the endeavors of modernizing Greece and instead supported traditional Orthodox values to defend creationism at a scholarly level. One such layperson is the respected businessman Ioannis Skaltsounis, who published several articles and books on materialism and the creation of man.[4] According to him, the most dangerous materialist was not Darwin but the German Ernst Haeckel: "Darwin proposed as probable the existence of an archetype for the origin of man and the monkey, and he several times admitted the logical weakness of his explanations. But Haeckel tried to define with mathematical precision the bestial origin of man."[5] Skaltsounis maintained that all living beings were created to the measure of humankind by a supernatural force; nothing could be engendered by itself. He could not accept university professors who supported materialist ideas because he feared that their prestige might legitimate these ideas in Greek Orthodox society. To prevent that from happening, he proposed the demolition of the University of Athens.

However, the intellectual environment at the science departments of the university was from the start rather favorably inclined to Darwin's ideas. In 1880 the course on evolutionary theory given by Professor Ioannis Zochios, professor of anatomy and physiology, was so successful that no university hall was large enough to hold those students who wished to attend. Although the Holy Synod of the Church of Greece protested against these courses, the rector defended his professor and merely recommended that Zochios be more prudent about offending the religious sentiments of his audience.

With the rise of socialism and communism, the church gave priority to the struggle against materialism supported by the new European revolutionary forces. For the official Greek Orthodox Church, these ideas were twice as dangerous: they were atheistic and came from the West. As evolutionist ideas could be used to furnish the arguments for atheism and materialism, the reactions from the church became stronger. Nevertheless, the institutional involvement of the church with the state, which made the Greek Church more dependent upon the government, ensured that the Holy Synod was always careful not to make decisions that might offend state institutions. The role of attacking the evolutionist professors of the University of Athens was thus undertaken by publications with close connections to but not run by the church, such as the journal *Anaplasis*. It was first published in 1887 and, with a circulation of forty-five hundred, was an important vehicle for communicating the opinion of the church on these matters. The Holy Synod recommended the journal to the clergy, which in turn recommended it to the local communities.[6] The journal was published by an association created in order to support the church in its new struggle against materialism:

> The association is composed of more than two thousand members of the civilized world and among them are the distinguished patriarchs of our Orthodox Church, the metropolitans of the Orthodox states, Greek and foreign scientists, celebrated theologians . . . It has the principal goal of safeguarding with all its strength our holy religion that is offered to us by God and our morality against all the enemies of Christianity that have reappeared, and above all against the pernicious materialism that has been rife for some time, and which, in the name of a new errone-

ous science, denies any spiritual existence and openly combats the great truths of Christianity.[7]

The first and only example of a trial against the introduction of evolution ideas in schools took place in 1914. It began with the introduction of new teaching methods in the Greek schools conducted by Professor Alexander Delmouzos, who had studied new pedagogical methods at Jena University. In 1908 Delmouzos was appointed director of the girls' school in Volos, where he encouraged critical thinking among the students. These novel ideas disturbed conservatives, who tried to get rid of what they considered dangerous anarchists by organizing demonstrations. These became so heated that the cavalry was forced to intervene to defend the school and the house of Delmouzos. The municipal council of Volos eventually decided to close the school in 1911 and prosecute Delmouzos and his collaborators. The trial took place in the town of Nafplion, with the prosecutors accusing Delmouzos and his collaborators of promoting atheism. The theory of evolution was cited: "In various periods, from September 1908 until the end of March 1911, in Volos and in Larissa, principally at the Workers' Foundation and the School for Girls in Volos, they [the accused], teaching out loud or with the aid of printed brochures, tried to proselytize in favor of so-called religious dogma, i.e., atheism. These actions are incompatible with the preservation of the political order, for they teach that God does not exist . . . , that man was created from monkeys, that God is a cucumber, etc."[8]

Neither the government nor the Holy Synod of the Church of Greece was directly implicated in the trial, and the accused were acquitted for lack of proof. However, the affair became an important political issue symbolizing the struggle between conservatives and progressives until the present. The propagation of socialist ideas after the first quarter of the twentieth century coincided with the propagation of materialist ideas. To conservative forces and the official church, scientific materialism was accordingly identified with subversive socialist and communist ideas. This was reinforced by the fact that most Greek intellectuals sympathized with the political left, with Christian scientists remaining a minority.[9]

Orthodox Scientists, Political Regime, and Materialism

In 1936 the parliamentary government was overthrown by the dictator Ioannis Metaxas, strongly influenced by Italian Fascism. Metaxas's regime privileged the church and the Orthodox traditions, despite its modernizing efforts in science and technology. In 1937 a fraternity of theologians called Zoë (Life) founded the

Christian Union of Scientists, which published the journal *Aktines* (Rays). The outbreak of World War II and then the occupation of the country by the German army in 1941–44 marked a break in local discussions about science and religion. Just after the war, in 1946, a civil war between communist and national armies broke out. In December of that year, in an effort to gather non-Marxist scientists, the Christian Union of Scientists issued the *Declaration of the Christian Union of Scientists,* addressed to "all Greeks" and signed by 222 people of science and letters. Most of them were university professors, some of whom had previously been attacked by those from church circles as enemies of Orthodoxy.[10]

Without explicitly condemning communism or dialectical materialism, the declaration endorsed the values of spirituality and work, and denounced materialism, Darwinism, monism, and Sigmund Freud. It emphasized the importance of science for society and pointed out that Greece since its independence, instead of basing its civilization on its Christian tradition, had drifted away by imitating everything that came from the West. The *Declaration* aimed to reconcile science and religion by citing recent discoveries that, according to the authors, were not in contradiction with Christianity. The Christian Union of Scientists asserted that a Supreme Being who defined the laws of nature created the world out of nonbeing and that physics did not deny this. The *Declaration* took on Haeckel and Ludwig Büchner, whose book *Force and Matter* (first published in 1855) was translated into Greek in 1910 and was published in several subsequent editions, and it accused them of trying to replace faith by science.

The goal of the *Declaration* was to show that science refuted evolution. The Christian Union of Scientists indicated that the transformative evolution of the French naturalist Jean-Baptiste Lamarck was not corroborated by the paleontological findings, since intermediate types had not been found. Darwin's natural selection could explain the disappearance or conservation of species, but not their creation. The *Declaration* argued that the latest findings showed a stability of species and that there was no scientific proof of the process by which new species appeared. In relation to the origin of humans, the *Declaration* made a distinction between Darwin and the popularization of his work by the "atheists" who used *Pithecanthropus erectus,* a hominin fossil found in Java in the 1890s, as proof of the affiliation of our species to the apes. The Swedish geneticist Nils Heribert-Nilsson, who contested evolutionary theory, was abundantly cited.[11]

The *Declaration* was a political response by Orthodox Christian scientists to the materialist ideas propagated by the Communist Party camp, which at the time was fighting hard for power. The civil war and the strong-arm republic that followed it caused a deep conflict and separation between the left and right camps lasting for almost thirty years. The main slogan of the seven years' dictatorship of the colonels from 1967 to 1974 was "Greece of Greek Christians." This overtly religious ideology promoting Orthodox propaganda in schools and universities affected education at all levels. Creationism and antievolutionism were part of the teaching in schools in the mandatory course on religion, which was more a sort of Orthodox catechism than an academic religious study.

During the colonels' era, Christian associations had a strong influence on the state and the Holy Synod. The association Zoë was strongly influencing the Holy Synod, controlled by the regime, which placed Ieronymos I as its head and replaced most of the former members. Members of Zoë also tried to influence the university students and were backed in this effort by the pro-regime councils, but without a great success. In the last years of the regime, Panagiotis Christou, a professor of theology at the University of Thessaloniki, became minister of education and religious affairs. After the fall of the colonels' regime, Christou was dismissed from the university as a collaborator of the Junta and devoted himself to studies of early church fathers.

In 1969, two years after the coup d'état, the Russian-born Theodosius Grigorovich Dobzhansky, a key figure in constructing the modern evolutionary synthesis, was invited to a conference organized by the Greek Anthropological Society. Dobzhansky, who had remained Orthodox, was upset by the personal attacks from Greek theologians, especially Marcos Siotis of the University of Athens. Siotis argued against the idea that humankind should have descended from brute ancestors because that would contradict the book of Genesis. Dobzhansky protested against these attacks. Both as a scientist and as a member of the Eastern Orthodox Church, he supported the views that humans were "created in God's image by means of evolutionary development." After the event, he observed, "Fortunately, the narrow-minded rigidity of the Greek section is not shared by the Eastern Church as a whole."[12]

The Change of 1974 and the New Trends in Greek Orthodoxy

In Greek history 1974 marks the change of the regime from dictatorship to democracy. The importance of this year is to be compared to the importance of the end of the Salazar and Franco regimes for Portugal and Spain, or of the fall of the communist regimes for the Eastern Europe countries.

The return of left-wing political refugees who had fled the country in two waves, in the 1950s and after the colonels' coup in 1967, hugely changed the political and intellectual environment. Left-wing intellectuals felt they were reestablished and that it was their turn to control academia. The Orthodox Church, especially the Holy Synod and the Christian associations, as close collaborators with the colonels' regime were discredited, particularly among young people. The church subsequently needed a rejuvenation to be reestablished among the young intellectuals. This was in part brought about by left-wing people who rediscovered the "values" of Orthodoxy, especially in contrast to Western Christianity. The "new Orthodox," as they have been called, tried to combine everything: asceticism, science, traditional values, modern way of life, and ecology. This movement rediscovered the esoteric approach of Christianity and a personal relation with God downplaying the importance of associations and institutions. The counterparts to this movement were the associations of the Orthodox Church, such as Zoë or Sôter, which had been created under the indirect influence of Protestantism and the official church, represented by the Holy Synod. The latter had been totally reformed. Its head Ieronymos, because he had been illegally named by the colonels, was dismissed and replaced by the moderate Serafeim, who worked to reconcile the church with the progressive political forces, notably the socialists and the communists.

This plurality of institutions and movements, characteristic of the Orthodox Church's practice of not relying on a single authority, translated into many and contradictory positions with regard to creationism. The situation became even more complicated by the attitude of the Holy Synod, which avoids making official statements on subjects related to science, including evolution. The Holy Synod of the Greek Church is constituted by the Metropolites, or bishops. It makes major decisions and elects the archbishop. All leanings are represented at this board from fundamentalists who believe in a six-day creation to those who under certain conditions would accept evolution. Various religious fraternities, for example, the aggressive fundamentalist Chrysopigi, back some of the groups within the Synod. Members of the Synod often express their own views publicly, ignoring the declarations of the archbishop on the subject.

The structure and the tradition of the Greek Orthodox Church allowing the presence and expression of many views make mapping the various attitudes toward creation and science difficult. Generally it can be simplified by dividing it into four main trends, from right to left. The "fundamentalists" are represented in the second half of the twentieth century by Avgoustinos of Florina but also include the fraternity Chrysopigi, while the "compromisers" are represented by Archbishop Ieronymos II. The "New Orthodox movement" includes many leanings, and the "New Orthodox theology" is mainly present in the faculties of theology. The most important of these groups is the compromisers, while during the 1980s the most dynamic group was the New Orthodox, and during the 1990s the fundamentalists. The new orthodox theology group remains a small minority.

The Holy Synod and Creationism

The direct involvement of the official Greek Church and the state since the foundation of Greece in 1830, and for the Orthodox Church as a whole since Constantine the Great, has historically invited the Holy Synod to moderation. The facts that the Ministry of Education is also in charge of the church, that all clerics are public servants, and that the Church possesses huge real estate holdings without being taxed are strong arguments for avoiding problems with those in political power. This is the main reason why the church has avoided taking an official position when a problem arises with science, including discussion about creation and evolution, and has instead put forward its friends and supporters or its most combative metropolites. Nevertheless, recently Christodoulos, archbishop from 1998 to 2008, departed from this cautious path and involved the Greek Church in political issues, such as the one concerning new identity cards, where the traditional mention of religion was abandoned.

Another example from recent history where the church has taken a more direct approach on topics usually avoided is a brief article by Archbishop Christodoulos in *Efimerios*, the journal of the Holy Synod of the Greek Church. This is a rare testimony about the official position of the Holy Synod, but it still does not constitute an official document of the church. In many ways remarkable, this text extends a great respect for the sciences but also takes a firm stance on the limits of scientific knowledge and the dangers of atheism, while evoking an Old Testament God teaching humanity humility through disaster:

> All of us have once heard that "God has created the world ex-nihilo." This is for some people a problem. Not all are ready to accept this answer to the question, "How has the world been created?" The problem of who has created and how he has created the world is a central problem of our life. We all know the great progress that science has done in this domain. Many sciences complete each other in a common effort to discover the principles of life. We are grateful to science for its efforts to reach the limits of knowledge and throw light in all the secrets of creation. We, believers, should not fear the progress of science; on the contrary, we should expect conclusions and proposals from it which strengthen our faith. Nevertheless, we do not ignore that in the past, and precisely in the preceding century [i.e., the nineteenth century], the distrust of scripture came from certain laboratories and lasted a long time. It was the period of the myth that science is omnipotent and

> can give answers to the main human queries . . . Then, when one believes in the omnipotence of man on earth, comes accidents such as *Challenger* or Soviet Chernobyl to demonstrate the weakness of man . . . Science is a holy gift, but within limits. It stands in between physics and metaphysics. With the means of observation, experiment, and mathematics, it tries to explore events that cannot be perceived. But its horizon is always limited. Nevertheless the query for atheist arguments in scientific results has not ceased even today to be a phenomenon, not of course in science laboratories, but in the imagination of some people who pretend that science has the status and the authority to decide on whether God exists or not.[13]

The big synthesis of divine providence, science, and humans' place in nature did include ideas about creation but had little specific to say on evolution. Christodoulos cited the British Orthodox bishop Kallistos Ware for saying, "God's love is expansive; it makes God to go out from himself and create things different from himself."[14] Ware himself is known not to have any issues with evolutionary theory and has found it to be fully supported by the scientific evidence. As long as one keeps the religious and scientific domains apart and accepts that they are addressing different kinds of questions and giving different kinds of answers, there are no problems. Equally, Christodoulos on this occasion expressed fairly moderate views about evolution. This was not, however, the outlook of the orthodox fundamentalist tradition he came from.

The "Right-Wing" Views on Creationism

Konstantinos Karoussos, who was elected metropolitan of Piraeus under the name of Kallinikos, collaborated with Christodoulos in founding the Christian fraternity Chrysopigi in 1973 during the colonels' regime. This fundamentalist fraternity fought against the teaching of evolution and materialism, and especially against Marxism. During the Cold War, the majority within the Greek Church feared that if Marxists prevailed, they would abolish church structures and support atheism. Kallinikos wrote two small books presenting his views about science: *Scientists Speak about God* and *Man from Monkey?* In the latter he presented the fraternity's views on creation, evolution, materialism, and Marxism:

> The enemies [of the Christians] of the first days were followed by heretics, false brothers, those who altered the truths of Faith . . . In recent years materialists and Marxists have appeared, who on the basis of so-called scientific results systematically fight to explain everything by materialist criteria and principles, denying the existence of God and the invisible Creation . . . Nowadays materialism has a

> party cover, because it is the foundation of Marxism. Marx was a materialist. He declared this himself . . .
>
> Marxism is not only an economic, social, and political system; it is a materialist and atheistic theory; its basis is historical materialism, and that is why it is contrary to the Christian idea of life . . . Now, the materialists strike mercilessly at anything that is related to God and the church. And they forge myths, like that of the origin of man from the monkey, which they propagate by any means possible.

Speaking on the theory of evolution, which Kallinikos thought identical with "man originates from monkey," he wrote: "Defenseless children are poisoned by this book [a schoolbook presenting evolution]. They learn to disrespect the teaching of the Holy Scripture, which says that man is a special creation of God, that he is the coronation of the Creation of God, that he belongs to two worlds, the celestial and this of earth, that he has been fashioned in God's image."

Kallinikos's view was that scientific theories on creation could not reach the prestige of the explanation of the Bible:

> Duane Gish, who holds a PhD in Biochemistry from the University of California, Berkeley, says: "The theory of evolution has not a greater prestige than that of the biblical explanation as far as it concerns the origin of species. If we compare the events of the versions (of Bible and Darwin) we can verify that the discourse of the Holy Bible has greater validity." . . . The prophet Moses enlightened by the Holy Spirit describes the cosmogony in the book of Genesis . . . There, we have the voice of the Creator of man. We have not hypothesis and theories and human imaginations. We have the testimony of the Holy Spirit, we have the apocalypses of God concerning the creation of man and the whole world . . . Some years ago, the great American physicist, Millikan, proclaimed that "if he would like to describe in a few words cosmogony, he would exactly copy the first chapter of the *Genesis* of Moses."[15]

Kallinikos went on to found the Piraeus Association of Scientists in 1993, dedicated to fostering nationalist and antievolutionary ideas. Christodoulos was appointed archbishop in 1998, and owing to his strong personality combined with powerful rhetoric talents, he enjoyed a great popular success. Backed by Chrysopigi, Christodoulos continued to promote fundamentalist ideas. Not helped by the case of the identity cards, he soon found himself at odds not only with the Ministry of National Education and Ecclesiastical Affairs but also with the socialist government, something that previous archbishops had tried to avoid. His premature death in 2008 and the succession by the moderate Ieronymos II marked the end of this particular church-state controversy. Ieronymos II, like most of his predecessors, has avoided interfering in any open way in matters of education and the sciences.

Fundamentalist views have also been expressed by traditionalist monks, who have been very popular among Orthodox people. The monk Paisios, for example, was a famous father of Mount Athos. He used to live in a skiti, an isolated home in the Holy Mountain, where he received visitors who came to seek religious answers. Paisios continued the practice of the ascetic monks of Byzantium, who gave simple answers based on Holy Scripture and the oral ascetic tradition. His views on evolution were in line with this tradition: brief and negative. Even supporters of evolutionary theory, he claimed, did not believe it, but only taught it in schools to turn children away from the church. Similar conservative views were held by others, but there were further factions within the Orthodox Church.

The "Moderate" Views and the "New Orthodox"

The moderate group avoided attacking evolutionary ideas directly, while endorsing an intellectual conception of evolution. Humans were by definition special but could evolve through faith to become closer to God. According to the metropolite of Nafpaktos (Lepante), Ierotheos, "In the church we are speaking about the evolution of man, not from monkey to man, but from man to God. And this ecclesiastic theory of evolution gives a meaning to life and satisfies internal problems and existential queries of man."[16]

With regard to the teaching of evolution in schools, the moderates put in a line of defense against it through the dogma that no nonproven theory could be taught. As the theory of evolution was thought to fall under that category, it could easily be dismissed. Furthermore, natural selection was argued to be against free will, a concept central to Orthodoxy. Violating Orthodoxy, natural selection was considered unacceptable. It came to a heated debate when evolution was introduced in the examination program in the grammar school system. One biologist, Georgios Hasouros, strongly opposed the mandatory teaching of evolution. He was concerned that what the children had learned of their own existences as the fruit of the infinite love and independent will of God would now be threatened by a random mutation and "blind and faceless natural selection." The Darwinian hypothesis, he claimed was a hypothesis at a religious level and should be considered illegal to teach at schools. He went on to state that teaching the theory of evolution was against the Greek Constitution forbidding the use of false information in schools and that it was even against the European

Constitution as an offense against human rights. Hasouros presented an unforgiving creationist view that went straight to the heart of Greek educational policy, appropriating arguments from rationalistic Catholicism to postmodernism to European law.[17]

The monk Seraphim Rose has also played a role in Greek theology's interpretation of evolution. Born Eugene Dennis Rose in San Diego, California, he adhered to the Russian Orthodox Church and founded the St. Herman of Alaska Orthodox Monastery in Platina, California. Rose was, and his writings continue to be, very popular among American Orthodox, but also among Russian and Greek Orthodox, especially the New Orthodox. Rose has been unyielding in his dismissal of evolutionary theory, a thing he considers even worse than heresy. Evolution is not just all that is wrong with Western society; it is all that is wrong with Western Christianity, a soft and corrupted form of Christianity invented by the devil to lure true Christians away from the right path. His Orthodox take on the evolution-creation debate in *Genesis, Creation and Early Man* dismissed the theory of evolution as a mere pseudoscience in colorful language, vivid metaphors, and unwavering antimodern views.[18]

For some in the New Orthodox, evolution is closely connected to Western modernity and is related to a mechanical selection theory that is contrary to the Orthodox definition of humans. To them as well as to the majority of Orthodox theology, what differentiates humans from animals is not logic but free will. In that sense, it is the personal relation with scripture and their understanding through spirituality and not philosophy that fits to Orthodox mysticism. Accordingly, another prominent representative of the New Orthodox movement, Stelios Ramfos, attacked the teaching of evolution in schools, seeing a direct link between " Western-fashion ideas" and the decay of a society that has lost its traditional values.

Orthodox Supporters of Evolutionary Theories

The faculties of theology play an important role in the Greek Orthodox Church, as they form the higher clergy that constitutes the Holy Synod. Lower clergy is usually formed by the schools of the church. Traditionally, in these faculties there have been moderate and conciliatory views concerning delicate questions such as creationism and evolution. In this group there have even been defenders of Darwinism. The best-known university theologians who have accepted evolution are Nikos Nisiotis, who was a professor of philosophy and religion at the Theological Faculty of the University of Athens; Ioannis Zizioulas, a metropolite of Pergamos, emeritus professor of dogmatic theology at the University of Thessaloniki, and chairman of the Academy of Athens; Nikos Matsoukas, who was a professor of dogmatic theology at the University of Thessaloniki; and Marios Begzos, a professor of theology at the University of Athens.

In his work *Prolegomena in Theological Theory of Knowledge*, Nikos Nisiotis defended evolution and mainly referred to the French Jesuit philosopher and inspirer of the "Nouvelle théologie" Pierre Teilhard de Chardin, whose opinions were condemned in 1950 by the Vatican but restored after the reforming Vatican Council II (1962–65). In the chapter "The Theory of Evolution and Its Point Omega," Nisiotis developed Teilhard de Chardin's idea that the universe evolves toward a maximum degree of complexity, the Omega point.[19] In *Creation as Holy Communion*, Ioannis Zizioulas argued that Darwinism was a slap in the face to the scholastic view that the image of God in man is its logic and intellect. Following this line of thought, he claimed that the Western church had failed to react positively to the Darwinian challenge. For Zizioulas, the most important aspect of being human was not logic but freedom, freedom he found contradicted in Darwinian evolution.[20]

In *Science, Philosophy and Theology in the Hexaemeron of Saint Basil*, Nikos Matsoukas noted that evolution in the Hexaemeron of Basil and mainly in the Hexaemeron of Gregory of Nyssa presented many similarities with evolutionary theories with respect to principles of thought. Creation goes forward in an evolutionary way through a natural order, he claimed, and he pointed out that the basis of the evolutionary theory and mainly the view of the unity between organic and inorganic fields could be found already in the Hexaemeron.[21]

Continuing on this conciliatory path, Marios Begzos used the occasion of the Darwin celebrations in 2009 to reflect upon the relation between science and religion. This had been problematic exclusively in Western Europe and within Roman Catholicism and Protestantism, Begzos claimed. For historical reasons, this had never been a problem for Greek Orthodoxy because it was asleep during the Ottoman regime and suffocated under the Russian tsar's theocracy. In Begzos's mind, there was no conflict and no problem left: "The controversy creationism-evolutionism on the scientific and theological levels is now considered as an out-of-date problem. Creation and evolution do not exclude each other, but they complete each other . . . The biologist is permitted to not accept Creation, but the theologian forbids himself to reject the theory of evolution for reasons of faith to Bible, which does not exclude the evolution of the Creation."[22] He concluded by quoting Dobzhansky: "Man has been created as the image of God through the evolutionary process." Evolution and creation might express two different perspectives, but they were representing the very same reality.

Creation and Evolution in School Textbooks

Biology was for the first time introduced to Greek education in 1931, taught in the fourth year of secondary school. Initially it was complementary to botany, zoology, and anthropology and was taught with these topics. The first biology textbook was published in 1933. Here it was straightforwardly stated that "natural selection is not valid" and that the problem of how the species change remains unsolved. This last statement was repeated in the subsequent textbook published in 1952. The concluding remarks of the second textbook went even further and beyond the mere critical evaluation of evolutionary theory in an openly creationist declaration: "The world is the work of the Creation." The third textbook was introduced in 1969 when biology became an independent field of study with its own course plan. This was under the dictatorship of the colonels in Greece, who claimed to be Greek and Orthodox. In this book, religious views were presented together with scientific ones. The book's conclusion was that we had to accept the existence of a God Creator, who supervised the extremely complex phenomena of evolution.[23]

After the fall of the colonels, an entirely new biology book was introduced to Greek schoolchildren. It was published in 1976, edited by Costas Krimbas, a famous Greek evolutionist (and later historian of evolutionary theories) who had worked with Dobzhansky. In this book as well as in the three subsequent textbooks that followed until the end of the 1990s, evolutionary theory was extensively and exclusively presented in an academic manner. Gone were the religious underpinning and justification of biological study. In the first two editions of Krimbas's volume for the senior year, it was stated that "man and the superior apes are close relatives concerning their genetic material and must have a common ancestor. This common ancestor should look like a monkey, if he was alive and examined by a specialist of mammal categories. This is the opinion of Simpson, the greatest paleontologist of our times." This paragraph was removed from subsequent editions without the author's consent and despite his protests.

In 1984, during the first Greek socialist government of Andreas Papandreou, the Ministry of Education and Ecclesiastical Affairs introduced a book with the title *The History of the Human Race*. It was written by the historian Lefteris Stavrianos and was aimed at secondary education. The book was soon attacked in the press by a wide range of people. The Holy Synod asked officially that the book be removed from schools. Questions were raised to the minister from both left and right political parties, and protests were organized by the official church and associations controlled by it, such as the association of *polyteknoi*, families with more than three children. These disparate groups of people all protested against the view that their ancestors were monkeys. Curiously, such a statement did not even appear in the book.

One of the leaders of the protest was the metropolitan of Florina Avgoustinos, Kantiotis, who ferociously fought for the book's retraction: "Do you know the conclusions of famous scientists, biologists, geneticists, embryologists, geologists, paleontologists, at the conferences in Chicago in 1980 and in Liverpool in 1982? If you don't, then we inform you that these men of science concluded that the origin of man is not yet demonstrated in a scientific manner. The theory of evolution was condemned, and the various genealogical trees, the results of the imagination of those who unreservedly support them, should disappear from the education books."[24]

Facing such harsh protests, the minister was forced to react. In his reply he carefully balanced support for scientific evidence and concern for his constituency: "Concerning Darwin's theory on the origin of humans, we note that in the book for the first year of grammar school there is a simple mention of the topic, as there has been in the past in the textbooks of anthropology and biology without the Permanent Holy Synod or other party having raised any protests. In any case, the author makes the distinction between men and 'humanoids,' but there is some confusion in this passage, as I have already written to the Greek Association of Theologians, and in a next edition it will be clarified."[25]

The protests not only were based on antievolutionary and anti-science sentiments but were deeply rooted in conservative nationalism. Accordingly, Stavrianos's book was also accused by a wide range of patriots for underestimating the role of Hellenism in the history of humanity. The book was eventually taken out of circulation in 1990. Remarkably, students who aimed to continue their education in medicine or biology had evolution as a mandatory topic in the two senior years, which never raised such protests. With the general Greek school reform in 1996, evolution was reintroduced in textbooks for primary and secondary education without major reactions. The long-standing religious and conservative response to evolution and education influencing Greek schoolchildren for generations since the interwar period thus seemingly came to an end.

Since the turn of the century, Greek schoolchildren have generally been fairly positive toward evolution. The majority has accepted it as an explanation for biological diversity and for the evolution of humans. Most schoolchildren, however, believe it is called a theory because of lack of evidence, while they have demonstrated a rather poor understanding of fundamental concepts in evolutionary theory. The same has turned out to be the case for biology teachers, who lack not only the knowledge to teach it properly but also a place on the curriculum to do so. Whereas the public debate was less visible compared to earlier decades when representatives of the church argued vehemently against evolution, slight changes occurred, making it even more difficult to have evolutionary theory part of the curriculum. Since 2000 only the textbooks for ninth grade and the senior year in secondary school have chapters on evolution. In both cases these are the final chapters. However, the teaching curriculum was not decided by the contents of the textbooks but rather by local school administrations. It was decided that evolution should not be among the topics for the general exams, and hence it almost disappeared altogether from biology classes. This might explain why students were positive toward evolutionary theory but had so little knowledge about it.[26] In the first decade of the twenty-first century, an influential conservative Greek creationism had turned into an effective covert antievolutionary educational practice.

NOTES

1. European Commission, *Biotechnology*, Special Eurobarometer (2010), 204, http://ec.europa.eu/public_opinion/archives/ebs/ebs_341_en.pdf.

2. Alexander Theotokis, *Γενικοί ζοολογικοί πίνακες ή Πρόδρομος της ελληνικής ζωολογίας* (Corfu, 1848). See Costas Krimbas, "Alexandre Theotokis, la notion de l'évolution et le premier texte de zoologie grecque," *Historical Review*, Institute of Neohellenic Research, 4 (2007): 191–197; Anna Sotiriadou, "Η εμφάνιση της θεωρίας της εξέλιξης των ειδών, δεδομένα από τον Ελληνικό χώρο" (The appearance of the theory of evolution of species: Data from the Greek space) (PhD diss., Thessalonica, 1990), 99–100; Charles Darwin, *On the Origin of Species*, 5th ed. (London: John Murray, 1869), 402; *The Descent of Man, and Selection in Relation to Sex* (London: John Murray, 1871), 1:197.

3. S. Sougras, *Η νεωτάτη του υλισμού φάσις, ήτοι ο Δαρουϊνισμός και το ανυπόστατον αυτού* (The newest phase of materialism, i.e., Darwinism and its nonfoundation) (Athens, 1876).

4. Ioannis Skaltsounis, *Ο άνθρωπος και ο υλισμός* (Man and materialism) (Milan, 1882); *Ψυχολογικαί μελέται* (Psychological studies) (1887); *Περί γενέσεως του ανθρώπου* (On the birth of man) (1893); *Θρησκεία και επιστήμη* (Religion and science) (Trieste, 1894); articles in the journal *Anaplasis* and the Greek newspaper of Trieste, *Ημέρα* (The day).

5. I Skaltsounis, *Θρησκεία και επιστήμη* (Religion and science) (Trieste, 1894), 207–208.

6. Sotiriadou, "*Η εμφάνιση της θεωρίας της εξέλιξης των ειδών, δεδομένα από τον Ελληνικό χώρο*," 142.

7. *Anaplasis* 119 (Apr. 1893): 1712.

8. On the trial, see Efi Gazi, *Πατρίς θρησκεία, οικογένεια* [Homeland, religion, and family]. *Ιστορία ενός συνθήματος (1880–1930)* (Athens: Polis, 2011).

9. Efthymios Nicolaidis, *Science and Eastern Orthodoxy* (Baltimore: Johns Hopkins University Press, 2011), 186–187.

10. Ibid., 188–190.

11. Διακύριξης της Χριστιανικής Ενώσεως Επιστημόνων (Declaration of the Christian Union of Scientists) (Athens, 1946).

12. Costas B. Krimbas, "The Evolutionary Worldview of Theodosius Dobzhansky," in *The Evolution of Theodosius Dobzhansky: Essays on His Life and Thought in Russia and America*, ed. Mark B. Adams (Princeton: Princeton University Press, 1994), 179–193. See also Francisco Ayala, "'Nothing in Biology Makes Sense Except in the Light of Evolution': Theodosius Dobzhanski, 1900–1975," *Journal of Heredity* 68 (1977): 3–10.

13. *Efimerios*, Oct. 2004, 10–12.

14. Kallistos Ware, *The Orthodox Way* (London and Oxford: Mowbrays/Crestwood; New York: St. Vladimir's Seminary Press, 1979), 52.

15. Kallinikos Karousos, *O άνθρωπος από τον πίθηκο; Απάντηση στην υλιστική άποψη* (Man from monkey? Reply to the materialist thesis) (Athens: Chrysopigi, 1987), 5, 10, 11, 28, 42. Note that although Gish earned a PhD in biochemistry at the University of California, Berkeley, he never taught there.

16. Ierotheos of Nafpaktos, *Οι Δεσποτικές Εορτές* (The festivals of Our Lord) (Nafpaktos: editions of the monastery Genethliou Theotokou, 1995), chap. 1.

17. Orthodox blog www.egolpion.com, Sept. 30, 2011, originally published in the journal *Orthros*.

18. Fr. Seraphim's letter to Dr. Alexander Kalomiris against the fallacy of theistic evolution, in Seraphim Rose, *Genesis, Creation and Early Man: The Orthodox Christian Vision* (Platina, CA.: St. Herman Press, 2000).

19. N. Nisiotis, *Προλεγόμενα εις την θεολογικήν γνωσιολογίαν* (Introduction to theological theory of knowledge) (Athens: Minima, 1986).

20. I. Zizioulas, *Η κτήση ως ευχαριστία* (Creation as Holy Communion) (Athens: Akritas, 1992).

21. N. Matsoukas, *Επιστήμη, φιλοσοφία και θεολογία στην Εξαήμερο του Μ. Βασιλείου* (Science, philosophy and theology in the Hexaemeron of Saint Basil) (Thessaloniki: Pournaras, 1990).

22. See www.vimaideon.gr, June 2, 2008: Αντιφωνο επιστημων, φιλοσοφιας, τεχνων και θεολογιας.

23. L. Prinou, L. Halkia, and C. Skordoulis, "The Inability of Primary School to Introduce Children to the Theory of Biological Evolution," *Evolution: Education & Outreach* 4 (2011): 275–285.

24. Kantiotis, metropolitan of Florina, encyclical 403/13-3-1985.

25. Prinou, Halkia, and Skordoulis, "The Inability of Primary School to Introduce Children to the Theory of Biological Evolution."

26. L. Prinou, L. Halkia, and C. Skordoulis, "The Inability of Primary School to Introduce Children to the Theory of Biological Evolution," *Evolution: Education & Outreach* 4 (2011):275–285, and "What Conceptions Do Greek School Students Form about Biological Evolution?," *Evolution: Education & Outreach* 1 (2008): 312–317.

CHAPTER NINE

Russia and Its Neighbors

INGA LEVIT, GEORGY S. LEVIT, UWE HOSSFELD, AND LENNART OLSSON

Creationism in Russia and in what is commonly called the "Russian-speaking world" exhibits some distinctive features. In this chapter, we outline the situation in Russia and two of its neighboring states, Ukraine and Belarus, and specify the role of the Orthodox Church and of Protestant movements in the growth of creationism in this region. We have singled out Ukraine and Belarus for inclusion in our discussion because, despite their increasing political and linguistic autonomy, these countries are tightly interwoven with Russia, both culturally and historically. After the dissolution of the Soviet Union near the end of 1991, all fifteen constituent republics became independent states. Eleven of them became members of the Commonwealth of Independent States, an amorphous political and economic alliance of former Soviet republics. The three Baltic States, Estonia, Latvia, and Lithuania, have become fully integrated into the European Union. Georgia has been pursuing a policy of confrontation with Moscow and sought collaboration with the European Union and with NATO. As a result, the political, cultural, and economic connections between the two countries have weakened considerably. After the South Ossetia or Russian-Georgian War in 2008, diplomatic relations were broken off, which also had an impact on more informal communications. Because, today, the Baltic States and Georgia have an entirely different cultural-political context, we do not include these countries in our discussion. For the same reason, we do not discuss the situation in the former Soviet republics that have a predominantly Muslim population, such as Uzbekistan and Turkmenistan.

The contemporary religious landscape in Russia, Ukraine, and Belarus is a product of long-term and short-term historical circumstances. From a long-term perspective, the major characteristic of these countries is their multiethnicity, which they inherited from tsarist Russia, accompanied by an extreme diversity of spoken languages and traditional cultures. This is especially true for contemporary Russia with its 142 major and 40 minor ethnic groups. For the majority of the population of the former Soviet Union, ethnic origin correlates with a certain kind of religiosity.[1]

From a short-term perspective, the most striking phenomenon is the explosive growth of religious communities after the breakdown of the Soviet Union. For example, if in 1989 most Russians, 75 percent, declared themselves to be atheists and only 17 percent claimed to belong to the Orthodox Church, twenty years later the situation was reversed; in 2009 73 percent of the population identified as Orthodox believers, and only 7 percent as atheists. According to the most recent study of the Russian Public Opinion Research Center (VCIOM), in 2010 75 percent of the population identified as Orthodox, 5 percent as adherents of Islam, 1 percent as Catholics, 1 percent as Protestants (of all kinds), 1 percent as Buddhists, 3 percent as independent believers (not belonging to any religious organization), and 8 percent as atheists. In addition, 5 percent of the population claimed to be unsure about its beliefs, about 1 percent claimed to belong to "other confessions," and 1 percent had difficulties answering the question.[2]

As for the "heterodox religiosity" (noninstitutionalized, eclectic religiosity, which includes beliefs in horoscopes, aliens, various "esoteric" teachings, etc.) according to a recent study Russia "follows the general pattern observed in Western countries: many elements of 'esoteric' or 'heterodox' religious worldviews are widely held in the population and possess relevance to practical actions." According to Demyan Belyaev, for many people there is no contradiction between "heterodox" religiousness and Orthodox Christianity, and they adhere to both.[3]

In Ukraine, as well as in Russia, the state authorities patronize traditional churches. According to a 2003 sociological study, 69.6 percent of Ukrainian citizens identified as believers. Remarkably, 40.7 percent of believers listed a particular religious confession but 29 percent did not. Those identifying as atheist were 16.1 percent of the population. The remaining 14.3 percent could not identify unequivocally with a particular organization. The largest Ukrainian religious organization is the Ukrainian Orthodox Church (Moscow Patriarchate), with 15.4 percent of the adult population. It is followed by the Ukrainian Orthodox Church (Kiev Patriarchate), which embraces 11.7 percent of the population; the Ukrainian Greek Catholic Church (UGCC), with 7.6 percent; and "Protestants" (Lutherans, Baptists, Pentecostals, Jehovah´s Witnesses, etc., those who are most active in propagating creationism), with only 2.4 percent of the population. The quantity of Muslims and Jews is less than 1 percent.

The situation in Belarus is less transparent, because of the authoritarian tendencies in the country. According to available data, religion plays a significant role for only 27 percent of the population. Paradoxically 58.9 percent claim to be believers, of which 82.5 percent belong to the Orthodox Church, 12 percent to the Catholic Church, 4 percent to Muslim religious communities, and 2 percent to various Protestant confessions.[4]

Creationism in Tsarist Russia

Creationism in the sense of direct intervention of religious institutions in the scientific enterprise has no historical roots in Russia. After the publication of *On the Origin of Species* in 1859 and especially of the German translation in 1860, evolutionary theory became rapidly accepted by most Russian biologists. In Russian historiography, Russia is commonly labeled "the second birthplace of Darwinism." However, although the theory became instantly popular in the Russian Empire, its dissemination was not met with massive clerical opposition. As Georgievsky and Khakhina put it: "The major specificity of the relationships between evolutionary theory and religion in Russia was the absence of open confrontation between them, which could lead to rigid *resistance to the development of science*." This is not to say that worldviews that were inspired by Charles Darwin's theory did not encounter any resistance in tsarist Russia. Still, even the Orthodox Church, the most powerful religious institution in Russia at the time, did not directly oppose evolution, and this for two reasons.[5]

First, in contrast to the Roman Catholic Church, the Russian Orthodox Church (ROC) had (and has) no institutional instruments to formulate a coherent concept opposing or supporting a theory of evolution (such as the famous Encyclical Letter of Pope Pius XII). The Sacred Synod of the ROC has no organs analogous to the Pontifical Academy of Sciences or the Vatican's Congregation for the Doctrine of the Faith. The theological claims of the patriarch within the ROC are merely "opinions," which do not reflect the official position of the church. Instead, the ROC has developed its relation with science through the mediation of the state. In 1804 state censorship before printing became compulsory for all publications in the empire. The Ministry of Religious Affairs and Public Education, established in 1817, controlled and determined the strategy of censorship in relation to both religious and secular literature, including scientific publications. In 1865 the state censorship law was weakened, and publishers obtained the right to publish voluminous (more than ten quires) and highly specialized scientific works without preliminary censorship, although it still posed significant restrictions. Darwin's *On the Origin of Species* was first translated into Russian by S. A. Rachinsky under the explicit approval of censorship officials and was published in 1864. *The Descent of Man*, however, was published only after serious difficulties with censorship. At around the same time, Carl Vogt's lectures on evolutionary anthropology were translated, as well as Thomas Huxley's *Man's Place in Nature*. By contrast, Ernst Haeckel's *Natural History of Creation* was translated into Russian but the entire print run (1,975 copies) was destroyed following an order from the Committee of Ministers (although the book was published again a year later). The reason for prohibiting the book was its disrespect toward the Bible and Christian teaching. Haeckel's *The Riddle of the Universe* was published twice already at the beginning of the twentieth century (1902 and 1906) and prohibited both times, because of its emphasis on the "animal origin of man." Censors openly admitted that scientists were allowed to read Haeckel in German and that the prohibition should first of all protect the youth from harmful ideas. In other words, censorship was directed against the popularization of Darwinism, rather than against strictly scientific publications.[6]

The second reason for the relatively mild clerical resistance to evolution was the very nature of early Darwinism in Russia. Russian scientists who worked on the basis of Darwin's theory were far less speculative than their British and, especially, German colleagues. For example, A.O. Kowalevsky and I. I. Metschnikov mainly conducted empirical studies and internal biological discussion. They were also very critical of Haeckel's speculations that ultimately resulted in a monistic, anti-Christian philosophy. As a result, tensions between the clergy and Russian biologists were not as strong and as well articulated as they were at times in Britain and Germany.

Creationism in Post-Soviet Russia

After the October Revolution in 1917, the "scientific philosophy" of Marxism became the official ideology of the growing socialist empire and religious opposition to any scientifically informed worldview disappeared from the cultural-political landscape. With the exception of the ideologically biased social sciences, scientific education in the late USSR was arguably one of the strongest worldwide and enjoyed governmental support at all levels of the educational system. The regime also invested in the popularization of science and antireligious propaganda (e.g., the society "Znanije"). Scientific creationism remained virtually unknown in the Russian-speaking world until the breakdown of the USSR in 1991. The dissemination of creationism in Russia and its neighboring countries began with the establishment of Western Protestant missionaries. Decades of atheist propaganda and anticlerical repression had turned the country into a virgin land in matters of religion. Combined with a general admiration for Western goods and values, theological naiveté rendered the population highly receptive and vulnerable to religious propaganda. In 2011, 23,848 religious organizations belonging to more than sixty different confessions were registered by the Ministry of Justice. Under the banner of religious freedom, various churches actively promoted creationism and found an audience that associated evolutionary theory with Soviet ideology rather than with empirical natural science. In the early 1990s, several "classical" writings of US creationism were translated into Russian and widely diffused. One of the first post-Soviet creationists, Dmitri Kouznetsov, made an attempt to promote the argument from design, but because of his criminal record in the United States and his marginal position within the mainstream religious and creationist movements, he is currently mostly remembered by anticreationists.[7]

In larger cities, creationist books were even distributed through central bookstores, such as the famous "House of the Book" in St. Petersburg. As a result, creationism soon became a "fact of social life," and Russia became a country with significant antievolutionist resistance within its educational system.[8]

Partly under the influence of the polemics surrounding this imported Protestant creationism, members of the ROC intervened in the debate between science and religion. In some debates, they engaged with scientists, as, for instance, Nobel Prize winner Vitaly Ginzburg, who claimed that religious beliefs are incompatible with science. However, until the end of the 1990s, Orthodox creationism remained primarily an issue of internal discussions within the ROC.[9]

In the early 2000s, the situation changed significantly when creationism was popularized by the mass media. One typical example is the case of Maria Schreiber, a schoolgirl from St. Petersburg. In 2006 a federal court in that city tried a case in which Schreiber demanded that the Ministry of Education permit an "alternative" to evolution to be taught in high school biology classes. As the newspaper *Gazeta.ru* reported from the court, the case revolved around a textbook, *General Biology* by Sergei Mamontov, in which the biblical creation story was called a "myth." Maria's father, Kirill Schreiber, urged her, through her lawyer, Konstantin Romanov, a distant relative of the last Russian tsar, to demand an apology from the author and from the Ministry of Education. Unexpectedly, Andrei Fursenko, the minister of education and science, expressed his support for the plaintiff by welcoming the teaching of "alternative ideas" in schools. The plaintiff suggested replacing Mamontov's textbook with an "Orthodox" biology textbook written by Sergej Vertjanov (aka Dr. Valschin, a physicist), in which the biblical story was presented as an alternative to evolution. Vertjanov's textbook is but one of a number of "Orthodox" biology textbooks currently available on the Russian market. However, none of these books reflects an official position of the ROC.[10]

Vertjanov's high school biology textbook stands as an example of the latest generation of creationist publications. The book is well illustrated and combines "Orthodox" interpretations with scientifically sound biological explanations. The structure of the textbook copies that of secular textbooks and corresponds to the "Educational Standard" of the Russian Federation. The textbook discusses cell biology, ecology, and genetics and touches on endosymbiosis theory and other topics that are commonly addressed in biology textbooks. However, the "Orthodox" nature of the book is evident in the preface, where the author claims that the origin neither of man nor of "ordinary biological species can be explained by chance processes." Along with technical details of mitosis and meiosis, chapters include statements on the impossibility of abiogenesis, the divine creation of the first man (Adam), the wonderful properties of DNA that should induce us to consider the existence of the Creator, and other creationist remarks. The textbook also includes a supplement with quotes of the Holy Fathers that are thought to relate to biological problems. Yet the most explicitly creationist parts of the book are chapters 10 and 11 (of the third edition), which are devoted to the origin of life and evolution. For example, section 49 begins with a discussion of the "unfoundedness of the evolutionary hypothesis."[11]

Generally, the authors of such "Orthodox" books copy the general pattern from Western creationist literature. They submit that there are no "transitional forms" in the paleontological record or that a "plan of creation" determines the course of evolution. Vertjanov, for instance, argues that the earth had been created in six days, seventy-five hundred years ago. Also, he claims that "contemporary science slowly comes to accept every word of the Holy Bible." In support of this claim, he makes "contributions" to demographic studies demonstrating that, in the pre-Flood era, human life expectation was about eight hundred years. It is worth noting that Vertjanov's textbook was subjected to criticism not only by scientists but also by Orthodox theologians. At present, there is still no official declaration of the Russian Orthodox Church toward evolutionary doctrine. The ROC stands divided on the subject of evolution. In one school of thought, "Orthodox evolutionists" interpret evolution as the realization of divine plans. The transitions from the lifeless to the biotic world and from animals to humans are thereby considered to be results of direct divine action. Another school, "Orthodox creationists," rejects evolution altogether, on the basis of theological and creation-scientific arguments. Although neither of these schools favors the theory of evolution by natural selection, Vertjanov clearly belongs to the latter. The Ministry of Education has never recommended his textbook. However, it has been used in private and state schools, for instance, in the Moscow private grammar schools "Jasenevo" and "Saburovo" and, as an experiment, in state school number 262.[12]

Another "expert" who was called to the stand as a witness in the Schreiber case by the plaintiffs was Professor V. B. Slezin, head of the Psychophysiology Laboratory of the well-known Psychoneurological Bekhterev Institute. In Russia, Slezin is known for his book *The Genocide of the White Race*, in which he speculates about the decline of the West and describes "experiments" that purportedly demonstrate the influence of prayer on the brain. In court, Slezin claimed that "Darwinism" blocks scientific progress, especially because this theory conflicts with the "fact" that the volume of the human brain had not increased during evolution.[13]

The defense pointed out that Mamontov's textbook does in fact mention creationist concepts, such as the ideas developed by the comparative anatomist Georges Cuvier in the early nineteenth century. It also argued that the textbook corresponds to the secular nature of the Russian educational system in that it does not contain religious teachings and that a scientific theory by its very nature cannot hurt religious sensibilities. On February 21, 2007, the court turned down Maria Schreiber's complaint. The federal court substantiated its decision with the argument that the scientific virtue of Darwinian theory, the role of brain increase, and the role of bipedalism should be discussed within science by scientific methods. Demonstrating a remarkable sense of humor, the court noted that "the very remoteness of events" in question (millions of years ago) prevented it from gathering the evidence necessary for making a legal decision about the truth or falsity of the theory. The scientific credibility of particular theories, such as Darwin's theory, they argued, cannot be determined by legal regulations.[14]

The secular media reported the Schreiber case mostly in ironic terms. For example, *Gazeta.ru* published an article entitled "Darwin in Slippers" as a play on Kirill Schreiber's promise that he would go to the European Court of Human Rights in Strasbourg in his slippers. Many supporters of the Schreiber case regarded this case as an artistic performance or a public relations stunt rather than a serious attack on evolutionary theory. Nevertheless, the discussions about biology textbooks catalyzed the Russian debate on science and religion in general, involving some of the highest officials of the ROC. For instance, His Holiness Alexy II, the former patriarch of Moscow and All Russia, stated in a lecture he delivered in the Kremlin that "those who want to believe that they are descended from apes, should do so, but they should not force their opinion upon others."[15]

Another significant media event was the "letter of 10 academicians" to President Vladimir Putin that was published on July 23, 2007, in which ten full members of the Russian Academy of Sciences, the most influential scientific institution in the Russian Federation, protested against an attempt to introduce the "Basics of the Orthodox Culture" as a subject in school curricula. In 2002 the Federal Ministry of Education had sent a letter to the education departments of the local governments with instructions on how to establish a new optional course in "the basics of the Orthodox Culture." The letter stated that this course should be taught at all levels of education, from elementary to high school and include such topics as the "Orthodox worldview," the "Orthodox way of life," "God and Creation," and "The Natural and Supernatural Worlds." To test pupils' knowledge, it was suggested that teachers pose questions such as "What did God create first?" Although this course prompted extremely sharp debates in Russian society, it was nevertheless established in many schools. For example, in 2003, 70 percent of schools in the Belgorod region (Central Federal District) adopted the new course as part of their curricula.[16]

The ten academicians, including two Nobel Prize winners, Vitaly Ginzburg and Zhores Alferov, not only argued that theology should not be confused with science but also pointed out that making such a course compulsory in a multiconfessional country would lead to national tensions. Indeed, the attempts of the ROC to gain a foothold in the school system are confronted not only by atheist movements and scientists but also by the Muslim communities of the Russian-speaking world. Nafigullah Ashirov, chairman of the Muslim Board for the Asian part of Russia, sharply criticized the plans of the Orthodox Church, siding with the scientists that this situation can lead to ethnic conflicts.

The academicians also protested against the attempts that had been made to add theology to the list of government-recognized scholarly disciplines. They saw both cases as a part of the dangerous process of desecularization of society: "The incorporation of the Church into a government body is an obvious breach of the Constitution. The Church has already infiltrated the army and now the media broadcast the blessings of new military equipment (battleships and submarines are now required to be blessed—which, alas! does not always help). Religious ceremonies attended by high government officials are also widely covered. These are all examples of the clericalization of this country."[17]

Both the "letter of academicians" and the Schreiber case attracted wider public attention to the growing creationism as a cultural-political problem and to the internal debates within the Orthodox Church.

Radical Orthodox Creationism

The ROC is divided between a liberal and a radical wing. The former is hostile to plain creationism and includes many well-educated priests who received their PhD-level scientific degrees from well-known institutions. For instance, Archpriest Kirill Kopeikin, the secretary of the St. Petersburg Theological Academy's Academic Council, formulates explicitly proscientific views in his publications and interviews and pleads that science and the Orthodox Church should be more closely connected with each other. He believes that only an appeal to the Christian tradition (and not to Eastern traditions such as Hinduism) "will help to clarify metaphysical preconditions and theological explications of modern European science." At the same time, he fully accepts biological evolution and contemporary cosmological models.[18]

Another example is Deacon Andrei Kuraev, who claims (in his paper "Orthodoxy and Evolution") that "there are neither textual nor doctrinal reasons to reject evolutionism." A tolerant position toward evolution is in the tradition of the Orthodox Church, he argues, while creationism is something new and introduced from outside. Kuraev thinks that creationism is in fact an attempt of some Western Protestants "to revive a pagan prejudice of identifying matter with passivity." Yet it should be noted that the acceptance of evolution by Orthodox evolutionists does not necessarily mean the acceptance of modern evolutionary theory. For instance, another Orthodox evolutionist, the priest Alexander Timofeev (who was trained as a paleontologist), argues that orthogenesis, the view that particular constraints determine phylogenetic paths, is a better explanation for evolution than Darwinian selectionism. In his view, there is biological evolution, but as a realization of divine intentions.

The radical wing of the ROC differs crucially from its liberal wing and appears to be much more intimate with modern Protestant creationism. Many authors see Satan as personally responsible for the introduction of evolution. Deacon Daniil Sysoev and Konstantin Bufeeff (the chief of the "Shestodnev" center) are arguably the most radical authors. They attack not only the scientific theory of evolution but also "orthodox evolutionism." Bufeeff characterizes Kopeikin's views as "shocking" and "Sadducean," playing in league with Catholic theologians, whom he characterizes as contemporary "Sadducees" as they follow Pierre Teilhard de Chardin and combine the idea of evolution with Christian belief. "None of the Orthodox Holy Fathers supported such views," Bufeeff summarized.[19]

Sysoev's recent publication "Who Is Like God? or How Long Was the Day of Creation?" delivers characteristic antievolutionary arguments at full length. This publication therefore deserves special attention. The first sentence of the title is a quote from Archangel Michael's cry "Who is like God?" when he stopped the rebel. Sysoev assumes that evolution is based on the idea that God is not almighty and therefore could not create the world as perfectly as he wished it to be. Correspondingly, evolutionary theory is a satanic theory. It is very sad, Sysoev claims, that even some members of the ROC have subscribed to the view that a day of creation is, in fact, not a day but is something like an epoch, that is, that they have begun thinking about biblical stories in metaphorical terms. Sysoev's book is structured as a description of a court case and consists of two major chapters: "The Evidence of the Prosecution" and "The Evidence of the Defense." Notably, Sysoev is not attacking evolutionary theory as such. His criticism is directed primarily against Orthodox thinkers (such as Deacon Andrei Kuraev and Archpriest Kopeikin) who plead against a literal interpretation of the Shestodnev (first chapters of Genesis).[20]

Generally, the radical creationists trace the origin of the idea of spontaneous generation back to myths of the appearance of gods out of initial chaos. The emphasis on the importance of the sun (characteristic for the life sciences) is expectedly reduced to sun worshiping, as with the sun gods of ancient Egypt. The concept of the affinity between humans and animals is explained by radical creationists as the persistence of totemic myths.[21]

Another characteristic feature of radical (both Protestant and Orthodox) creationism is the negation of the "scientific character" of evolutionary theory. They argue that such a "vague hypothesis" is rigidly defended as a "theory" and that some "progressive scientists" have already shown its weaknesses. These claims are often accompanied by accusations of immorality. Sysoev even suggests that some evolutionists demand the legalization of cannibalism, although he does not support this claim with any references.[22]

Most of these authors make evolutionism responsible for profligacy, fascism, adultery, theft, dictatorship of the proletariat, and capitalist rivalry. Allegedly, evolutionism, also contributes to the increase in abortions and the legalization of euthanasia and in-vitro fertilization. In addition, some have pointed out that the growth of evolutionism is a sign of the nearing doomsday.[23]

Many radical orthodox creationists have developed a kind of homemade epistemology for criticizing the methods of evolutionary theory. For example, Sysoev devotes fifty pages of his book to undermining the idea of the objectivity of science and to the claim that science is in fact based on the same foundations as creationism but hypocritically denies its own foundations. The only way for humans to survive, he says, is to bring science back to its original roots, that is, to theology.[24]

The major problem however, is that the denial of creation leads to the rejection of the major dogmas of Christianity and, correspondingly, to the impossibility of salvation. For the radical creationists, evolutionary theory is not only a doubtful scientific theory but the essence of temptation for contemporary European civilization. It is therefore impossible to argue with them by means of scientific arguments. They are, however, open to theological discussion.

Theological Discussions about Evolutionary Theory

The central question in the theological discussions about evolutionary theory is whether the idea of organic evolution is at all compatible with orthodox theology. A conference devoted to the problems of orthodox-theological education, which took place in Moscow (October 29–30, 1999) provided an opportunity to each of the competing sides to clearly formulate its views. The conference revealed that within the ROC the position of the radical creationists is much stronger than that of the evolutionists. Several reasons account for their dominance.

First of all, neither the Bible nor the writings of the Holy Fathers mention anything about evolution. It is, however, still up for debate whether this means that the existence of evolution is rejected. Also, the discussion about evolution touches on a general theological question concerning the interpretation of biblical claims. In fact, to a significant extent, the debate within the Orthodox Church narrows down to a philological discussion about the meaning of particular Hebrew words and quotes from Genesis. Traditionally, exegetics has remained mostly undeveloped within Orthodox theology, especially when compared to Catholic theology. At the beginning of the twentieth century, an influential Russian theologian, N. N. Glubokovsky, wrote that no systematic scholarly interpretation of the Bible existed within the Orthodox tradition. Under the Soviet regime, which promoted atheism, this situation did not change. Recently, however, the theological discussions have become increasingly "scientific," owing to the efforts of the evolutionists.[25]

The antievolutionists base their specifically Orthodox theological argumentation on the "opinions" of the Holy Fathers and Orthodox righteous men (*pravedniki*). At present, compendiums of antievolutionary quotes by Orthodox righteous men such as the very influential hesychasts, members of a mystic movement in the Orthodox Church, St. Theophan the Recluse and St. Ignatius Brianchaninov, circulate among creationists. Both men expressed explicitly antievolutionary views. St. Theophan the Recluse claimed that people who believe in the animal origin of humans do not distinguish between soul and spirit: if we admit, he argued, that the very essence of man consists of his spirituality, the theory of evolution becomes meaningless (1, п. 106, с. 100). He also unequivocally demanded that the teachings of Darwin and Ludwig Feuerbach should be anathemized. Brianchaninov held similar views. The Creator, he claimed, designed the visible and invisible world ex nihilo and did this by his word only. New creatures were created by his word as well. The act of creation was not a work since the perfect and almighty Lord did not need to work in order to make his creation. Everything occurred perfectly corresponding to his thought and word from the very beginning.

Another example of early anti-Darwinian thought in the ROC is that of St. John of Kronstadt, a Russian Orthodox archpriest. St. John explained that both the "undereducated" and the "overeducated" do not believe in a personal, just, almighty and eternal God but in "blind evolution" without any divine creative power, and therefore idolize human reason. To a large extent, today, Orthodox creationists repeat these arguments of the fathers and righteous men, while at the same time trying to respond to the challenges of contemporary science.[26]

The major internet platform of contemporary Orthodox creationists is the portal www.shestodnev.ru, but a lot of creationist material can be found also on www.pravoslavie.ru. The Shestodnev society (formally The Centre of Missionary and Outreach Shestodnev) is headed by the archpriest Konstantin Bufeeff, who has a PhD in geology and mineralogy. Bufeeff asserts that Alexy II, the former patriarch of the Russian Orthodox Church between 1990 and 2008, had thoroughly read the society's publications and approved of its efforts "to defend the patristic teaching of creation." In 2009 Shestodnev published its "proceedings" exclusively devoted to the theological analysis of contemporary science ("Proceedings N5," available online). The conference was opened by Hieromonk Damascene of the St. Herman of Alaska Monastery in Platina, California, who emphasized that, "although this conference may seem like a 'mustard seed,' its importance is far-reaching, and will prove to be even more so in the years ahead. Orthodox Christians from all over the world benefit from the defense of our common Faith that is being undertaken in Russia."

The proceedings are remarkable in the sense that they reveal a new tactic of the creationists, who now attempt to disprove evolutionary theory by citing outstanding natural scientists of the recent past. In one paper, for instance, Bufeeff invokes the authority of one of the most outstanding Russian/Soviet scientists, Vladimir Ivanovich Vernadsky, to back up his attack on evolution. Probably, Bufeeff singles out Vernadsky for two reasons. First, Vernadsky is indeed one of the greatest Russian naturalists. He is regarded as one of the founders of modern geochemistry and biogeochemistry and was a pioneer of radiogeology. He is also regarded as a pioneer of genetic mineralogy and as an outstanding crystallographer. Renowned for his encyclopedic knowledge, he is also considered to be one of the great thinkers in the history of science. However, the most valuable contribution of Vernadsky to modern science is his grandiose theory of the biosphere and of living matter. The author of contemporary Gaia theory, James Lovelock, wrote: "We [Lovelock and Lynn Margulis, coauthor of the Gaia theory] discovered him to be our most illustrious predecessor." Elsewhere, Margulis wrote: "Indeed, Vernadsky did for space what Darwin had done for time: as Darwin showed all life descended from a remote ancestor, so Vernadsky showed all life inhabited a materially unified place, the biosphere." Accordingly, Vernadsky is one of the most influential Russian thinkers. At present, there are about one thousand published works about Vernadsky.[27]

The second reason why Bufeeff chose Vernadsky to underwrite his antievolutionary argument is that Vernadsky developed the concept of the eternity of life. Indeed, Vernadsky coined a peculiar space-time theory to prove the thesis of the cardinal difference between living and inert matter and, hence, that biological processes could not be deduced from physical and chemical laws. If impassable boundaries did not separate living from nonliving matter, Vernadsky argued, we would observe abiogenesis, the origination of living matter from nonliving matter, occurring regularly in the biosphere, which is not the case. According to Vernadsky, living matter is a regular, nonsporadic, perpetual phenomenon in the universe, the origin of which cannot be explained in terms of biogeochemistry, the science he founded.

Bufeeff, however, misinterpreted Vernadsky's theory and argued that Vernadsky "proved the failure of evolutionary theory." In support of this claim, Bufeeff reviewed Vernadsky's publications for quotations that could serve his purposes and aligned these with biblical citations, thereby denying Vernadsky's intentions. Vernadsky's objective was to create an all-embracing theory of evolution, in which the entire biosphere is regarded as an evolving bio-inert body. His criticism of traditional Darwinism resulted from his interest in the evolution of the global ecosystem and had nothing to do with a creationist rejection of evolutionary theory.

Misinterpretations of scientific literature have become very common among Russian creationists. For instance, they use publications of Scott Gilbert, pioneer in the field of evolutionary developmental biology, to disprove Ernst Haeckel's biogenetic law and, in the same breath, the very idea of evolution. These examples illustrate that contemporary radical-creationist Orthodox theology consists of an amalgam of citations from indisputable religious authorities and selected claims by outstanding scientists.[28]

Protestant Creationism in Belarus and Ukraine

In the Western parts of the former Soviet union, particularly Belarus and Ukraine, the influence of the Protestant and "new age" Christian communities is especially strong. For instance, the community of the "evangelical Christians-Baptists," with its center in Belarus (www.rogdestvo.by), attracts creationists from both Belarus and the Ukraine. On the community's website one can read an interview with Vladislav Olkhovsky, head of the laboratory of nuclear processes of the National Academy of Sciences of the Ukraine. Olkhovsky is convinced of the reality of creation and claims that his belief is supported by scientific evidence. If we would change any of the four known forces (gravity, the weak force, the strong force, and the electromagnetic force) only slightly (1% to 3%), life on earth would be impossible. This scientific discovery, Olkhovsky argues, perfectly correlates with biblical statements such as: "For this is what the Lord says: he, who created the heavens, he is God; he who fashioned and made the earth, he founded it; he did not create it to be empty, but formed it to be inhabited" (Isaiah 45:18). He also claims that no one can prove the theory of abiogenesis and that no evidence exists in support of the theory of macroevolution. Evolution cannot explain the complexity of the living world, but he believes that microevolutionary processes are possible.

The very structure of Olkhovsky's arguments, his attempts to appeal to the authority of well-positioned scientists, and his opposition to macroevolution are typical of the imported Protestant creationism in Russia and Ukraine. Creationism is influential in the Protestant churches of the Ukraine, Belarus, and Russia, and, unlike the situation in the Russian Orthodox Church, there is no visible conflict between "radical creationists" and "evolutionists." Almost all books discussing scientific issues that are published by Protestant communities are written by creationists.

Ukraine's creationists became active right after the breakdown of the USSR. For example, the Christian Center for Science and Apologetics (Crimea), a "nondenominational ministry to the subcultures of intellectuals unreached by the Gospel," was founded in 1991 by Sergei Golovin. In the subsequent years, Golovin became one of the leading young-earth creationists in Ukraine. Crimea is active not only in the Ukraine but also in Russia. Being one of the largest creationist centers in the Ukraine, it organizes both online and offline events. For example, it offers summer camps for children between eight and fifteen years old. These camps, with the title "Shestodnev," are structured in accordance with the six-day creation story, but also include a seventh and eighth day, which stand for God's rest and the fall and salvation. The center also organizes the lecture series for adults called "Lessons of Evolution," which emphasizes that evolutionary theory is a form of religious belief.[29]

Scientific creationism initially came to Russia in the form of translations of texts written by Western Protestant creationists and members of the intelligent design movement. Because the most important creationist arguments are of a universal antiscientific nature, they are easily converted into any cultural context and were therefore able to influence Orthodox creationists, who utilized them in their doctrinal attack on secular education. At the same time, only Russian Protestant creationism exists as part of a transnational creationist network. US creationists often visit creationist organizations in Russia and neighboring countries and support their activities. For example, by the beginning of the 1990s, the publishing house Protestant was actively supported by American creationists. Arguments of the Protestant creationists are directed toward the broadest possible audience without confessional barriers.[30]

Russian Orthodox creationism differs in this respect from its Protestant counterpart. Encouraged by the successes of the Protestant creationists and by the growing influence of the Orthodox Church in Russia, the Orthodox creationists strengthened their efforts to give Russian education confessional colors, thereby changing the educational landscape. They apply two parallel tactics in hopes of achieving this goal. The first tactic is to try to make religious education, with an Orthodox bias, part of the *compulsory* curriculum. The course "The Basics of the Orthodox Culture" for ordinary schools is an example of this tactic. The second tactic consists of intervening into areas of science that have a bearing on the development of a modern worldview. The production of new "Orthodox" science textbooks and the participation in the Maria Schreiber trial are examples of this second tactic. Although Orthodox creationism has been strongly influenced by the methods and arguments of Western scientific creationism, it is relatively autonomous in relation to transnational creationist movements. The major reason is that the polemics within the Orthodox tradition are primarily based on statements of the Holy Fathers and proceed on a purely theological level. Protestant scientific creationism has developed neither methodological instruments nor arguments, which could be used within Orthodox polemics of this kind. This does not mean, however, that Russian Orthodox creationists are entirely disconnected from their American and European Orthodox and Protestant fellows. Russian-speaking Orthodox creationists have borrowed their theological arguments to a significant extent from the writings of the American Hieromonk Seraphim Rose of the Russian Orthodox Church Outside Russia, who cofounded the St. Herman of Alaska Monastery in Platina, California. Russian Orthodox creationist workshops have also welcomed prominent figures of Western intelligent design and creationist movements. For example, a workshop organized by Shestodnev (2005) and cochaired by Konstantin Bufeeff brought together antievolutionists from several countries and backgrounds, including Siegfried Scherer from Germany, Hugh Owen and Hieromonk Damaskin from the United States, Guy Berthault from France, and Sergej Vertjanov from Russia.[31]

The question of how influential both Protestant and Orthodox creationism are in contemporary Russia is a controversial one. In the absence of reliable sociological data, it remains a matter of contention among researchers. Arguably the best-known Russian anti-creationist, the paleontologist Kirill Eskov, claims that "the popularity of creationism is strongly exaggerated by journalists thoughtlessly PRing various forms of deviant behavior, creating an illusion of its mass character, and then falling prey to this illusion."[32]

By contrast, the historian of science Mikhail Konashev believes that "contemporary neocapitalist, neoliberalist, neoclerical Russia became nearly a world leader of the antievolutionist campaign."[33] Konashev provides statistics according to which 24.4 percent of the Russian population maintains that evolutionary theory is proven, 24 percent supports the theory of creation, and a staggering 34.5 percent maintains that contemporary science is unable to explain the origin of man.[34] If these data are correct, then Russia, over the past twenty years, has gone through a dramatic transformation and can no longer be considered a country dominated by scientism.

NOTES

1. Dm. Bogojavlenskij, "Сколько народов живет в России?," *Население и общество. Институт демографии Государственного университета - Высшей школы экономики* 4 (2008): 319–320; http://demoscope.ru/weekly/2008/0319/tema01.php.

2. Natalya Zorkaya, "Православие в безрелигиозном обществе," *Russian Public Opinion Herald* 2 (2009) 65–84; http://wciom.ru/index.php?id=268&uid=13365.

3. Demyan Belyaev, "'Heterodox Religiousness' in Today's Russia: Results of an Empirical Study," *Social Compass* 58 (2011): 353.

4. The number of religious communities in the Republic of Belarus (as of Jan. 1, 2010), Office of the Commissioner for Religions and Nationalities, www.belarus21.by/ru/main_menu/religion/relig_org/new_url_1949557390.

5. A. B. Georgievsky and L. N. Khakhina, *Развитие эволюционной теории в России* (Saint Petersburg: RAN, 1996), 9, 147.

6. L. M. Dobrovolsky, *Запрещенная книга в России* (Moscow: Vsezojuznaja knizhnaja palata, 1962), 232; G. V. Zhirkov, *История цензуры в России XIX-XX вв. Учебное пособие* (Moscow: Aspekt Press, 2001); Carl Vogt, *Человек и его место в природе*, vols. 1–2 (Saint Petersburg: Gaideburov, 1866); Ernst Haeckel, *Естественная история миротворения.* (Saint Petersburg: Demakov, 1873).

7. Ronald L. Numbers. *The Creationists: From Scientific Creationism to Intelligent Design*, expanded ed. (Cambridge, MA: Harvard University Press, 2006), 413–416.

8. www.gks.ru/bgd/regl/b11_13/IssWWW.exe/Stg/d1/02-13.htm; Duane T. Gish, *Ученные креационисты отвечают своим критикам* (Saint Petersburg: Biblija dlja vsech, 1995); Ken Ham, C. Snelling, and C. Wieland, *Книга ответов* (Saint Petersburg: Biblija dlja vsech, 1994); Reinhard Junker and Siegfried Scherer, *История происхождения и развития жизни* (Saint Petersburg: Kairos, 1997); Henry Morris, *Библейские основания современной науки* (Saint Petersburg: Biblija dlja vsech,1993); Henry Morris, *Начало мира* (Moscow: Protestant, 1993); Henry Morris and Martin Clark, *Ответ в Библии* (Moscow: Protestant, 1993); D. Petersen, *Открывая тайны творения* (Saint Petersburg: Biblija dlja vsech, 1994); M. B. Konashev, "Эволюционная теория и нео-модернизация России," in *Научное, экспертно-аналитическое и информационное обеспечение национального стратегического проектирования, инновационного и технологического развития России*, ed. J. Pivovarov (Moscow: INION RAN, 2010), 77–83.

9. E.g., V. L. Ginzburg, "Вера в Бога и научное мышление," *Poisk* 29–30 (1998): 479–480.

10. Inga Levit, "Evolutionstheorie und religiöses Denken in der zeitgenössischen orthodoxen Theologie," in *Netzwerke: Verhandlungen zur Geschichte und Theorie der Biologie*, vol. 12, ed. M. Kaasch et al. (Berlin: VWB, 2006), 233–247; *Gazeta.ru*, Oct. 27, 2006; http://antidarvin.com; News agency Rosblat, Jan. 3, 2007.

11. S. Vertjanov, *Общая биология*, 3rd ed. (Moscow: Svjato-Troitzkaja Lavra, 2012), 4.

12. Ibid., 224; S. G. Mamontov, "Вера и наука. Рецензия на учебник С. Вертьянова—«Общая биология» для 10—11 классов Москва Свято-Троицкая Лавра 2005," http://isps.ru; Inga Levit, "Теория эволюции и современная православная теология," in *В тени дарвинизма: Альтернативные теории эволюции в 20-м веке*, ed. Georgy S. Levit et al. (Saint Petersburg: Fineday Press, 2003), 149–155; M. Zeleznova, "В начале было тесто," *Russian Newsweek* 38 (2005), 68.

13. V. Slezin, *Геноцид белой расы. Кризис эпохи* (Saint Petersburg: AST, 2010).

14. http://humanism.su/ru/articles.phtml?num=000398.

15. *Die Presse.com*, Feb. 6, 2007.

16. "Open Letter to the President of the Russian Federation Vladimir V. Putin from the Members of the Russian Academy of Sciences," http://scepsis.ru/eng/articles/id_8.php; "Letter from the Ministry of Education to the Local Education Departments," Oct. 22, 2002, no. 14-52-876.

17. "Open Letter to the President of the Russian Federation Vladimir V. Putin from the Members of the Russian Academy of Sciences."

18. K. Kopeikin, *Наука и религия на рубеже третьего тысячелетия. Актовая речь на торжественном заседании посвященном 200-летию Санкт-Петербургской Духовной Академии* (Saint Petersburg, 2009); http://svitk.ru/004_book_book/15b/3407_kuraev-pravoslaviya_i_evolyuciya.php; www.pravmir.ru/kreacionizm-ili-evolyuciya/.

19. D. Sysoev, *Кто как Бог? Или сколько длился день Творения* (Moscow: Izdatelstvo Zentra Ioanna Kronschtatskogo, 2003), 3–6; K. Bufeeff, "Ересь эволюционизма," in *Шестоднев против эволюции*, ed. D. Sysoev (Moscow: Palomnik 2000); S. Schubin, "Ложь «православного эволюционизма»," in ibid.; D. Sysoev, "Эволюционизм в свете православного учения," in ibid., http://shestodnev.ortox.ru/publikacii/view/id/11327.

20. http://creatio.orthodoxy.ru/kkB/index.html.

21. Bufeeff, "Ересь эволюционизма"; Schubin, "Ложь «православного эволюционизма»"; Sysoev, "Эволюционизм в свете православного учения"; D. Sysoev, *Летопись начала* (Moscow: Izdatelstvo Stretenskogo Monastzrja, 1999); Sysoev, *Кто как Бог? Или сколько длился день Творения*; V. N. Trosnikov, "Научна ли научная картина мира," *Novij mir* 12 (1989): 257–263.

22. Sysoev, *Летопись начала*.

23. N. Kolzunskij, "Зеленый свет," 2001, www.creatio.orthodoxy.ru.

24. Sysoev, *Кто как Бог? Или сколько длился день Творения*, 62.

25. N. N. Glubokovskij, *Церковный вестник* 50–51 (1909): 1575–1582; A. I. Jurchenko, "Начало библейского богословия от «чайников»—«чайникам» от богословия," 2002, www.textology.ru.

26. http://pravbeseda.ru/library/index.php?page=book&id=301; Ju. Maksimov, "Богословские аспекты проблемы согласования православного и эволюционного учений о происхождении человека," in *Шестоднев против эволюции*, ed. D. Sysoev (Moscow: Palomnik 2000), 129–138.

27. www.shestodnev.ru/PravOsm005/Greeting2009eng.htm; James Lovelock, "The Biosphere," *New Scientist* 1517 (1986): 51; Lynn Margulis and Dorion Sagan, *What Is life?* (New York: Simon & Schuster, 1995).

28. See, e.g., A. Khomenkov, "Неоправданные стереотипы: о некоторых стратегических ошибках в современном креационнном мышлении," 2005, www.goldentime.ru/hrs_text_007.htm.

29. See, e.g., S. N. Golovin, *World-Wide Flood: Myth, Legend or Reality?* (Simferopol: Christian Center for Science and Apologetics 1999); http://scienceandapologetics.org/.

30. www.icr.org/article/creationism-russia/.

31. www.creatio.orthodoxy.ru/english/rose_genesis/index.html.

32. K. Eskov, "Популярность креационизма сегодня сильно преувеличена," Feb. 29, 2012, www.chaskor.ru/p.php?id=3036.

33. M. B. Konashev, "Эволюционная теория и культурно-идеологическое состояния российского общества во второй половине XIX–XXI вв," in *Социальный диагноз культуры российского общества второй половины XIX–начало XXI вв: Материалы всероссийской научной конференции*, ed. Kozlovskij (Saint Petersburg: Inersocis, 2008), 134–139.

34. M. B. Konashev, "Дарвин и религия," *Chelovek* 5 (2009): 22–37.

CHAPTER TEN

Turkey

MARTIN RIEXINGER

Although only 3 percent of Turkey's land area falls within Europe, and although more than 99 percent of its population of 73.7 million are at least formally Muslim, there are good reasons to include a case study on this country in a book on the history of creationism in Europe. The predecessor of the Turkish Republic, the Ottoman Empire has belonged to the European system of powers since its emergence. With the Tanzimat (Reform) era in the 1830s a process began, during which the institutions of the sultanate, which were based on Islamic and Byzantine traditions, were replaced by new ones designed after European models. This process accelerated after the coup/revolution in 1908–9 by the İttihad ve Terakki Cemiyeti (Committee for Unity and Progress, commonly known in the West as Young Turks) and finally culminated in the early phase of the Turkish Republic founded by Mustafa Kemal Atatürk (1881–1938) in 1923, when equal rights for women, Western clothing, the Latin alphabet, surnames, the Christian week rhythm, and the Gregorian calendar were adopted. After World War II Turkey joined NATO and the Council of Europe, and it was invited to join the European Economic Community (EC) as far back as 1959. An offer for membership was refused by the leftist Evevit government in 1978, but in 1987 the Özal government submitted a membership request to the EC. And although the multiparty democracy established in Turkey after World War II surely has had major shortcomings—it was, for example, interrupted by coups in 1960, 1971, and 1980—the Turkish political system differs markedly from the Arab successor states of the Ottoman Empire that developed into one-party dictatorships, authoritarian monarchies, or fragile democratic systems torn apart by civil war.[1]

In spite of the westernization and secularization of the political system, the majority of the population remained deeply committed to Islam. For several decades this was neutralized by two factors. First, Atatürk's successor, İsmet İnönü, already laid the basis of a state-controlled Diyanet İşleri Reisliğī, later Başkanlığı (Directorate for Religious Affairs), which developed from an organ of supervision into a means of religious support under leaders of a different political orientation. Second, the center-right parties integrated segments hostile to westernization into the political system. Hence support for parties emerging from the Islamist Millî Görüş (National Vision) remained restricted to about 10 percent of the electorate. From the 1990s onward, however, their share of the votes grew dramatically. In 1995 Necmettin Erbakan's Refah Partisi (Welfare Party) emerged as the major party of a coalition government, and in 2002 his former party fellow Recep Tayyip Erdoğan came to power with his Adalet ve Kalkınma Partisi (AKP, Justice and Development Party).[2]

Religious movements that did not join ranks with the Islamists nevertheless engaged in a cultural opposition that could be articulated relatively freely since the political liberalization that had begun in 1946. This deserves particular attention in our context, because one major aspect of this cultural opposition to westernization is the rejection of the theory of evolution.[3]

Islamically motivated opposition to the theory of evolution can be traced back to the last decades of the Ottoman Empire, but it did not become a major issue before the 1970s. From the 1990s onward, the Islamic creationist discourse that had emerged in Turkey gained popularity in Muslim communities elsewhere, above all among Muslims migrants in the West. Thus it became one of the most important forms of creationism in several European countries.

Evolution in the Late Ottoman and Early Republican Period

Although several Ottoman Turkish authors presented the theory of evolution already in the 1870s and 1880s, it did not become a major issue until some secular intellectuals from the Imperial Medical School, including Baha Tevfik and Abdullah Cevdet, used the theory of evolution for antireligious propaganda. They belonged to the Young Turk opposition to the autocratic rule of Abdülhamit II, who legitimized his rule in Islamic terms. These authors did, however, not propagate the theory of evolution on the basis of Charles Darwin's writings. Instead they referred to the reformulations of Ernst Haeckel and the German "Vulgärmaterialists" Carl Vogt and Ludwig Büchner. Certain distinctive positions held by these authors—such as the eternity of matter, spontaneous generation, and the notion that the different human races have descended from different ape species—became important for the discourse on the theory of evolution in Turkey.[4]

Mustafa Kemal, the founder of the Turkish Republic, and his political allies had belonged to the Young Turk movement. So it came as no surprise when the theory of evolution was integrated into the secondary school curricula in the 1920s. The outstanding document for this is the course book *Türk Tarihin Ana Hatları* (Main lines of Turkish history), which describes life as completely governed by the laws of chemistry and physics. The authors assert that current life forms have developed from more simple ancestors and that humans represent the highest stage of development. In addition, the theory of evolution was used to legitimize a racist concept of history in which a superior role is attached to the Turks (the term was quite generously attached to a large variety of Eurasian peoples). This particular racist element was given up in the 1940s, but until the 1970s the theory of evolution was an undisputed part of the Turkish biology curricula; even the most contentious aspects, the struggle for survival and the descent of man, were presented in the textbooks.[5]

Religious Opposition to the Theory of Evolution

The Qurʾan does not contain a detailed account of creation, but in several places there are references to the creation of Adam from earth, clay, and argil (6:2, 15:26, 28, 33; 23:12; 32:7–9; 37:11; 55:14). Furthermore, gaps in the Qurʾanic account are filled with sayings ascribed to Muḥammad (*ḥadīth*), so that the premodern Islamic position on the Flood, the time frame of creation, and the creation of species do not differ substantially from the Christian ones. Hence the theory of evolution confronted traditional Islamic concepts with challenges similar to those encountered by Christianity.[6]

The foundation text for an Islamic variant of creationism was written in the Arabic provinces of the Ottoman Empire where the theory of evolution had been presented by Christian academics. Ḥusayn al-Jisr al-Ṭarābulusī, an educational reformer from Tripoli in what is today Lebanon, observed that the educational gap between the Christian and the Muslim populations in this Ottoman provincial center was widening. Hence he dedicated his efforts to the establishment of secular schools and the propagation of secular knowledge. At the same time he struggled to defend the core doctrines of Islam against ideological challenges arising from the very same secular knowledge he propagated. In 1881 he wrote the tract *al-Risāla al-ḥamīdīyya fi taḥqīq al-diyāna al-islāmiyya wa-haqqiyyat al-sharīʿa al-muḥammadiyya* (Treatise dedicated to Sultan Abdülhamit II to demonstrate the truth of the Islamic religion and the Muḥammadan Sharīʿa) in order to single out acceptable and undesirable elements. Translations of this treatise into Turkish, Tatar, and Urdu ensured that its influence did not remain restricted to the Arab world and religious scholars. Al-Jisr explains the theory of evolution and describes it as a rationally tenable but unproven hypothesis, whereas several verses in the Qurʾan explicitly state that Adam was specially created. Thus it is not permitted to interpret these verses allegorically. In addition, he presents dozens of examples of astonishing features in animals and plants that are supposed to substantiate the argument from design. The insistence on design and teleology has remained a central aspect of Islamic creationism until the present.[7]

The first comprehensive attack on the theory of evolution in Turkish was formulated by a religious conservative intellectual in the early republic. İsmail Fenni (Ertuğrul since the introduction of surnames in 1935) published his *Maddiyun mezhebi izmihlali* (Dissolution of the materialist school/religion) in 1928. In contrast with later antievolutionist literature, this book is remarkably free from malicious polemic. For example, his attempt to refute Lamarckism and Darwinism is preceded by an exposition of both concepts in a neutral tone. He uses various arguments to delegitimize evolutionist concepts. Sticking to the ideal-type concept of species, he claims that any deviation (*inhıraf*) from the ideal type has to be considered a sign of decay. In addition, he quotes arguments from anti-Darwinist literature in French, and he follows the claim of the American naturalist Louis Agassiz that the fossil record does not provide proofs of Darwin's assertion that life forms from lower geologic strata are less complex than life forms from higher strata. Hence life must have come into existence in different forms. Furthermore, he ridicules the assertion that apes and humans are related. In the appendix to his work, he stresses that hitherto the Darwinists have been unable to explain the mechanism underlying evolution and disputes the value of fossils described as missing links.[8]

Only after adducing philosophical and scientific arguments does İsmail Fenni attack the theory of evolution from a religious point of view. He juxtaposes the hypothetical character of the theory of evolution to the unambiguous statements on the creation of man in both the Qurʾan and the Hadith. From religion he turns to morals. Quoting Gustave Le Bon, he stresses that men have to be aware of the deep gaps that separate them from animals. Only then will they be able to direct their efforts to sublime goals; otherwise they will be caring only for their physical necessities.[9]

Whereas Fenni rejects the theory of evolution, he refers to radioactivity in order to disprove the concept that matter is eternal. Thus, he was one of the first authors to use modern physics in an effort to delegitimize nineteenth-century materialism. In fact, this aspect of his work was more influential in the beginning: antimaterialist polemics in the 1940s were dominated by references to radioactivity, the theory of relativity, or quantum physics, whereas hardly any attention was paid to evolution.[10]

Fenni was not the only Turkish critic of evolution in the first half of the twentieth century. Ömer Nasuhi Bilmen, a conservative religious scholar who managed to make a career in the official religious institutions owing to his nonconfrontational approach—he became head of the Directorate for Religious Affairs in 1960—attacked the theory of evolution in his catechism, first published in 1947 and often republished since. He employs arguments such as the impossibility of spontaneous generation, and he asks why the alleged transformation of life forms cannot be observed at present. According to Bilmen, Adam came into being independently from other creatures, and the biological perfection of man proves that he has been designed. As in the case of Fenni, Bilmen's refutation of Darwinism was part of a general attack against materialism.[11]

The Role of Said Nursi

The cases of Fenni and Bilmen show that the theory of evolution was rejected in religious circles, but they also demonstrate that it was not a major issue before 1970. In order to understand the shift that occurred at this time, we have to turn to a scholar who never explicitly mentioned Darwin and evolution.

Said Nursi was born to a family of Kurdish peasants in southeast Anatolia, most likely in 1876. Already at an early age he was renowned for his religious learning, but unlike most religious scholars in the eastern provinces he took a great interest in secular learning. This resulted in his plan to found a university in the city of Van that could combine religious and secular education. However, he failed to gain official support for this project. His attitude toward the Young Turk Revolution in 1908–9 was ambivalent. On the one hand, he saw the necessity of a thoroughgoing reform of the Ottoman Empire to end the autocracy of Abdülhamit II. On the other hand, he abhorred the Young Turks' positivist and materialist ideology. After he had returned from captivity in Russia during World War I, he stayed in İstanbul and supported the resistance that Mustafa Kemal organized against the occupation by Greek and Entente troops. In 1923 he was called to the new capital, Ankara, in order to provide religious support for the new government.[12]

However, when it became obvious that secularization was the objective of Mustafa Kemal and his followers, Said Nursi withdrew to Van and dedicated himself to worship. After being falsely accused of participating in an insurrection motivated by both Kurdish separatism and resentment against secularization in 1925, he was arrested and spent the next two and a half decades in prison or in banishment in central and western Turkey. In this situation, he formulated his theological, ethical, and political ideas that were collected as the *Risale-i Nur* (Treatise of the [Divine] Light). He was freed one year after the Democratic Party of Adnan Menderes (executed in 1961 after the 1960 coup) had come into power in 1950. Said Nursi supported the Menderes's center-right policies because his government put an end to restrictions on religious practice and because of his policy of pro-Western alignment. Said Nursi died on March 23, 1960, two months before the first coup in the history of modern Turkey that aimed unsuccessfully at the restoration of undiluted Kemalist secularism.[13]

Said Nursi did not formulate a program of radical political opposition to Kemalism but attempted to undermine its materialist intellectual foundation instead. His attitude to modern science was rather inconsistent. For example, he defended modern astronomy and praised it as an example of divine harmony observed in nature. Nevertheless, he clung to an affirmative interpretation of Muhammad's heavenly journey (*miʿrāj*), which is based on a spherical concept of the cosmos. A large part of the *Risale-i Nur* is dedicated to the affirmation of a concept originating from classical Sunni theology (*kalām*) according to which all events are directly created by God. In one chapter of his most famous work *Sözler* (The Words), he attempts to refute that matter can organize itself in more complex forms. For this purpose, he formulates the parable of a pharmacy in which a storm has raged so that all the different tinctures and pastes have been spilled. Nobody would expect life to emerge in the course of such an event. This is supposed to prove that a conscious creator, and not chance, has brought all life forms into being.[14]

Although neither Darwin nor evolution is mentioned, Said Nursi's intention is obvious: he praises al-Jisr and also mentions in passing that all species are perfect and without predecessors and successors. The omission of an explicit reference might be explained by the intention to immunize a rural and small-town audience, whose members might have never heard of the theory of evolution, without making them curious with a detailed treatment.[15]

The Rise of Islamic Creationism in the 1970s and 1980s

In the 1960s Said Nursi's followers, the Nurcus (disciples of the [divine] light), dedicated themselves primarily to the propagation of the *Risale-i Nur* in print and in ceremonial lectures called *ders* (lesson). However, by the end of the decade some authors began to adapt his teachings to current problems. One of them was Fethullah Gülen (born 1938 or 1941). Originating from a village in the conservative province of Erzurum, he was posted after his studies as a preacher (*vaiz*) of the Directorate for Religious Affairs in İzmir and Edirne, two of the country's most westernized cities. In İzmir he addressed in particular the students of the local university. In this period left-wing Kemalists and Marxists dominated the campuses, so that many students with a conservative outlook, often those with a modest background, felt discriminated against. In 1973 Gülen delivered a series of lectures directed against the theory of evolution that were recorded and then copied and distributed on audiotape.[16]

In the following years, other Nurcus joined the bandwagon. The most noteworthy was Âdem Tatlı, a biologist who translated writings of leading American creationists such as Duane Gish. Anti-Darwinism was from then on propagated in various tracts and in many articles in the Nurcu magazines *Köprü* (The bridge), *Zafer* (Victory), and *Sızıntı* (The leakage).[17]

Throughout the 1970s the creationist propaganda produced by the Nurcus was reinforced by two authors from other religious currents: Haluk Nurbaki, a physician and contributor to right-wing magazines, and Zekeriya Beyaz, a member of the Nationalist Movement Party (Milliyetçi Hareket Partisi) and an employee of the Directorate for Religious Affairs. Their arguments were essentially the same as those of the Nurcus, with the exception that Beyaz introduced the allegation that Darwinism was part of a worldwide Jewish conspiracy.[18]

The arguments brought forth against the theory of evolution in the Turkish context can be separated into three partially overlapping categories: scientific, ethical, and political. To the first category belongs the objection that the complexity of organisms and single cells necessitates a designer, the defense of the idealist concept of species implying that mutations mean deterioration and that species are immutable, the functionality of all organisms in general and their beneficial nature for humans in particular disproving "the survival of the strongest," the demonstration of the inability of the evolutionists to explain the origin of life, the misrepresentation of Darwinian evolution as saltationism, the emphasis on an unbridgeable gap between humans and apes, actual and alleged forgeries of missing-link fossils, the low degree of mathematical formalization, counterarguments by renowned scientists (often misquoted), and finally the allegation that the theory of evolution is not falsifiable. The argumentation is supposed to undermine certain aspects of the theory of evolution, while attempts to provide detailed explanations of alternative creationist positions, like the creation of Adam as a chemical procedure in a kind of laboratory, remain isolated phenomena.[19]

On the religious and ethical level, the Islamic creationists appeal to the consensus of monotheists with regard to creation, and they claim that evolution is a concept derived from pagan myths. Moreover, they denounce the principle of competition as immoral, which is occasionally combined with the allegation that an evolutionist perspective deprives nature of its aesthetic value.[20]

The boundary between the ethical and the political arguments is diffuse. The Islamic creationists point to the emergence of evolutionary ideas in the Age of

Enlightenment to denounce its atheist origin. In the context of Turkish history, they stress that the Young Turks were the first to propagate the theory of evolution, which then became the basis of Kemalism. Moreover, they identify the struggle for survival with racism and fascism on the one hand and capitalism on the other. Their main target is, however, Marxism, to which they ascribe a particular affinity for the theory of evolution.[21]

This last point can easily be explained by the position of Nurcus during the political confrontation that haunted Turkey throughout the 1970s. Since the late 1960s Bülent Ecevit had turned the elitist Kemalist Cumhuriyet Halk Partisi (Republican Peoples' Party) into a center-left party able to challenge the mass appeal of Süleyman Demirel's hitherto dominant center-right Adalet Partisi (Justice Party), which the Nurcus supported. The latter had to cope with the emergence of an Islamist and a nationalist party to its right. The paramilitary organizations of these two parties as well as a wide array of Marxist factions were prone to violence and thus responsible for bringing Turkey to the brink of a civil war. Although the Nurcus positioned themselves on the right, they kept aloof from violent activities. In the same way as their founder had considered the intellectual confrontation with Kemalism as his task, they concluded that they should undermine Marxism with arguments targeting materialism.[22]

A decisive shift for the development of Islamic creationism in Turkey came about with the military coup of September 12, 1980. Whereas the 1960 and the 1971 coups aimed at reinstalling secularism, the junta headed by Kenan Evren propagated a "depoliticized" version of Islam in order to gain popular support. The first elected government by Turgut Özal's ANAP (Motherland Party) enforced this tendency from 1982 onward. Özal himself as well as many deputies and cabinet members hailed from the former Islamist Millî Selâmet Partisi (National Salvation Party). In this period conducive to religious conservatism, the followers of the aforementioned Fethullah Gülen were able to found a media empire comprising the daily *Zaman* (Time), now Turkey's most successful newspaper, and the TV channel Samanyolu (The Milky Way).[23]

Among Özal's followers with an Islamic background was the secretary of education, Vehbi Dinçerler, who issued a white book on the dangers emanating from the theory of evolution prepared by Tatlı. The document reiterated the arguments known from the 1970s. The report did not result in the total deletion of the theory of evolution from biology textbooks, but the most contentious aspects, the struggle for survival and the descent of man, were omitted, while creation and Lamarckism were introduced as alternative concepts. In addition, malicious polemics characterized the treatment of Darwinism in the new compulsory subject *din ve ahlak bilgisi* (religion and ethics). The Directorate for Religious Affairs joined the campaign by publishing anti-Darwinist literature. In the 1980s opposition to Islamic creationists remained weak. Thus, meetings of academics in protest against government-sponsored creationist propaganda had no major impact on the public.[24]

The Global Impact of Turkish Islamic Creationism in the 1990s

During the 1990s, at a time when polemics against the theory of evolution were toned down at the official level, Turkish creationism became a global phenomenon. Responsible for this was Adnan Oktar, an interior designer by education, who began to address the high-society youth of İstanbul with religions sermons in the late 1970s. His four main topics were an ultra-idealistic critique of materialism, Darwinism in particular; eschatology, in particular the imminent advent of the Mahdi (a savior figure in Islamic eschatology); glorification of the Ottoman Empire; and anti-Semitic and anti-Masonic conspiracy theories, including Holocaust denial. In the 1980s he also began to publish books with similar content under the pseudonym Harun Yahya. Although Oktar received a lot of press coverage, he remained a minor figure on Turkey's religious scene, and his movement was considered by some observers in the mid-1990s to be doomed to vanish.[25]

This, however, turned out to be a complete misjudgment. In 1999 Oktar discovered the opportunities the internet provided, and so he produced online versions of his tracts, which were subsequently enhanced with interactive and multimedia facilities. At the same time he began to have them translated into various European and Central Asian languages, as well as into Indonesian and Malay. Thus he gained enormous popularity, especially in the Western diaspora and Southeast Asia. However, his success was restricted to creationism and some related subjects, such as the Big Bang as proof for the *creatio ex nihilo* and archaeological evidence for the Qurʾanic reports on the prophets before Muhammad. Eschatology seems to have been a nonseller, and his advocacy for Holocaust denial until 2002 brought his German supporters into the argument.[26]

With regard to the theory of evolution, Oktar draws his inspiration mostly from the Nurcu writings of the 1970s and 1980s. However, he has professionalized the creationist enterprise considerably. His books are printed on high-quality paper, and they are constantly updated with references to new antievolutionist publications from the West.[27]

Oktar holds Darwinism responsible for both communism and fascism, and by referring to a single anti-Turkish remark by Darwin in a private letter in

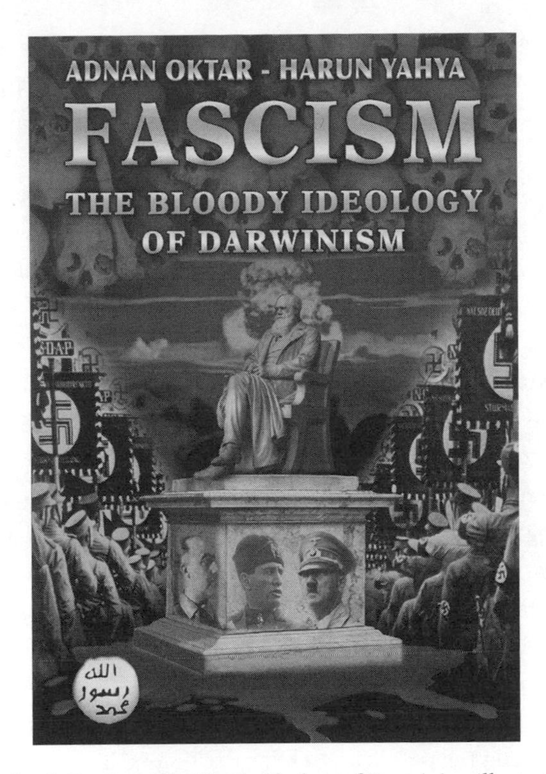

The cover of the book *Fascism: The Bloody Ideology of Darwinism* illustrates Adnan Oktar's assertion that Darwin is responsible for the crimes of Hitler and Mussolini. The book was originally published in Turkish. © Harun Yahya.

1881, in which he claimed, "The more civilised so-called Caucasian races have beaten the Turkish hollow in the struggle for existence," Oktar concludes that anti-Turkish racism was Darwin's main motivation. Moreover, he applied 9/11 to his creationist agenda. In his book and website *Islam Denounces Terrorism*, he asserts that Darwinism was responsible for the attacks because it promotes an "ideology of conflict." Apparently this struck a chord among Muslims in the diaspora, who were eager to disassociate Islam from these events.[28]

Oktar did not associate himself with any particular Islamic movement in Turkey. Nevertheless, the support of the Islamist Millî Görüş movement helped him to further his cause in Turkey and among the Turkish diaspora. In 1990 he institutionalized his activities with Bilim Araştırma Vakfı (BAV, Science Research Foundation), which built systematic contacts with American anti-Darwinist organizations, such as the young-earth creationist Institute for Creation Research in San Diego, as well as the Seattle-based Discovery Institute, which advocates intelligent design. He also invited major Western creationists such as Duane Gish and John Morris to several conferences in Turkey. Now, lecturers trained by the Science Research Foundation hold conferences in Turkey, and they are also often invited to speak at conferences of Islamic organizations abroad. In 2006 he organized exhibitions of his "Creation Museum," which provides alleged evidence for the existence of contemporary species after extremely long periods, in many Turkish cities and at conferences of Islamic organizations in other countries. On one of the most important aspects of Oktar's activities, their funding, no reliable information has been available until now.[29]

In Turkey, Adnan Oktar's activities have met little resistance until recently. First, amateurish websites were designed by some opponents, and recently adversaries have countered him with weblogs. Moreover, magazines such as *Bilim ve Gelecek* (Science and future) and *Bilim ve Ütopya* (Science and utopia) counter creationist allegations. The latter magazine does, however, belong to Doğu Perinçek's İşçi Partisi (Workers Party). This might lead to the impression that the defense of the theory of evolution belongs to the agenda of the elitist and antidemocratic ideology of *ulusalcılık*, an eclectic left-wing nationalism based on elements from Kemalism, Marxism, and either Pan-Turkism or Eurasianism (the idea of an alliance of Slavs and Turkic people against the West). The influence of the İşçi Partisi on the public is out of proportion with its meager electoral appeal documented by its 0.4 percent in the 2007 elections, because its supporters are overrepresented among the academic elite and owing to the talent of its leader Perinçek to draw attention to himself. One of Oktar's most outspoken opponents, the physician-biologist Ümit Sayın, is associated with the *ulusalcı* movement. In February 2008 he and Perinçek were apprehended in connection with the alleged "Ergenekon conspiracy," which aimed to create tension by bomb assaults on public institutions and celebrities in order to create an atmosphere favorable for a coup against the AKP government. Both were indicted in August 2008, and Sayın was sentenced to four years in prison, whereas Perinçek received a life sentence.[30]

Since the AKP came in power in late 2002, Adnan Oktar has found political support for his activities. For example, shows of his "Creation Museum" have been supported by local authorities. After the reintroduction of the primacy of evolutionary theory during Social-Democratic rule in the 1990s, creationism is now again presented on a par with the theory of evolution in biology textbooks, and educators who teach the theory of evolution have come under pressure. Moreover, Richard Dawkins's website has been blocked in Turkey, because he exposed a picture from Harun Yahya's massive and widely distributed *Atlas of Creation* as a fraud, and the editor of the magazine of the Turkish Academy of Sciences was sacked, because she had published a cover article on the occasion of the Darwin year in 2009. These developments led to a stronger reaction against creationism. Academics now organize themselves to counter what they see as religious infringement in the teaching of biology, and they alert the international public to this development.[31]

The impact of Islamic anti-Darwinism from Turkey in the West is not restricted to Muslim migrants and converts. Mustafa Akyol, an erstwhile assistant to Oktar who was sent to the Discovery Institute to forge ties with the intelligent design movement in the mid-2000s, began to work independently as a freelance writer and was excluded from Oktar's organization. In Turkey he tried to popularize intelligent design outside Islamic circles, while he looked for support among American conservatives, whose Muslim darling he became. Thus he declared Darwinism responsible for jihadist terrorism since it causes a moral revolt against the West in the Islamic world. In May 2005 he was invited to testify before the Kansas State Board of Education, the Caucus for Education. He had been invited by some members of the Republican majority who intended to introduce intelligent design as an alternative to the theory of evolution that would deserve equal time in biology classes. Akyol's task as a Muslim was to dispel the suspicion that the theory of intelligent design was Christian theology cloaked in the terminology of biochemistry. Thus it would be possible to circumvent the constitutional provision banning the teaching of religion in public schools. However, his efforts were for naught, because in August 2006 the local legislators had already been defeated in the Republican primaries in Kansas. Akyol himself renounced intelligent design in favor of the anthropic principle in late 2009. Nevertheless, Akyol's desertion is most probably the reason why Oktar stopped endorsing intelligent design.[32]

American Christian and Turkish Islamic Creationists: Strange Bedfellows?

Several aspects of the Turkish creationist discourse show that it is part of a more general cultural struggle by religious circles in Turkey against westernization in general and materialist ideologies associated with leftism in particular. In this context, the pro-American attitude that some religious groups in Turkey, the Nurcus in particular, exposed until the 1980s deserves attention, because in this respect the situation in Turkey differs markedly from that of other Muslim countries.

When the Nurcus aligned themselves politically with the center-right in the late 1940s, they also supported the move away from Kemalist neutralism to NATO-membership, which was completed, but not initiated, by the Menderes government. Up to the present, the participation of some of their members in the Korean War is one of their favorite subjects. Therefore, Nurcus and some Islamic publishers friendly to them drew a very favorable picture of America. In particular, they contrasted the great role attached to religion in American public life to the suppression of religious practice by the Kemalists, which they equated to the treatment of religion under communism. The most extensive example of this attitude is to be found in the popular novel *Minyeli Abdullah* by Hekimoğlu İsmail from 1967, where the protagonist praises the dominance of free enterprise and the omnipresence of religion in American public life. The emphasis on religion gained new strength in the late 1970s and the 1980s because of the strong religious element in the rhetoric of Presidents Carter and Reagan.[33]

In this context, the idea to turn to American Protestant fundamentalists to borrow from their extensive anti-Darwinist literature seemed plausible. However, in addition to ideological affinity, a contingent detail also played a role. The demand to prove the historicity of the Deluge motivated several American creationists to travel to Turkey in order to search for remnants of Noah's Ark on the accessible Turkish part of Mount Ararat. These expeditions received press coverage in Turkey and apparently popularized creationist concepts in Turkey. However, although the Turkish Islamic creationists took over many arguments from their American Protestant role models, they never subscribed to the concept of a young earth, because the Qur'an mentions no time scale for the creation. They accept the time frame of the geological consensus, and some of them argue that man appeared early in the Tertiary.[34]

The Rejection of Evolution and the Internalization of Islamic Norms

Certain experiences made by religious-minded circles in Turkey and the ideological conclusions the Nurcus drew from them also have to be taken into consideration as explanation for the emergence of creationism in Turkey. Furthermore, it might provide a clue to its popularity among Muslim minority communities in the West.

Even in religious circles in Turkey, the demand for the reinstallation of the Islamic legal system (*sharia*) is a minority issue. An Islamic way of life has become a matter of choice instead of something to be imposed by law. Thus the most radical politics of secularization in the Islamic world have not led to a vehement counterreaction but to a particular form of adaptation. The Nurcus play a pivotal role in this process, because their founder had declared already in 1909 that the *sharia* is "99 percent of ethics, while only 1 percent is law and hence the business of the state."[35]

In the case of the Nurcus this reflects not only an illusion-free attitude toward the feasibility of an Islamic order but also an ethical choice, because according to their view moral behavior can result only from conviction, not from force. This has implications for one's worldview, since in order to take the correct ethical decision, one has to be aware of one's position in the order of the world, which is determined not by random evolutionary processes but by a purposeful and harmonious order of creation. This aspect is particularly conspicuous in a type of pseudodocumentary novels that have become popular in recent years. They report the conversion of people who lead an unethical and unfulfilled life (as communists or youths with behavioral problems) until they become aware that God has created a harmonious order. Therefore, they renounce the theory of evolution, begin to pray, and start a meaningful life.[36]

Moreover, the emergence of their ideology has to be considered not only in a political context but also with regard to the social background of most Nurcus. The Nurcu movement is well represented in the provinces between the westernized coast and the conservative heartland and, with regard to social stratification, among people from a provincial background who owe their worldly success to the modern secular educational system. Like upwardly mobile migrants with a strong religious background, they have experienced a shift that causes ambivalence: success on the one hand, exposure to a sinful society on the other. The Nurcu movement offers a program of adaptation on the political level that goes hand in hand with a strong effort to preserve the purity of one's belief in a hostile environment. Islamic creationism as an attack on the foundations of Western materialism is one part of this project of self-assertion.[37]

In Turkey creationist views are far more common than in most other European countries. According to a 2005 Eurobarometer poll, 51 percent of the Turkish population rejects human evolution, whereas only 27 percent accepts it, and a survey conducted by Ipsos in 2011 revealed that no less than 60 percent of the Turks identify as creationists. Even among science students, the number of those accepting the theory of evolution is not significantly higher (27.9 percent), although among that sample those who are undecided dominate with about 51.4 percent.[38]

This low rate of acceptance of the theory of evolution demonstrates that creationist views respond to deeply held convictions of large parts of the population, so the strong Turkish opposition to evolution cannot be explained simply as a cloned form of American creationism. Instead, one has to acknowledge that arguments from American creationists were appropriated in order to lend support to an opposition to the so-called materialist theory of evolution that can be traced back to the late nineteenth century. With regard to both this long historical memory and the more recent adaptation of American models, the opposition to the theory of evolution in Turkey shows—Islamic concerns notwithstanding—striking similarities to that in Russia and Greece.[39]

NOTES

1. Feroz Ahmad, *The Making of Modern Turkey* (London: Routledge, 1993).

2. Hakan Yavuz, *Islamic Political Identity in Turkey* (Oxford: Oxford University Press, 2003), 207–264.

3. Ibid., 144–205.

4. Remzi Demir and Bilal Yurtoğlu, "Unutulmuş bir Osmanlı düşünürü Hoca Tahsîn Efendî'nin *Târîh-i tekvîn-i hilkat* adlı eseri ve Haeckelci evrimciliğin Türkiyeye girişi," *Nüsha* 1 (2001): 166–197; Şükrü Hanioğlu, "Blueprints for a Future Society: Late Ottoman Materialists on Science, Religion and Art," in *Late Ottoman Society: The Intellectual Legacy*, ed. Elisabeth Özdalga (London: Routledge Curzon, 2005), 28–116; Cemal Güzel, "Türkiye'de Maddecilik ile maddecilik karşıtı görüşler," in *Modern Türkiye'de siyasi düşünce*, vol. 8: *Sol*, ed. Murat Gültekingil (İstanbul: İletişim Yayınları, 2007), 49–66, 60–61.

5. *Türk Tarihin Ana Hatları* (İstanbul: Kaynak Yayınları, 1996), 33–38, 47; Jens Peter Laut, *Das Türkische als Ursprache? Sprachwissenschaftliche Theorien in der Zeit des erwachenden türkischen Nationalismus* (Wiesbaden: Harrassowitz, 2000), 2–11; Nazlı Öztürkler, "Türkiye'de evrim eğitimin sosyolojik bir değerlendirmesi" (Yüksek lisans thesis, Ankara Üniversitesi, 2005), 76–77; Deniz Peker, Gulsum Gul Cömert, and Aykut Kence, "Three Decades of Anti-evolution Campaign and Its Results: Turkish Undergraduates' Acceptance and Understanding of the Biological Evolution Theory," *Science & Education* 19 (2010): 739–755, 740.

6. Martin Riexinger, "Islamic Opposition to the Darwinian Theory of Evolution," in *Handbook of Religion and the Authority of Science*, ed. Olav Hammer and James R. Lewis (Leiden: Brill, 2011), 484–509, 484–485.

7. Ḥusayn al-Jisr al-Ṭarābulusi, *al-Risala al-ḥamīdiyya fi taḥqīq al-diyāna al-islāmiyya wa-haqqiyyat al-sharīʿa al-muḥammadiyya*, ed. Khālid Ziyāda (Ṭarābulus: Jarrush Press, n.d.), 188–209, 237–258; Johannes Ebert, *Religion und Reform in der arabischen Provinz: Ḥusayn al-Ǧisr aṭ- Ṭarābulusī (1845–1909): ein islamischer Gelehrter zwischen Reform und Tradition* (Frankfurt am Main: Lang, 1991); Taner Edis, *An Illusion of Harmony: Science and Religion in Islam* (Amherst, NY: Prometheus Books, 2007), 151–158.

8. İsmail Fenni, *Mâddiyûn mezhebi izmihlali* (İstanbul: Orhaniye matbaası, 1928), 77–86, 89–92, 109, 694–698.

9. Fenni, *Mâddiyûn*, 112–114.

10. Ibid.; Mustafa Şekip Tunç, "Madde ve Ruh 5: (Aynştayn) bize ne kazandırdı?," *Büyük Doğu* 29 (Apr. 28, 1944): 4; Mustafa Şekip Tunç, "Madde ve Ruh 6: Dünyamızın sonu," *Büyük Doğu* 30 (May 5, 1944): 4.

11. Ömer Nasuhi Bilmen, *Büyük İslâm İlmihali* (İstanbul: Bilmen Yayınevi, 1990), 128, 172, 188, 192–196; Ahmet Selim Bilmen, *Ömer Nasuhi Bilmen: hayatı, eserleri, anılar* (İstanbul: Bilmen Basımevi, n.d.).

12. Şükran Vahide, *Islam in Modern Turkey: An Intellectual Biography of Bediuzzaman Said Nursi* (Albany: State University of New York Press, 2004).

13. Yavuz, *Islamic Political Identity*, 174.

14. Said Nursi, *Isharat al-i'jaz fi mazann al-ijaz* (İstanbul: Sözler Yayınevi, 1999), 138, 224; Said Nursi, *Sözler* (İstanbul: Işık Yayınları, 2004), 767–768; Said Nursi, *Âsâ-yı Mûsâ* (İstanbul: Yeni Asya Neşriyat, 2005), 142.

15. Said Nursi, *İşaratü'l-İcâz* (İstanbul: Yeni Asya Neşriyat, 2001), 144–145; Said Nursi, *Mektubat* (İstanbul: Yeni Asya Neşriyat, 2002), 195, 207.

16. Fethullah Gülen, *Yaratılış gerçeği ve evrim* (İstanbul: Nil Yayınları. 2003); Fethullah Gülen, *Evrim anaforunda gerçek*, audio CD (2005).

17. Duane Gish, *Fossiler ve Evrim* (İstanbul: Cihan Yayınları, 1984).

18. Haluk Nurbaki, *Verses from the Holy Qur'an and the Facts of Science* (Delhi: Kitab Bhavan, 2002); Zekeriya Beyaz, *Darwinizm'in yıkılışı* (İstanbul: Sağduyu Yayınları, 1978), 174, 179, 216; Lutz Berger, "Ein Türkischer Reformtheologe? Zekeriya Beyaz zwischen Tradition und Politik," *Welt des Islams* 45 (2005): 74–107.

19. Beyaz, *Darwinizm'in yıkılışı*, 15, 91, 131–136; Şemseddin Akbulut, *Darwin ve evrim teorisi* (İstanbul: Yeni Asya Vakfı, 1985), 6–8, 12, 21–28, 39–43, 50–54, 84–88; Mehmed Yıldız, "Yaratılışta tesadüfün hissesi var mıdır?," in *Merak ettiklerimiz*, ed. Âdem Tatlı (İstanbul: Cihan Yayınları, 1990), 21–24; Fethullah Gülen, *İnancın Gölgesinde* (İstanbul: Işık Yayınları, 2003), 1:23; Cemil Şahinöz, *Wer bist du? Die Reise des Menschen* (İstanbul: Nesil Yayınları, 2005), 23–29, 47–49; Selim Uzunoğlu and İrfan Yılmaz, *Alternatif biyolojiye doğru* (İzmir: Türkiye Öğretmenler Vakfı, 1995), xxii, 19–23, 222–226, 239–242; Abdüllatif Metin, "Klasik madde kavramı ile hayatı izah denemesinde karşılaşılan sınırlar," preface to Arthur Nelson Field and John N. Moore, *İlmî gerçekler ışığında Darvinizm* (İstanbul: Otağ Yaynları, 1978), 33–34, 39–42; Âdem Tatlı, *Evrim ve Yaratılış* (Konya: Toplum Kitabevi, 1992), 20–28, 70–71, 118–119, 133–134, 139–140, 153; Fethullah Gülen, *Asrın getirdiği tereddütler* (İstanbul: Işık Yayınları, 2003), 4:231–232; "Piltdown Adamı," *Köprü* 34 (Dec. 1979): 5–7; "Evrim Teorisi hâlâ neden tutuluyor," *Zafer* 72 (Dec. 1982): 7; Nurbaki, *Verses*, 48; İsmet Hasenekoğlu, *Evrim terorileri ve mutasyonlar* (İstanbul: Yeni Asya Yayınları, 1977), 126; İhsan İnal, "Ve bindörtyüzyetmiş insan," *Sızıntı*, Aug. 1986, 415–422; Ömer Said Gönüllü, "Evolüsyon ve evrim," *Sızıntı*, Sept. 1994, 344–347; Bahri Dayıoğlu: "İlk insan nasıl yaratıldı?" *Köprü* 22 (Sept. 1990): 21–34.

20. Beyaz, *Darwinizm'in yıkılışı*, 54–58; Dayıoğlu, "İlk insan nasıl yaratıldı?"; Akbulut, *Darwin*, 7, 102–105; Tatlı, *Evrim*, 61–62; Gülen, *Yaratılış*, 57; Ali Toker, "Zekâ mı sevk-i ilahi mi?," *Köprü* 42 (Mar. 1980): 16–17; M. Uslu, "Yardımlaşma," *Sızıntı*, July 1981, 25–26; Zekeriya Altuner, "Hayat bir mücadele mi?," in *Merak ettiklerimiz*, ed. Âdem Tatlı (İstanbul: Cihan Yayınları, 1990) 120–131, 125; Şahinöz, *Wer bist du?*, 43–47.

21. Akbulut, *Darwin*, 7; İzzet Akyol, "Darwinizmden Kemalizme," *Köprü* 72 (Aug. 1987): 42–44; "Yüz yıllık Tartışmazı," *Köprü* 86 (May 1985): 5; "Türkiye'de islâmî uyanış," *Köprü* 53 (Jan. 1986): 26–28; Bünyamin Duran, "Irkçılık, Milliyetçilik ve Müslümanca

Bakış," *Köprü* 52 (Autumn 1995): 11–30, 18–20; Metin Karabaşoğlu, *Camide dans var* (İstanbul: Karakalem, 1999), 29–30; Şahinöz, *Wer bist du?*, 51–53; Metin, "Klasik madde kavramı," 9–10; "Evrim teorisi sosyal hayata ne getirdi?," *Köprü* 81 (May 1985): 12; Uzunoğlu and Yılmaz, *Alternatif biyoloji*, 17–18; Beyaz, *Darwinizm'in yıkılışı*, 9; Gülen, *Evrim anaforu*; Gülen, *Asrın getirdiği tereddütler*, 4:233–234; Âdem Tatlı, "Yaratılış mı Evrim mi?," *Zafer* 73 (Feb. 1983): 6–9; "Darwinizmle Marksizm birbirinin destekçisidir," *Köprü* 81 (May 1985): 10.

22. Hugh Poulton, *Top Hat, Grey Wolf and Crescent: Turkish Nationalism and the Turkish Republic* (London: Hurst, 1997), 162.

23. Yavuz, *Islamic Political Identity*, 69–75; Christine Jung, *Islamische Fernsehsender in der Türkei: Zur Entwicklung des türkischen Fernsehens zwischen Staat, Markt und Religion* (Berlin: Klaus Schwarz, 2003), 181–196.

24. Türkiye Cumhuriyeti Milli Eğitim Gençlik ve Spor Bakanlığı, *Evrim Teorisi hakkında Rapor Özeti* (Ankara: Milli Eğitim Basımevi, 1985); Türkiye Cumhuriyeti Milli Eğitim Gençlik ve Spor Bakanlığı, *Liseler için Din kültürü ve ahlâk bilgisi* (İstanbul: Milli Eğitim Basımevi, 1994), 12–13; Öztürkler, "Türkiye'de evrim eğitimi," 56, 91, 109–110, 116–118, 134; Peker, Cömert, and Kence, "Three Decades," 741; Hüseyin Aydın, *İlim, felsefe ve din açışından yaratılış ve gayelik (teleoloji)* (Ankara: Diyanet İşleri Başkanlığı, 1986).

25. Martin Riexinger, "The Islamic Creationism of Harun Yahya," *ISIM-Newsletter*, Oct. 11, 2002, 5; Martin Riexinger, "Propagating Islamic Creationism on the Internet," *Masaryk University Journal of Law and Technology* 2:2 (2008): 99–112, 103; Ronald L. Numbers, *The Creationists: From Scientific Creationism to Intelligent Design*, expanded ed. (Cambridge, MA: Harvard University Press, 2006), 421–427; Ruşen Çakır, *Ayet ve Slogan: Türkiye'de islami oluşumlar* (İstanbul: Metis, 1995) 246.

26. www.creationofuniverse.com; www.kavimlerinhelaki.com/index2.html; www.perishednations.com/index2.html, all accessed June 19, 2012; Landesamt für Verfassungsschutz Baden-Württemberg, *Verfassungschutzbericht 2003* (Stuttgart: Landesministerium des Innern, 2004), 76–78; Riexinger, "Islamic Opposition to the Darwinian Theory of Evolution," 496–498.

27. Riexinger, "Propagating Islamic Creationism," 104–106.

28. Letter from Charles Darwin to William Graham, July 3, 1881, Darwin Correspondence Project, accessed Nov. 8, 2013, www.darwinproject.ac.uk/entry-13230; www.harunyahya.org/evrim/hy_darwinizmininsanligagetirdigibelalar/belaları.html; www.harunyahya.com/disaster1.php; www.harunyahya.org/evrim/hy_Turk_Dusmanligi/dtd1.html; http://harunyahya.com/en/books/735/Islam_Denounces_Terrorism/chapter/1659, all accessed June 19, 2012.

29. Riexinger, "Propagating Islamic Creationism," 106–108; www.yaratilismuzesi.com/; www.fossil-museum.com/, both accessed June 19, 2012; Numbers, *The Creationists*, 422–425.

30. http://web.archive.org/web/20041130175840/www.geocities.com/evrimkurami/; "Harun Yahya Çarpıtmaları" (Harun Yahya's treacheries), http://kodoman.wordpress.com/; "Evrim karşıtı yaratılışçı iddialara cevaplar" (Answers to creationist claims), http://web.archive.org/web/20101115183029/http://yaratiliscilaracevaplar.wordpress.com/, all accessed June 19, 2012; Martin Riexinger, " 'Turkey, Completely Independent!' Contemporary Turkish Left-Wing Nationalism (*ulusal sol / ulusalcılık*): Its Predecessors, Objectives and Enemies," *Oriente Moderno* 90 (2010): 353–395, 368–379; "Üniversite hocalarına Ergenekon gözaltısı," *Zaman*, Feb. 23, 2008, www.zaman.com.tr/haber.do?haberno=655572;

Erdal Kılınç, "Ergenekon soruşturması üniversiteye uzandı," *Milliyet*, Feb. 23, 2008, www.milliyet.com.tr/2008/02/23/yasam/axyas01.html; "Ergenekon'un Gizli Tanığı Ümit Sayın Tahliye Oldu," Jan. 31, 2010, http://bianet.org/bianet/toplum/119759-ergenekonun-gizli-tanigi-umit-sayin-tahliye-oldu, all accessed June 19, 2012; "İşte Ergenekon Davası'nda verilen cezalar," *Hürriyet*, Aug. 5, 2013, www.hurriyet.com.tr/gundem/24461529.asp.

31. Betül Kotan, "Bakan Çelik: Yaratılış aynen kalacak," *Radikal*, Mar. 5, 2006, www.radikal.com.tr/haber.php?haberno=180408; "AKP'li vekil 'Adnan Hoca'nın kitabını dağıttı," *Radikal*, Feb. 27, 2007, www.radikal.com.tr/haber.php?haberno=214156; Peker, Cömert, and Kence, "Three Decades," 742–743; Harun Yahya, *Atlas of Creation*, vol. 1 (Istanbul: Global Publishing, 2006); Riazat Butt, "Missing Link: Creationist Campaigner Has Richard Dawkins' Official Website Banned in Turkey," *Guardian*, Sept. 19 2008, www.guardian.co.uk/world/2008/sep/19/religion.turkey; Nicholas Birch, "Turkey: Scientists Face Off against Creationists," www.eurasianet.org/departments/insight/articles/eav052407.shtml; "Evolution, Science, Education" Symposium, İstanbul, May 2006, http://web.archive.org/web/20070617001137/; www.uniaktivite.net/ajanda/ajanda.asp?uid=49&id=3044, all accessed June 19, 2112; Aykut Kence and Ümit Sayın, "Islamic Scientific Creationism: A New Challenge in Turkey," *Reports of the National Center for Science Education* 19:6 (1999): 18–20, 25–29; Aykut Kence, Mehmet Somel, and Nazlı Somel, "Turks Fighting Back against Anti-evolution Forces," *Nature* 445 (2007): 147.

32. Riexinger, "Propagating Islamic Creationism," 109–111; www.akyol.org; www.thewhitepath.com; Mustafa Akyol, "Under God or Under Darwin? Intelligent Design Could Be a Bridge between Civilizations," *National Review*, Dec. 2, 2005, www.nationalreview.com/comment/akyol200512020813.asp; Tony Ortega: "Your OFFICIAL program to the Scopes II Kansas Monkey Trial," www.pitch.com/kansascity/your-official-program-to-the-scopes-ii-kansas-monkey-trial/Content?oid=2177607; Ralph Blumenthal, "Evolution Opponents Lose Kansas Board Majority," *New York Times*, Aug. 2, 2006; Mustafa Akyol: "Darwin Year Is Not the Year of Atheism," www.hurriyet.com.tr/english/opinion/11154116.asp?yazarid=301&gid=260, all accessed June 19, 2012; Numbers, *The Creationists*, 426.

33. Hekimoğlu İsmail, *Minyeli Abdullah* (İstanbul: Timaş Yayınları, 2004), 164–165; Filiz Çırpçı, "Amerikan usulü lâiklik," *Köprü* 81 (Feb. 1984): 7–8; "Reagan: Hakiki ve derin bir imana sahibim," *Yeni Nesil*, Aug. 3, 1985.

34. Daniel Martin Varisco, "The Archaeologist's Spade and the Apologist's Stacked Deck: The Near East through Conservative Christian Bibliolatry," in *The United States and the Middle East: Cultural Encounters*, ed. Abbas Amanat and Magnus Bernhardsson (New Haven: Yale University Press, 2002), 57–116; Numbers, *The Creationists*, 159, 287, 315; Seyyit Irmak, "Nuh Tufanı," *Köprü* 37 (Apr. 1980): 15–16; Mehmet Dikmen, "Hazret-i Nuh," *Köprü* 37 (Apr. 1980): 19–20; Safvet Senih and T. Çelikbilek, "Tufan gerçeği ve Nuh'un gemisi," *Sızıntı*, July 1983, 203ff.; Tarık Çelik, "Bir kere daha Tufan," *Sızıntı*, July 1986, 232–233; Taşkın Tuna, "Dünyanın yaşı," *Köprü* 52 (Jan. 1981): 18–19; Uzunoğlu and Yılmaz, *Alternatif biyoloji*, 61–73; İhsan İnal, "Hayatın izleri," *Sızıntı*, Apr. 1985, 55–57; Gülen, *Asrın getirdiği tereddütler*, 3:34.

35. Said Nursi, *Tarihçe-i Hayatı* (İstanbul: Yeni Asya Neşriyat, 2002), 59; Günter Seufert, *Politischer Islam in der Türkei: Islamismus als symbolische Repräsentation einer sich modernisierenden muslimischen Gesellschaft* (Stuttgart: Steiner, 1997), 183, 262.

36. Halit Ertuğrul, *Kendini arayan adam* (İstanbul: Nesil Yayınları, 2005); Halit Ertuğrul, *Düzceli Mehmet* (İstanbul: Nesil Yayınları, 2005).

37. Yavuz, *Islamic Political Identity*, 162.

38. European Commission, *Europeans, Science and Technology*, Special Eurobarometer 224, Wave 63.1 (Brussels: European Commission, 2005); Jon D. Miller, Eugenie C.Scott, and Shinji Okamoto, "Public Acceptance of Evolution," *Science* 313 (2006): 765–766; Ipsos Global @dvisory, "Supreme Being(s), the Afterlife and Evolution," released Apr. 25, 2011, www.ipsos-na.com/download/pr.aspx?id=10669; Peker, Cömert, and Kence, "Three Decades," 743–753.

39. Taner Edis, "Cloning Creationism in Turkey," *Reports of the National Center for Science Education* 19:6 (1999): 30–35; Nicolaidis and Levit, Levit, Hoßfeld, and Olsson, chapters 8 and 9 in this volume.

CHAPTER ELEVEN

Catholicism

RAFAEL A. MARTÍNEZ AND THOMAS F. GLICK

In the historiography of the controversies over evolution, no phenomenon called "Catholic creationism" appears explicitly. Ronald Numbers, in his classical work of 1992, *The Creationists,* mentions Catholicism (in America) but once, alluding to an attempt to establish a group "in favor of creation and against evolution," which soon dissolved for lack of members. In Europe especially, the common view of creationism is that it has been a typically American phenomenon, closely allied to a focus on the Bible as a unique source of religious authority and to biblical literalism, neither of which has any real counterpart in Catholic theology, historically or currently. Nevertheless, some recent events suggest that the situation may have changed. In 2009, while discussing "the myth that holds creationism to be an exclusively American phenomenon," Numbers now includes an explicit reference to Catholic creationist groups or activists, even though the Catholicism of such persons is in no way presented as relevant to their creationist perspectives.[1]

We perceive nowadays in scientific, intellectual, and political spheres, a diffuse concern about the penetration of creationist tendencies in religious circles quite distinct from the evangelical fundamentalism that gave rise to creationism, in particular in the Catholic world. The question is made even more complex in regard to intelligent design. Some episodes have gained notoriety in traditionally Catholic countries, and even have commanded international attention. In July 2005 the archbishop of Vienna, Christoph Cardinal Schönborn, published a letter in the *New York Times* in which he put forth a position that was largely perceived as supporting intelligent design and questioned "the supposed acceptance—or at least acquiescence—of the Roman Catholic Church when they [Schönborn refers here to those whom he calls "the defenders of neo-Darwinian dogma"] defend their theory as somehow compatible with Christian faith." During the "Darwin Year," 2009, along with celebrations of evolution, there were other activities clearly contrary to it, which were in some cases organized by groups of traditional Catholic orientations. Opposition to evolution, which for a long time was not openly evoked, now seems to attract an increasing number of thinkers as well as common people.[2]

However, it has become commonplace for scientific or cultural media in Catholic countries to accuse religious leaders of creationism or antievolutionism. Similar accusations have appeared in political circles. For example, in June 2007 a report prepared for the Parliamentary Assembly of the Council of Europe asserted, "The Catholic Church has clearly demonstrated for a very long time that it is creationist." This assertion was, however, nuanced by a reference to recent positions, seemingly more favorable to evolution. According to the same report, the Catholic Church has been "more discreet and almost remained aloof on this issue," starting with the Second Vatican Council. And only with the well-known declaration of John Paul II on October 23, 1996, in which he asserted that evolution "is more than a hypothesis," had the church finally accepted Darwin's theory in a positive way, although he also declares that "several movements still defend creationism as a dogma." Even if these observations are not in any way conclusive, they do demonstrate a growing acceptance in public opinion of the existence of a certain current of creationism in the Catholic world.[3]

As such, two issues need to be distinctly addressed in order to clarify the issue of creationism in Catholicism: the perception of Catholicism as somehow creationist, and its recent creationist (or near-creationist) tendencies. The first point depends, at least in part, on the peculiar history of the reception of evolution in the Catholic Church, marked by a sharp internal debate about the theological consequences of Darwin's theory that took place at the end of the nineteenth century. Although Catholic theology progressively accepted evolution, it often treated the subject with suspicion, sometimes involving theological stances that could be considered antievolutionary.

Early Reaction of Catholic Theology to Evolution: Opposition and Polemics

From the publication of *On the Origin of Species* in 1859 through the first years of the twentieth century, theological responses to evolution were often polemic. Even though the central authorities of the Catholic Church did not officially condemn evolutionism, clashes occurred in various places, including Rome, where particular individuals and groups tried to block the diffusion of evolutionary theories. For a long time, many theologians were openly hostile to Charles Darwin, in consonance with earlier stances on other "transformist" theories, such as that of Jean Baptiste Lamarck, and in general opposition to the "dialectical" postulates of nineteenth-century German idealism.[4]

However, to apply the term *creationism* to this period might be in some way anachronistic. Catholic antievolutionism of the late nineteenth century was primarily directed at pro-evolution sentiments within the church itself. Catholics favorable to evolution included Catholic scientists working in the biological field, such as St. George Jackson Mivart, but also scientists working in other disciplines, philosophers, and theologians. Theologians who opposed evolution paid no attention to natural science; they referred, at best, to a few anti-Darwinian scientists who mainly worked at the turn of the century, during the "eclipse of Darwinism."[5]

Theological, often summary, discussions of evolution could be typically found in a genre of theological textbooks called *De Deo Creante* (About God the creator). These books normally devoted very little attention to the characteristics of the physical and biological worlds. The study of creation was mainly based on revelation: the biblical text, magisterial teachings of the church (i.e., the teachings of the popes and church councils), and the witness of the tradition of the church. Rational or philosophical arguments were examined metaphysically, discussing for example the type of potentiality required by matter in the creative process. Discussions of the concrete way by which creation was produced were rare. The physical world and living animal and plant species (if one excludes the problem of the six days of creation, which had long been considered metaphorically) played an even smaller role. The creation of the human being always took center stage.

The main theological question was: Did God create the body of Adam (i.e., of the first human being) directly, or was it permissible to accept his provenance from preexisting organic matter, which had itself originated by natural laws? This would open the possibility of interpreting the history of humankind as the result of evolution. The debate focused on the body because, from the perspective of Christian doctrine, the soul, or human spiritual principle, is necessarily created by God. Authors favorable to evolution agreed on this point, without seeing in it any contradiction. Dalmace Leroy explicitly argued that the traditional assertion of the direct creation of the first man by God was compatible, from the perspective of traditional Catholic anthropology, with the evolutionary origin of humankind: for it is only at the moment in which man attains a spiritual dimension that he is properly human, and, inasmuch as the spiritual dimension transcends the biological plane and can be caused only by God, it is perfectly possible to assert that God created Adam directly, even if the human body was the result of biological evolution.[6]

Most theologians, however, opposed the compatibility, albeit for very diverse reasons. The most radical voices attributed the existence of the first human being exclusively to supernatural creative action and thus precluded any natural dynamism. Matthias J. Scheeben, for instance, one of the most important Catholic theologians of his time, wrote: "It is heresy to pretend that man, insofar as concerns his body, 'is descended from monkeys' as a consequence of a progressive change registered in forms, including the supposition that in the complete evolution of man's form, God has simultaneously created a soul." His opposition to evolution is however unclear, as he distinguishes between physical explanation, on the one hand, and supernatural or theological explanation, on the other. Talking in general about the biblical account of creation, he declares that "the explanation that the Hexameron [the narrative of the six days of creation], even though a true explanation of the origin of the cosmos, does not constitute a proper physical commentary. If it did, it would have to attribute the origin of its constitutive parts to natural causes; rather, it is related to divine actions."[7]

Another strongly antievolutionist thinker was Tommaso Zigliara, a leader of the neo-Thomist movement. He taught philosophy and theology in the Dominican house of Santa Maria sopra Minerva, in Rome, and was close to Pope Leo XIII, who named him cardinal in 1879. Zigliara was very influential in the Roman Curia till his death in 1893. His approach is quite different from Scheeben's, as he bases his critique of evolution on the similarity he perceives between evolution and Hegelian dialectic. To accept evolution would mean to accept that nature possesses, like the Hegelian absolute, its own dynamism that gives rise to everything real, and that is the equivalent of pantheism. Zigliara expresses these views in a report about a book published by a Catholic priest, Raffaello Caverni, professor of physics and mathematics, who favored evolution. In his critique, Zigliara suggests a possible solution: to agree that such a dynamism has been, in some way, infused by God. But that, in his opinion, would open the gates to Hegelian pantheism.[8]

Such radical attitudes, however, were grounded not on a literal interpretation of scripture or in an attempt to substitute the direct intervention of God by a natural process. Moreover, they were rare. Most theologians, even while maintaining a nonevolutionary vision of the physical world, did not oppose the plausibility of evolutionary processes. They left the door open: to adopt some evolutionary positions need not necessarily be contrary to faith. Camillo Mazzella was another of the most influential theologians in the Roman Curia. A Jesuit, professor of theology in the United States and afterward in Rome, cardinal from 1886, and successor of Zigliara at the head of the Congregation of Studies, he held that, while it might be more consistent with Catholic theology to maintain that God created the body of Adam directly, it is not a theological requirement. He therefore accepts that one can recognize the existence of other causal dimensions when speaking of creation, which in practice means the intervention of natural forces or dynamics.

Fighting against Evolution

Mazzella played an important role in the main polemics over evolution in the Vatican at the end of the nineteenth century. He had been prefect of the Congregation of the Index from 1889 to 1893 and continued as a member of the congregation until his death in 1900. He was surely the most renowned theologian among the other cardinals. From 1894 to 1899 the Index considered two cases related to evolution: the books of Dalmace Leroy, *L'évolution restreinte aux espèces organiques*, denounced in the summer 1894, and John A. Zahm, *Evolution and Dogma*, denounced in 1897. During the previous decades, several antireligious and anti-Catholic authors who espoused evolution had been condemned, but the main reason was always their religious views, and the issue of evolution was never central to the debate. But now, Leroy and Zahm were explicitly defending evolution as fully compatible with Catholicism, and, for that reason, discussions necessarily centered on evolution. These two cases offer deeper insight into the main antievolutionist authors, their reasons and arguments, and their "agenda."

The two actions were not part of an organized campaign: the case against Leroy was owing to an inconsequential complaint, but it stimulated an intense debate among the Roman theologians who advised the Vatican congregations. In the case of Zahm, other issues, related to the situation of the Catholic Church in the United Status, were in play, a situation that was picked up by some European "premodernist" movements amid a climate of conflict between "conservatives" and "liberals." In any case, the maneuvering of the parties provided an important proving ground for the strengths and weaknesses of those in favor of, and against, evolution.

The attitude of the Vatican consultors (*consultores*), upon examining the works implicated, was quite negative, if considered in the whole. Nevertheless, most recognized that the question was, at least, debatable and acknowledged that there were no decisive theological reasons for rejecting evolution. In some cases, some important officials supported positions favorable to evolution. Marcolino Cicognani, secretary of the Congregation of the Index, wrote in a personal memorandum for his statutory audience with the pope concerning the congregation's decision on Leroy: "His system is that God has created everything, that he governs with his Providence, and in the successive development of organisms he makes use of created forces, or secondary causes, always guided by his Providence . . . This opinion was not found contrary to either theology or faith; and of this book it was said: *dimittatur* [that the case be dismissed]." One of the consultors, Bishop Ernesto Fontana, clearly acknowledged this possibility: "Supposing, as the author supposes, that the creative act may have placed in the individuals of the primitive types aptitudes and potentials as embryonic, to be developed, evolution is not impossible." As a result of recognizing divine action as transcendent, operating at a "metaphysical" level (the level of "being"), there was always an implicit respect for the natural level: faith or theology cannot make decisions about the empirical, natural world.[9]

Nevertheless, opposition to evolution was strong among the consultors. Luigi Tripepi wrote at the end of 1894 a long (fifty-four printed pages) report on Leroy's book. As was typical of this genre, he began by pointing out that evolution was only a hypothesis and "is nowadays abandoned and refuted as false and absurd by the same rationalists and unbelievers who previously supported it." His central argument, however, is taken from theology, where he follows, sometimes literally, Mazzella's *De Deo Creante*. The main point discussed is whether Catholic doctrine requires the direct creation of human beings by God. It is clear that this cannot be considered as a *revealed truth*. Tripepi tries to show, however, that theologians and church authorities have always argued for direct creation; to deny it would be rash. In interpreting biblical texts about God's action, Tripepi always follows the most literal sense; for that reason he has to accept that the intervention of secondary causes (evolution) in the origin of plants and animals cannot be excluded, as the scripture tell us of the plants and animals "that the waters might bring forth" or "the earth might bring forth." However, he prefers the special creation of each species, which from that moment would be fixed and immutable.[10]

Leroy's book was also examined by a Dominican consultor, Enrico Buonpensiere, who delivered a very adverse judgment. Buonpensiere's arguments are not too different from those of Mazzella or Tripepi: evolution appears to be, first at all, false in empirical science, because of the "inexorable law of hybridization." But he adds a metaphysical argument: "In ontology, the essence of any object is an immutable type, that is, incapable of any change [*evoluzione*], whether toward the greater or toward the lesser." Apparently, Buonpensiere denies any value to empirical science. When examining Zahm's book three years later, he rejected his arguments "inasmuch as they lack both philosophical and traditional theological arguments." At the end of his report, Buonpensiere presents his own theological thesis, attempts to demonstrate it, and proposes that the Congregation of the Index condemn the contrary doctrine: "It is Catholic doctrine to state that God has made Adam, immediately and directly from the mud of the earth." For all that, his proposal was never considered. His proactive attitude seems to have been an exception: antievolutionist authors seemed to prefer a more passive, reactive stance. Although doctrinal censure was common in the Catholic Church (reaching a high point during the crisis of "modernism," in the following decades), evolution does not seem to ever have constituted a priority.[11]

Perhaps the only other proactive attitude was that of *La Civiltà Cattolica*. For a long time this influential journal had been publishing critiques of evolution: in 1877, a harsh review of Caverni's book, written by Francesco Salis Seewis, for example, and a long series of articles published between 1878 and 1880 by Pietro Caterini. In 1897 Salis Seewis published also an unfavorable review of the Italian edition of *Evolution and Dogma*. But it was only after Zahm's case exploded that *La Civiltà Cattolica* and its director, Salvatore Brandi, played a more open role. The Congregation of the Index had decided to condemn the work but not to publish its decision until Zahm's reaction was known. As had happened before in the case of Leroy, the condemnation was never published. Zahm's supporters, among whom were a cardinal or two and part of the American hierarchy, did everything they could to halt the decree, finally obtaining the pope's agreement.[12]

In the end, antievolutionist efforts did not succeed in obtaining any public condemnation of evolution. Attempts to oppose evolution on theological grounds never gained traction or definitive results. Indeed, antievolutionists operating within, or close to, the Vatican could be defeated *politically*, whatever the theological issues may have been. In the following decades, as the understanding of biological evolution grew, antievolution stances steadily lost strength, even if they never disappeared. They always ran up against a theological horizon marked intrinsically by the distinction between the natural and the supernatural, which left the door open to the recognition of the intrinsic autonomy of the physical and biological dimension. From the 1920s to the 1940s, explicit support for the validity of evolution and its compatibility with Christian doctrine had become habitual in Catholic intellectual circles. This is the case of the works of Vittorio Marcozzi in Italy, Ernst Messenger in the United States, or Henry de Dorlodot in Belgium. Later on, in 1950, Pius XII asserted, in the encyclical *Humani Generis,* that nothing impedes Catholics from considering the hypothesis of evolution insofar as it refers to the human body. This is most likely the only explicit papal affirmation of evolution before the various speeches and messages of John Paul II. It was not an endorsement of evolutionary theory, but from then on voices against evolution, at least in a theological context, seemed to have disappeared till very recent times.[13]

To this picture there was apparently only one significant exception: the *monitum* (warning) issued to Catholic seminaries by the Holy Office in 1962 concerning the ambiguities and doctrinal errors in the theological and mystical works of the most renowned Catholic evolutionist, Pierre Teilhard de Chardin. Although the note was perceived as an antievolutionary statement, it explicitly excludes any judgment of Teilhard's *scientific* work.[14]

Catholic Antievolutionist and Creationist Groups

During the second half of twentieth century, Catholic theology, following Karl Rahner's anthropological turn, lost interest in questions related to natural sciences. As a consequence, references to evolution were very scarce among theologians. Critiques of evolution were more common among some Catholic philosophers, such as Etiènne Gilson or Robert Spaemann. However, they mainly criticized some aspects of common interpretations of evolution, as naturalism or antiteleologism, while they remained generally open to the acceptance of the fact of evolution as well as transcendental interpretations, in the line of so-called theistic evolutionism.[15]

Only in recent decades have antievolutionary attitudes, mixed with creationism and intelligent design, reappeared. Some small but radical antievolutionary groups advanced alleged scientific criticisms to evolution. For instance, the Cercle d'Ètudes Scientifique et Historique (CESHE), a small and not too active group, continued the work of Fernand Crombette, a French autodidact who offered an account of ancient history and cosmology based on biblical literalism and supporting geocentricism. Its most active member seems to be Guy Berthault, an expert on sedimentology who argues that empirical facts about sediments contradict evolution.[16]

Another small group was founded in 1971 by John G. Campbell as the Counter Evolution Group and published a newsletter entitled *Daylight* from 1977. After Campbell's death in 1986, Anthony Nevard reorganized the group, initially as an English branch of the CESHE but from 1991 as the Daylight Origins Society. Activities have continued, with some interruption, mainly through the publication of an expanded version of *Daylight*. Some of the contributions are close to young-earth creationism; others just criticize different aspects of evolution. Religious opinions are often linked to traditionalist groups, and theological reflections are limited generally to a literal reading of the Bible and defense of the antievolutionist sense of the doctrinal declarations of church authorities. The group maintains regular contacts with other, non-European, Catholic and non-Catholic creationist groups, particularly with the Kolbe Center for the Study of Creation, an American Catholic creationist group based in Virginia, founded in 2000 and directed by Hugh Owen.[17]

These groups tend to collaborate very strongly: articles written by their collaborators are usually published in the other groups' media. However, their reflection in public opinion is scant. Attempting to stimulate wider attention, the Kolbe Center organized a symposium in Rome on October 24–25, 2002. Among the speakers was an Austrian bishop, Andreas Laun. His speech, fully contrary to evolution, accepted however "the theological possibility of evolution." According to him, and following the general ideas presented by many of the speakers, evolution just happens to be false from the scientific point of view. Among the other speakers were Guy Berthault and Dominique Tassot, from the CESHE; Robert Sugenis, another supporter of geocentricism; Robert Bennet; Maciej Giertych; and Joseph Seifert.[18]

Six years later, again in Rome, in preparation for the Darwin Year, a similar conference was organized by Roberto de Mattei, an Italian historian well known not only for his antievolutionist opinions but also for his traditionalist views that have several times put him at odds with theologians in questions related to the history of the church and the Second Vatican Council. Among the speakers were Guy Berthault, Dominique Tassot, Hugh Owen, Maciej Giertych, and Joseph Seifert. Again, as one of the speakers noted, "the media did not care to notice this session very much."[19]

The proceedings of this symposium once again reveal a mixed set of opinions. The volume does not present strong creationist statements, as had been usual in other publications from Kolbe or *Daylight*; instead, the focus is on certain critiques of science long championed by fundamentalist creationists—for example, results from the fossil record are falsified by wrong assumptions linked to stratigraphy (Berthault), the carbon-14 dating method is incorrect (Pontcharra), and thermodynamics would make impossible the scenario foreseen by today's cosmology. Seifert's criticisms are presented at the philosophical level. According to him, some "limited evolution" would be possible from the rational point of view; however, he considers that most current evolutionary theories offer only illogical and unproven hypotheses.[20]

One of the main speakers in these two meetings was Maciej Giertych, who is known because of his political activity. Since 2004 he has been a member of the European Parliament, where, on October 11, 2006, he organized a public hearing on the teaching of evolution in European schools. There were three invited speakers: one of them was Guy Berthault; the other two were Hans Zillmer, a German engineer and author of several books criticizing evolution, and Joseph Mastropaolo, an American human physiologist who is an active antievolutionist linked to creationism and intelligent design. The antievolutionist attitude of all the speakers (Giertych himself was the chair) was clear; however, the meeting did not consider creationism or intelligent design but only criticisms of evolution based on scientific arguments.[21]

The meeting was received by the press and other Parliament members as a creationist attempt to stop evolution in schools. As a result, a report on the dangers of creationism in education was commissioned by the Parliamentary Assembly of the Council of Europe. A first draft reported by Guy Lengagne on June 2007 was returned for revision, and a new version presented by Anne Brasseur was approved by the Parliament Assembly on October 4, 2007.[22]

Giertych strongly denies any creationist or intelligent design affiliation, and he does not use any argument based on his religious views. On the other side, he also complains about what he calls "a total blindness" for scientific criticism of evolution among Catholic theologians who accept the full compatibility between evolution and Catholic doctrine.

Present Tendencies versus Intelligent Design

If Catholics rarely endorse young-earth creationism, intelligent design seems to have had a greater impact during recent decades. Rational access to God has always been a fundamental tenet of Catholic faith, and intelligent design has been sometimes interpreted as a strong version of this belief, as it tries to prove the existence of God from empirical experience. Two American Catholics played a prominent role in its development. Young-earth creationist Dean H. Kenyon, professor emeritus in biology at San Francisco State University and adviser to the Kolbe Center, authored *Of Pandas and People* (1989), the first intelligent design textbook. The biochemist Michael Behe also collaborated in the textbook. His book *Darwin's Black Box* (1996) tried to provide the movement with scientific legitimacy. However, Catholics were also among the most active critics. Kenneth R. Miller, the plaintiff lead expert in the *Kitzmiller v. Dover* case, has written and lectured extensively against it, as have also Francisco Ayala, John Haught, and Stephen Barr.[23]

In Europe, some Catholic public figures and intellectuals have responded quite favorably to intelligent design. For instance, in 2005 the Dutch minister Maria van der Hoeven proposed to introduce its teaching into the schools. In Spain, France, Belgium, Poland, and Germany, other prominent Catholics have made statements in support of this movement or that at least could be interpreted as such. Catholic intellectuals concerned with materialistic interpretations of evolution, especially if they have conservative tendencies, tend to approach positively the arguments prompted by Behe and others. On the other hand, advocates of evolution, especially if they endorse naturalistic views, tend to interpret theological statements as expression of "hidden" creationism.[24]

The most representative case, however, is that of Cardinal Schönborn. His initial statements on evolution stimulated a polemic among Catholic scientists, philosophers, and theologians, which prompted Cardinal Schönborn to explain his ideas with greater precision.

A letter by Schönborn published on July 7, 2005, contains several statements that clearly reveal a close relation with intelligent design. While he accepts that there is some truth in evolution ("Evolution in the sense of common ancestry might be true"), he also attacks what he calls the neo-Darwinian dogma: "an unguided, unplanned process of random variation and natural selection." Against it, the letter stresses "that by the light of reason the human intellect can readily and clearly discern purpose and design in the natural world, including the world of living things." Quoting several statements from John Paul II and Benedict XVI, Schönborn then identifies the classical, philosophical, and theological concept of finality "with final cause, purpose or design."[25]

Ronald Numbers suggests that the cardinal had the support of Pope Benedict XVI, arguing that in the mid-1980s he had claimed that advances in microbiology and biochemistry had demonstrated rational design and had once used the expression "intelligent project" referring to the universe. Schönborn's article elicited strong reactions from several Catholic scientists, such as Francisco J. Ayala and George Coyne. The main reaction to Schönborn was, however, at the philosophical level. Stephan M. Barr criticized the concept of chance the cardinal referenced when speaking of an "unguided, unplanned process of random variation." Schönborn's response was published in the same journal. He stressed that his arguments were not to be understood as scientific but as rational or philosophical, and he apologized for the misunderstanding: "It seemed that, right or wrong, my original essay was all about *science*, about real, tangible, factual knowledge of the material world. But now I admit to be speaking in the language of natural philosophy."[26]

That was the first of a series of statements in which the cardinal tried to clarify his thought. From October 2005 to June 2006 he presented at Vienna Cathedral a series of catechetical lectures under the general title "Evolution and Creation," in which he does not criticize evolution from the scientific point of view, although he does stress its limits. First, he distinguishes "belief in creation" from "creationism" and explicitly rejects the latter as linked to biblical literalism and the "young-earth" thesis. His criticisms are directed against what he calls "evolutionism" or "evolution ideology": the attempt to consider the evolutionary process (a blind process led only by chance) as the only and total explanation of reality. That would lead to materialism, excluding God from our view of reality. But obviously, "meaning" can be linked to "purpose" and "design." One of Schönborn's main points is that reason can discover purpose in nature and, through it, the existence of God as creator. That clearly makes him sympathetic to intelligent design, although he notes—without stressing it too much—an important difference: the search for purpose should not be expected from scientific methodology; it is the result of philosophical reasoning.[27]

Cardinal Schönborn's position is in some way characteristic of a general tendency in the Catholic intellectual milieu that was also present more than a century ago: a reaction against materialistic and naturalistic interpretation of biological evolution, pointing out on one hand the alleged limits of the theory and, on the other, the capacity of human reason to supersede those limits on the philosophical and theological level, all of which end up at positions close to intelligent design and creationism. That position has been answered by Catholic scientists, who see evolution as not requiring this kind of materialistic interpretation and who therefore have no difficulty in identifying themselves as evolutionists and believers at the same time. Such ingenuous antimaterialism has been the target of criticisms from philosophical and theological points of view, as it confuses the physical and theological levels. Although it is not conceivable that near-to-intelligent-design attitudes would succeed in Catholic theology, the present situation makes clear that creationist tendencies are the result of the identification of evolution with its ideological interpretations (just as happened in Darwin's day), and the lack of a more profound comprehension of the implications of Catholic doctrine for the relationship between natural causes and divine action.[28]

ACKNOWLEDGMENT

We thank Stefaan Blancke for his insightful comments.

NOTES

1. Ronald L. Numbers, *The Creationists* (New York: Knopf, 1992), chap. 15; Ronald L. Numbers, "Myth 24: That Creationism Is a Uniquely American Phenomenon," in *Galileo Goes to Jail, and Other Myths about Science and Religion*, ed. Ronald L. Numbers (Cambridge, MA: Harvard University Press, 2009), 215–223.

2. Christoph Schönborn, "Finding Design in Nature," *New York Times*, July 7, 2005. To Schönborn's letter there was an interesting and vigorous reaction. In church circles he was attacked from both the "right" (Schönborn had ceded too much to evolution) and the "left" (he had misunderstood randomness). For the former, see John F. McCarthy, "Reviewing Cardinal Schönborn's Stand on Evolution by Chance or by Purpose," *Living Tradition: Organ of the Roman Theological Forum*, May 2008; for the latter, Patrick H. Byrne, "Evolution, Randomness, and Divine Purpose: A Reply to Cardinal Schönborn," *Theological Studies* 67 (2006): 653–665.

3. Parliamentary Assembly of the Council of Europe, Report "The Dangers of Creationism in Education," Doc. 11297 (June 8, 2007), n. 75.

4. Mariano Artigas, Th. F. Glick, and Rafael A. Martínez, *Negotiating Darwin: The Vatican Confronts Evolution, 1877–1902* (Baltimore: Johns Hopkins University Press, 2006).

5. Peter J. Bowler, *The Eclipse of Darwinism: Anti-Darwinian Evolution Theories in the Decades around 1900* (Baltimore: Johns Hopkins University Press, 1983).

6. Cf. Marie-Dalmace Leroy, *Lettre à M. L'Abbé A. Farges* (Paris, Oct. 1898); "Comptes-rendus à J. Guibert, *L'origine des espèces*," *Revue Thomiste* 7 (1899): 735–741.

7. Matthias Joseph Scheeben, "Handbuch der katholischen Dogmatik. Teil III: Schöpfungslehre," in *Gesammelte Schriften*, ed. Wilhelm Breuning and Franz Lakner, 3rd ed. (Freiburg: Herder, 1961), 5:115, 160–161.

8. Raffaello Caverni, *De' nuovi studi della Filosofia. Discorsi a un giovane studente* (Florence: Carnesecchi, 1877).

9. ACDF (Archives for the Congregation for the Doctrine of the Faith), Index, Protocolli, 1894–1896, f. 88; ACDF, Index, Protocolli, 1894–1896, f. 123.

10. ACDF, Index, Protocolli, 1894–1896, f. 125, p. 9; Camillo Mazzella, *De Deo Creante*, 4th ed. (Rome: Forzani, 1896); ACDF, Index, Protocolli, 1894–1896, f. 125, pp. 43–44.

11. ACDF, Index, Protocolli, 1894–1896, ff. 117–118, p. 3; ACDF, Index, Protocolli, 1897–1899, f. 180, pp. 8 and 45.

12. Francesco Salis Seewis, review of *De' nuovi studi della Filosofia. Discorsi di Raffaello Caverni a un giovane studente, La Civiltà Cattolica*, 10th ser., 4 (1877): 570–580; and 10th ser., 5 (1878): 65–76; Cf. Artigas, Glick, and Martínez, *Negotiating Darwin*, 156ff.

13. Vittorio Marcozzi, *Le origini dell'uomo* (Rome: AVE, 1944); Vittorio Marcozzi, *Evoluzione o creazione?* (Milan: Casa Editrice Ambrosiana, 1948); Ernst Ch. Messenger, *Evolution and Theology: The Problem of Man's Origin* (New York: Macmillan, 1932); Henry de Dorlodot, *Le darwinisme au point de vue de l'orthodoxie catholique. Premier Volume. L'origin des espèces* (Bruxelles and Paris: Vromant 1921).

14. Cf. Georges Chantraine, "Evolution According to Teilhard de Chardin," in *Biological Evolution: Facts and Theories*, ed. G. Auletta, M. Leclerc, and R. A. Martínez (Rome: G&B Press, 2011), 613–644.

15. Etiènne Gilson, *From Aristotle to Darwin and Back Again: A Journey in Final Causality, Species and Evolution* (Notre Dame, IN: University of Notre Dame Press, 1984); Robert

Spaemann, *Die Frage Wozu? Geschichte und Wiederentdeckung des teleologischen Denkens* (Munich: Piper, 1981).

16. Ronald L. Numbers, *The Creationists: From Scientific Creationism to Intelligent Design*, expanded ed. (Cambridge, MA: Harvard University Press, 2006), 411; www.ceshe.fr/index.htm, accessed April 2, 2014.

17. The magazine is distributed through the webpage www.daylightorigins.com/, with free access to the newsletters published until 2006, when the publication was discontinued. It started again in 2010, in both print and e-book formats. www.kolbecenter.org.

18. Andreas Laun, "Evolution and Creation—Theological Considerations," Oct. 24, 2002, www.kolbecenter.org.

19. Maciej Giertych, "Investigating Evolution in Rome," Kolbe Center, Dec. 10, 2008. Two other meetings were held in Rome on the occasion of the Darwin Year, one organized by the Pontifical Academy of Sciences ("Scientific Insights into the Evolution of the Universe and of Life," October 29–November 3, 2008), the other by the Pontifical Gregorian University, in collaboration with the Pontifical Council for Culture and some other Pontifical Universities ("Biological Evolution: Facts and Theories," Mar. 3–9, 2009). Both of these meetings were favorable to evolution and were widely covered by Italian and international media. See the proceedings: Werner Arber, Nicola Cabibbo, and Marcelo Sánchez Sorondo, eds., *Scientific Insights into the Evolution of the Universe and of Life* (Vatican City: Pontifical Academy of Sciences, 2009), and Gennaro Auletta, Marc Leclerc, and Rafael A. Martínez, eds., *Biological Evolution: Facts and Theories* (Rome: G&B Press, 2011).

20. Roberto De Mattei, ed., *Evolucionismo: il tramonto di una ipotesi* (Siena: Cantagalli, 2009).

21. Maciej Giertych, "Teaching on Evolution in European Schools," www.kolbecenter.org/images/stories/pdf_files/giertych_teaching_evolution.pdf. See Mastropaolo's webpage "Biology vs. Evolution," www.josephmastropaolo.com; for Giertych, see also Borszyk, chapter 7 in this volume.

22. Parliamentary Assembly of the Council of Europe, "The Dangers of Creationism in Education," Resolution 1580 (2007).

23. www.kolbecenter.org/advisors/; Percival Davis and Dean H. Kenyon, *Of Pandas and People: The Central Question of Biological Origins* (Richardson, TX: Foundation for Thought and Ethics, 1989); Michael J. Behe, *Darwin's Black Box: The Biochemical Challenge to Evolution* (New York: Free Press 1996); Kenneth R. Miller, *Finding Darwin's God: A Scientist's Search for Common Ground between God and Evolution* (New York: Harper, 2002); John F. Haught, *God after Darwin: A Theology of Evolution* (Boulder, CO: Westwiew Press, 1999); Francisco J. Ayala, *Darwin's Gift to Science and Religion* (Washington, DC: Joseph Henry Press, 2007); Stephen M. Barr, *Modern Physics and Ancient Faith* (Notre Dame, IN: University of Notre Dame Press, 2006).

24. See Lepeltier, Catalá-Gorgues, Blancke, Flipse and Braeckman, Kutschera, and Borczyk, chapters 1, 2, 4, 6, and 7 in this volume.

25. Christoph Schönborn, "Finding Design in Nature," *New York Times*, July 7, 2005.

26. Numbers, *The Creationists*, 395–396; Numbers refers to: Benedict XVI, *"In the Beginning . . .": A Catholic Understanding of Creation and the Fall* (New York: T&T Clark, 1995); originally published in German in 1986, reporting his sermon in Munich Cathedral in 1981; *Audience*, Nov. 7, 2005, www.vatican.va/holy_father/benedict_xvi/audiences/2005/documents/hf_ben-xvi_aud_20051116_it.html; Stephen M. Barr, "The Design of Evolu-

tion," *First Things* 156 (Oct. 2005): 9–12; Christoph Schönborn, "The Designs of Science," *First Things* 159 (Jan. 2006): 34–38.

27. Christoph Schönborn, *Ziel oder Zufall? Schöpfung und Evolution aus der Sicht eines vernünftigen Glaubens* (Freiburg im Breisgau: Herder 2007); English translation: *Chance or Purpose? Creation, Evolution and a Rational Faith* (San Francisco: Ignatius Press 2007). Here we quote the English translation.

28. Martin Rhonheimer, "Neodarwinistische Evolutionstheorie, Intelligent design und die Frage nach dem Schöpfer. Aus einem Schreiben an Kardinal Christoph Schönborn," *Imago Hominis*, 14 (2007): 47–81, and "Teoria dell'evoluzione neodarwinista, Intelligent design e creazione. In dialogo con il cardinal Christoph Schönborn," *Acta Philosophica* 17 (2008): 87–132. For an analysis of the *rhetoric* of Schönborn's letter, see José Antonio Díaz Rojo, "El caso Schönborn: un cambio retórico en la postura católica ante la evolución biológica," *'Ilu. Revista de Ciencias de las Religiones* 14 (2009): 33–58, who stresses Schönborn's decontextualization of John Paul II's 1996 statement.

CHAPTER TWELVE

Intelligent Design

BARBARA FORREST

Since the late 1980s, American creationism has included the concept of *intelligent design*, a variant of creationism that has been shaped by creationists' consistent defeats in American federal courts. The Discovery Institute (DI), a Seattle, Washington, think tank, promotes intelligent design according to its "Wedge strategy," which is outlined in what is now widely known in the United States as the "Wedge document." Although setbacks have forced DI to alter its methods, the goals of the Wedge strategy remain largely intact, including the establishment of "an active design movement in Israel, the UK and other influential countries outside the US." Although DI did not meet its 2003 deadline for this goal, it is belatedly turning more toward Europe, where its allies are helping to implement its plans.[1]

The Wedge: Background and Content

In 1987 the US Supreme Court ruled in *Edwards v. Aguillard* that teaching creationism in government schools violates the US Constitution, which prohibits government establishment of religion. Immediately afterward, several creationists who were working independently became collaborators, rebranded creationism as "intelligent design," and coalesced around law professor Phillip E. Johnson. In 1996 DI established the Center for the Renewal of Science and Culture (now Center for Science and Culture), with this core group as fellows and Johnson as adviser. Here they wrote the Wedge document, a 1998 fundraising appeal that outlines a plan to advance intelligent design into America's cultural and educational mainstream.[2]

DI's efforts to implement the Wedge strategy in the United States were dealt a severe blow in December 2005 when Judge John E. Jones III (Middle District of Pennsylvania) ruled in *Kitzmiller et al. v. Dover Area School District* that teaching intelligent design in government schools is unconstitutional. Two DI fellows testified as witnesses on behalf of the Dover, Pennsylvania, school board, which had adopted a 2004 policy to promote intelligent design as science in a local school. Perhaps anticipating defeat, DI held its first European intelligent design conference, "Darwin and Design: A Challenge for 21st Century Science," in Prague in October 2005. American intelligent design creationists now travel to Europe frequently.[3]

Spanning twenty years (1998–2018), the Wedge strategy outlines the mission of DI "to replace materialistic explanations [in science] with the theistic understanding that nature and human beings are created by God." The wedge metaphor reflects DI's aim to "wedge" the supernatural into the public understanding of science. Moreover, DI's goal of promoting "a science consonant with Christian and theistic convictions" reveals intelligent design's distinctly Christian foundation. Unsurprisingly, its stated ambitions "to see intelligent design theory as an accepted alternative in the sciences and scientific research being done from the perspective of design theory" and "to see intelligent design theory as the dominant perspective in science" have not been realized. However, DI has managed to "build up a popular base of support among our natural constituency, namely, Christians," by means of religious "apologetics seminars." Additionally, the "integration of design theory into public school science curricula" is always within reach in the United States.[4]

Because DI is committed to influencing government school science instruction, the Wedge document calls for "teacher training" and "possible legal assistance in response to resistance." The organization has made some progress on this front in the United States. In 2008 DI and a politically powerful Christian group in Louisiana persuaded the state legislature to enact the Louisiana Science Education Act. Based on DI's own model legislation, the law permits science teachers to use creationist instructional materials in government schools. In 2012 Tennessee adopted a similar law, also through the influence of DI and a politically oriented Christian group.[5]

Intelligent Design in Europe

DI also seeks to affect science education abroad. With its assistance, European intelligent design supporters are implementing major goals of the Wedge strategy, especially in the United Kingdom. They hold intelligent design events in "significant academic settings" ("Academic Conferences"), publicly debate prominent pro-evolution figures ("Cultural Confrontation & Renewal"), promote intelligent design via the media ("Publicity & Opinion-Making"), and distribute intelligent design materials for use in government schools (to "rectify ideological imbalance in . . . science curricula & include design theory"). Furthermore, they are involved in "alliance-building, recruitment of future scientists and leaders, and strategic partnerships with think tanks, social advocacy groups, educational organizations and institutions, churches, religious groups, foundations, and media outlets."[6]

The European effort differs in one remarkable respect: in Europe, DI has forged a prominent, public partnership with young-earth creationists. In the United States, intelligent design proponents William Dembski and Phillip Johnson have based intelligent design (a form of old-earth creationism) on the Gospel of John instead of Genesis, thus keeping young-earth creationism at arm's length and alienating young-earth creationists whose support DI needs. However, DI has not shied away from its young-earth connections abroad. In 2004 Johnson toured the United Kingdom with Andrew Snelling of the American young-earth creationist organization Answers in Genesis. Sponsored by the Elim Pentecostal Church, the tour was organized by Peter Loose, a member of the Elim Church National Management Board and other pro-creationist organizations.[7]

Nothing better illustrates the European focus of DI and its partnership with European young-earth creationists than the staff roster at its Washington State "research" center, the Biologic Institute, and the editors of its online journal, *BIO-Complexity*, which "aims to be the leading forum for testing the scientific merit of the claim that intelligent design is a credible explanation for life." The Biologic Institute's fifteen "researchers" include three European young-earth creationists: British mechanical engineer Stuart Burgess, Finnish chemist Matti Leisola, and British operational research professor Colin Reeves. *BIO-Complexity*'s thirty-member editorial team, led by Leisola, includes ten Europeans from Finland, the United Kingdom (England and Scotland), Germany, Croatia, Ireland, and the Czech Republic. Nine are creationists or sympathetic to intelligent design concepts and as many as six may be young-earth creationists. Stephen Meyer, DI fellow and director of the Center for Science and Culture, considers *Bio-Complexity*'s editorial board "very prestigious."[8]

Leisola, who appears in young-earth creationist documentaries and writes for young-earth publications, epitomizes this European partnership between intelligent design and young-earth creationism. The *Journal of Creation* reports that "Leisola regularly speaks on intelligent design in various Finnish Universities. He has translated or edited into Finnish several intelligent design books, and has been involved in organizing university lecture tours for a number of speakers including [young-earth creationist] Professor [A. E.] Wilder-Smith in 1979." In 2004 DI fellows Paul Nelson and Richard Sternberg spoke at Helsinki University of Technology (now part of Aalto University), where Leisola teaches.[9]

This chapter concentrates on Western European countries with which DI has a documentable relationship, such as attendance at European intelligent design events and links to pro-intelligent design websites. DI's Intelligentdesign.org website links to "Worldwide Intelligent Design Organizations" in Scotland, Italy, England, Spain, Finland, Denmark, and Germany. Although some of these "organizations" may be merely websites, the list provides an outline of the Wedge strategy in specific countries.[10]

Intelligent Design in the United Kingdom

A 2009 Ipsos Mori poll revealed that 54 percent of UK respondents favored teaching creationism, including intelligent design, along with evolution. This level of public support explains why DI has mainly focused on the United Kingdom. There are also other incentives. First, there is no language barrier. Intelligent design material can be disseminated via websites, broadcasts, and public lectures. Second, DI and its UK supporters have alluded to the absence of an American-style separation of church and state in the United Kingdom, which may reduce legal challenges to advocating the teaching of intelligent design in government schools. British young-earth creationist David Tyler, who writes an online column for an American website that partners with DI to promote intelligent design, told the BBC that "intelligent design . . . should be taught in [British] schools." Third, people in the United Kingdom appear to be less familiar with intelligent design. Reflecting on a London lecture he delivered, Meyer described the British as "not quite so paranoid" about equating intelligent design with "young-earth creationism," adding that "it's kind of nice to be able to start fresh with an audience that doesn't think they already know what you're saying before you say anything." Fourth, DI's prospects of promoting intelligent design as science in the United States have probably peaked, as Johnson admitted after *Kitzmiller*: "I . . . don't think that there is really a theory of intelligent design at the present time to propose as a comparable alternative to the Darwinian theory." In 2012 Dembski, too, confessed, "Whereas a decade ago I was all gung-ho about intelligent design becoming the new reigning paradigm that would replace conventional evolutionary theory, I no longer have that optimism."[11]

Nonetheless, in the United Kingdom, two creationist organizations with direct ties to DI, Truth in Science (England) and the Centre for Intelligent Design (Scotland), promote intelligent design as genuine science. Biologic Institute staff member Stuart Burgess serves on Truth in Science's Council of Reference, whose members are "Christians with traditional Biblical beliefs." *BIO-*

Complexity editorial team member Norman Nevin is president of the Centre for Intelligent Design, which DI calls "our British sister organization." The center's website registrant is the Pentecostalist Peter Loose.[12]

Truth in Science

Burgess cofounded Truth in Science in 2004 to get creationism into UK schools: "Our aim is to compliment the work of existing Creation groups by targeting education in particular." In September 2006, the organization distributed to all UK secondary schools an intelligent design "resource pack" containing DVDs from Illustra Media, a film company that works closely with DI. Truth in Science asserts that "our packs discuss intelligent design, and not Creationism." The same month, DI, alluding to the United Kingdom's closer church-state relationship, announced Truth in Science as "a new player in the United Kingdom in the debate over how best to teach evolution"—a player that, "because of the different education and policy environment in the UK, versus that of the United States, . . . endorses teaching both the criticisms of evolution and the scientific theory of intelligent design."[13]

In December 2009, Truth in Science distributed DI's textbook, *Explore Evolution*, to "all UK school libraries where biology is taught at advanced level," along with a letter from board member and young-earth creationist Andrew McIntosh announcing plans for teacher training seminars. Using the same euphemisms that DI now uses to promote intelligent design in the United States after the *Kitzmiller* ruling, McIntosh recommended the book for teaching the "strengths and weaknesses of modern evolutionary theory" and "the evidence for and against evolution."[14]

Centre for Intelligent Design

According to the British Centre for Science Education (a pro-science group modeled after the National Center for Science Education in the United States), the Centre for Intelligent Design "has closely aligned itself with the strategy of the Discovery Institute." It is therefore reprising aspects of DI's Wedge strategy. For instance, as DI did in the early stages of its promotion of intelligent design in the United States, the Centre for Intelligent Design publicly confronts pro-evolution scientists in the United Kingdom. Between 1997 and 2000, DI invited pro-evolution scientists to conferences at American universities for what the Wedge document called—unbeknown to these invitees—"major public debate[s] between design theorists and Darwinists"; DI's intention was that "the attention, publicity, and influence of design theory should draw scientific materialists into open debate with design theorists, and we will be ready." Similarly, to publicize the January 2010 UK release of the intelligent design movie, *Expelled: No Intelligence Allowed*, Dembski and Meyer debated, respectively, biologist Lewis Wolpert and chemist Peter Atkins—both atheists—on Premier Christian Radio's *Unbelievable* radio program, which also publicized Michael Behe's November 2010 UK tour.[15]

The Centre for Intelligent Design also makes use of university settings to create an impression of academic legitimacy, just as DI has done in the United States. As part of the center's debut, Behe's UK tour included the cities of Oxford and Cambridge. Ostensibly showcasing the respectability of intelligent design by lecturing in "significant academic settings," as the Wedge strategy calls for, Behe actually spoke in lesser-known schools and churches. According to the British Centre for Science Education, "the tour never went anywhere near any science departments or any science organisations." Yet another conference was held in Cambridge in July 2012, this time featuring Meyer and sponsored by the Tyndale Fellowship, which promotes "research into philosophical aspects of the historical Christian faith." In 2000 DI held similar lectures at Yale University; however, such events are not sponsored by the universities. Their purpose is mainly to attract religious followers and financial donors who will promote intelligent design. Tellingly, the Elim Pentecostal group that sponsored the Johnson-Snelling tour was also involved in Behe's. (In 2011, the Centre for Intelligent Design used Elim's conference center for an intelligent design "Summer School" and a conference featuring Jay Richards, another DI fellow. In September 2012, the center was the setting for another conference featuring Biologic Institute director Douglas Axe, who emerged as an intelligent design proponent in the United States after conducting research in England for a number of years.)[16]

Finally, like DI, the Centre for Intelligent Design cultivates publicity, religious alliances, and political connections. In 2011 Meyer delivered the "Inaugural Lecture of the Centre for Intelligent Design, UK" in London. The center's director, Alastair Noble, announced that "90 invited guests included leading scientists, philosophers, Parliamentarians, educationalists, theologians, lawyers, and representatives of the media and business sectors." DI has cultivated similar American allies, including conservative Christian politicians and a nationwide network of evangelical organizations. Former US senator Rick Santorum is a supporter (others have included current Speaker of the US House of Representatives John Boehner). Another is Focus on the Family, a politically powerful evangelical organization with thirty-six state affiliates called "Family Policy Councils."[17]

Through the Centre for Intelligent Design, DI gains similar UK supporters. For instance, conservative Christian lawyer and Scottish MP Lord Mackay of Clashfern, who was Lord Chancellor under prime ministers Margaret Thatcher and John Major, is apparently Santorum's UK counterpart. Mackay, who advocates basing Scottish law on biblical principles, earned prominent mention in DI's *Evolution News and Views* blog for hosting Meyer's London lecture. As the British Centre for Science Education has documented, Noble "is also Educational Consultant to (and former Education Officer of) CARE [Christian Action Research and Education] in Scotland." With offices in London, Glasgow, Belfast, and Brussels, CARE is a politically oriented evangelical organization that, according to the British Centre, "aims to get Christians elected to political office and use the gospel to 'transform' society." CARE's activities demonstrate that it is the UK equivalent of Focus on the Family. Its website also states that the organization "monitors and contributes to the process of influencing education policy and practice."[18]

The Centre for Intelligent Design clearly aims to influence UK science education. Noble, who has many connections in the Scottish educational system, declared the intelligent design debate's entry into UK schools "inevitable." In CARE's "Keeping Faith in Schools" video, Noble asserts that "it's not hard to see . . . where Christians should be exerting influence—schools." In an October 2011 interview with DI, Noble, complaining about efforts to ban intelligent design statutorily from British science classes, observes that "we haven't had a court case like Judge Jones [in *Kitzmiller*], but I suppose we may have one of these before much longer." Noting that Britain lacks "formal separation of church and state," he contends that Britain's tradition of "free academic inquiry" should permit students to "talk about [either] intelligent design . . . or creationism within the science class." While the center is "not actively asking for intelligent design to be taught in the school curriculum," according to Noble, "pupils should have the freedom to ask those questions." This "academic freedom" approach mirrors DI's American strategy. Indeed, the center's website features an "Academic Freedom" statement.[19]

Intelligent Design on the Continent

DI's Intelligentdesign.org list of "Worldwide Intelligent Design Organizations" shows its continental allies, although most lack the organizational level of the UK supporters. The exception is Germany, where former DI fellow and microbiologist Siegfried Scherer, a young-earth creationist, promotes intelligent design and young-earth creationism through his organization Wort und Wissen

(Word and Knowledge). As noted, DI also has Finnish supporters, Leisola being the most prominent. Both the Finnish website Älykkään Suunnitelman (Intelligent Design) and the Danish website Intelligent Design reflect DI's influence.[20]

In Italy, geneticist Giuseppe Sermonti has been a minor ally. As editor in chief of the lower-tier journal *Rivista Biología*, he published an article by Jonathan Wells, a DI fellow, enabling DI to tout an intelligent design publication in "an internationally respected biology journal." The Italian website Progetto Cosmo (Project Cosmos) features little more than DI talking points, books, and articles.[21]

In Spain, businessman Felípe Aizpún promotes intelligent design on the ¿Darwin o Diseño Inteligente? (Darwin or Intelligent Design?) website. He partners with a small group of Hispanic intelligent design supporters, Organización Internacional para el Avance Científico del Diseño Inteligente (International Organization for the Scientific Advancement of Intelligent Design), which lists leading DI fellows as collaborators.[22]

European Prospects

Prospects for the advancement of intelligent design in Europe depend on cultural, political, legal, and demographic factors. State-established Christian churches notwithstanding, Europeans generally are markedly more secular than Americans; consequently, creationism is not culturally entrenched there as in the United States. Moreover, European pro-science activists have access to more than a decade of scholarly and scientific criticism of intelligent design. European governments have also refused to back education ministers who tried to promote it. In addition—to DI's displeasure—the Council of Europe's (nonbinding) 2007 resolution on "The Dangers of Creationism in Education" explicitly includes intelligent design. Most recently, England's Department for Education banned teaching creationism, including intelligent design, in government-funded "free schools" (similar to American charter schools), despite requiring that such schools "teach religious education, and provide for a daily act of worship." Operating in a very different constitutional environment from that of the United States, UK authorities can nonetheless prevent creationism from entering science education by administrative means via ministerial control of curricula in state-funded schools.[23]

Nonetheless, America offers a lesson for Europe: creationism dovetails closely with right-wing politics. The Council of Europe recognizes this: "Today creationist ideas are tending to find their way into Europe and their spread is affecting quite a few Council of Europe member states . . . The war on the theory of evolution and on its proponents most often originates in forms of religious extremism closely linked to extreme right-wing political movements. The creationist movements possess real political power." One European Parliament member stated, "Worryingly, religion is also increasingly making its presence felt in . . . the European Union—even though the EU was designed as a strictly secular project . . . We are witnessing the emergence of the European equivalent to the 'religious right' in the US." Hence, creationists could eventually benefit from a continued upswing in right-wing conservatism in Europe.[24]

The Discovery Institute's establishment of the Center for the Renewal of Science and Culture in 1996 was not accidental. DI exploited the political momentum of converging right-wing religious and economic conservatism, which began in the United States in the 1980s, crested in the mid-1990s, and has filtered down to conservative-dominated state legislatures, enabling DI to amplify its efforts through religious lobbying organizations. Intelligent design is among a spectrum of "culture war" issues that include opposition to stem cell research, same-sex marriage, and climate science. In the Wedge document, DI links its antievolutionism to its views on "sexuality, abortion and belief in God" and even "free enterprise." The "Mission and Program" of the institute's new Center on Wealth, Poverty, and Morality—launched in March 2012 chiefly to promote the moral virtues of free-market capitalism—remarkably resembles the Wedge strategy. *Indivisible*, the book coauthored by Jay Richards, a DI fellow, as part of the center's debut, transitions seamlessly from denouncing evolution to promoting capitalism.[25]

Similar developments could happen in Europe. The rise of the Christian right, allied with conservative politics, may create an environment in which intelligent design can establish a foothold in the European mainstream.

ACKNOWLEDGMENTS

The author thanks the Humanist Society Scotland, Paul Braterman and the British Centre for Science Education, and Robert Morris, Honorary Senior Research Fellow, Constitution Unit, University College London, for their kind assistance.

NOTES

1. Molleen Matsumura and Louise Mead, "Ten Major Court Cases about Evolution and Creationism," National Center for Science Education, accessed Jan. 8, 2012, http://ncse

.com/taking-action/ten-major-court-cases-evolution-creationism. See also Roger Downey, "Discovery's Creation," *Seattle Weekly*, Feb. 1, 2006, www.seattleweekly.com/2006-02-01/news/discovery-s-creation.php/; Discovery Institute, "The Wedge," accessed Jan. 8, 2012, www.antievolution.org/features/wedge.html.

2. "Establishment of Religion," *Findlaw.com*, accessed Mar. 11, 2012, http://caselaw.lp.findlaw.com/data/constitution/amendment01/02.html#1. See also Paul Nelson, "Life in the Big Tent: Traditional Creationism and the Intelligent Design Community," *Christian Research Journal* 24:4 (2002): 3, accessed Jan. 13, 2012, www.equip.org/PDF/DL303.pdf; Barbara Forrest and Paul R. Gross, *Creationism's Trojan Horse: The Wedge of Intelligent Design* (New York: Oxford University Press, 2007), 19.

3. *Kitzmiller et al. v. Dover Area School District* (2005), Middle District of Pennsylvania, accessed Jan. 13, 2012, www.pamd.uscourts.gov/kitzmiller/kitzmiller_342.pdf. See also "First European Conference on Intelligent Design Will Feature Scientists Presenting Evidence for Design from Biology, Paleontology, and Astrophysics," PR Newswire, accessed Jan. 13, 2012, www.thefreelibrary.com/First+European+Conference+on+Intelligent+Design+Will+Feature...-a0137725873.

4. Discovery Institute, "The Wedge."

5. Ibid. See also Barbara Forrest, "It's Dèjá Vu All Over Again: The Intelligent Design Movement's Recycling of Creationist Strategies," *Evolution: Education and Outreach* 3 (2010): 173–175, doi 10.1007/s12052-010-0217-1. See also Discovery Institute, "Model Academic Freedom Statute on Evolution," accessed Jan. 13, 2012, www.academicfreedompetition.com/freedom.php; National Center for Science Education, "'Monkey Bill' Enacted in Tennessee," Apr. 10, 2012, http://ncse.com/news/2012/04/monkey-bill-enacted-tennessee-007299.

6. Discovery Institute, "The Wedge."

7. Nelson, "Life in the Big Tent." See also Forrest and Gross, *Creationism's Trojan Horse*, 294; William A. Dembski, "Signs of Intelligence: A Primer on the Discernment of Intelligent Design," *Touchstone* 12:4 (1999): 84; A. Halloway, "Elim-Backed Antievolution Tour Fills Churches as Law Professor and Geologist Unite against the Theory," *Direction*, Jan. 2005, 18, www.darwinreconsidered.org/media/p18-23Screen.pdf; "Andrew A. Snelling, B.Sc. (Hons), Ph.D.," *Answers in Genesis*, accessed Mar. 11, 2012, www.answersingenesis.org/home/area/bios/a_snelling.asp; Tour Sponsors, "Darwin Reconsidered," accessed July 20, 2012, www.darwinreconsidered.org/tour_sponsors.html; "Peter Loose," British Centre for Science Education, accessed Mar. 2, 2012, www.bcseweb.org.uk/index.php/Main/PeterLoose. The Netherlands is an exception, however; see Blancke, Flipse, and Braeckman, chapter 4 in this volume.

8. "Purpose and Scope," *BIO-Complexity*, accessed July 20, 2012, http://bio-complexity.org/ojs/index.php/main/about/editorialPolicies#purposeAndScope. See also Glenn Branch, "The Latest 'Intelligent Design' Journal," *Reports of the National Center for Science Education* 30:6 (2010), accessed Mar. 11, 2012, http://ncse.com/rncse/30/6/latest-intelligent-design-journal; Forrest, "It's Déjá Vu All Over Again," 176–178; "People," Biologic Institute, accessed July 20, 2012, http://biologicinstitute.org/people/; John Pieret, "The Pros from Dover," *Thought in a Haystack Blog*, Aug. 7, 2009, http://dododreams.blogspot.com/2009/08/pros-from-dover.html; John Lynch, "Three (YEC) Amigos Join the Biologic Institute," *A Simple Prop Blog*, Aug. 7, 2009, http://blog.jmlynch.org/2009/08/07/three-yec-amigos-join-the-biologic-institute/. For young-earth creationists Burgess, Reeves, and John Walton, see British Centre for Science Education, "Who's Who in the UK Cre-

ationist Movements," Dec. 22, 2010, www.bcseweb.org.uk/index.php/Main/WhoIsWho. Concerning Leisola, see Lynch, "Three (YEC) Amigos Join the Biologic Institute," and Pieret, "The Pros from Dover." Concerning Peter Imming, see Wort und Wissen, accessed July 20, 2012, www.wort-und-wissen.de/index2.php?artikel=info/i06/1/i06-1.html. Concerning Siegfried Scherer, see Forrest and Gross, *Creationism's Trojan Horse*, 275. For intelligent design supporters, such as Wolf-Ekkehard Lönnig, see Ulrich Kutschera, "The German Anti-Darwin Industry," *Reports of the National Center for Science Education* 28:1 (2008), accessed July 20, 2012, http://ncse.com/rncse/28/1/german-anti-darwin-industry. Concerning Norman Nevin, see "About CID," Centre for Intelligent Design, accessed July 20, 2012, www.c4id.org.uk/index.php?option=com_content&view=article&id=166&Itemid=26. Concerning Jirí Vácha, see Jirí Vácha's Personal Homepage, accessed Mar. 2, 2012, www.jvacha.com/articles_eng.html. See also Stephen Meyer, interview by David Boze, *ID the Future*, podcast audio, Dec. 9, 2011, http://intelligentdesign.podomatic.com/entry/2011-12-09T16_51_29-08_00.

9. Pauli J. Ojala and Matti Leisola, "Haeckel: Legacy of Fraud to Popularise Evolution," *Journal of Creation* 21:3 (2007): 110, accessed July 20, 2012, http://widgets.new.digg.com/newsbar/topnews/haeckel_legacy_of_fraud_to_popularise_evolution. See also Carl Wieland, "In Charles Darwin's Footsteps," Creation Ministries International, Apr. 22, 2009, http://creation.com/charles-darwin-voyage-movie; Nelson's and Sternberg's presentations on the Finnish ID website, Älykkään Suunnitelman, at "In English," accessed July 20, 2012, www.intelligentdesign.fi/sivut/in-english/; Leisola's faculty listing at "Contact Information of Personnel," Department of Biotechnology and Chemical Technology, Aalto University, accessed Mar. 4, 2012, http://chemtech.aalto.fi/en/department/personnel/.

10. "Worldwide Intelligent Design Organizations," Intelligentdesign.org, accessed Feb. 3, 2012, www.intelligentdesign.org/resources.php.

11. Jessica Shepherd, "Teach Both Evolution and Creationism Say 54% of Britons," *Guardian*, Oct. 25, 2009, www.guardian.co.uk/science/2009/oct/25/teach-evolution-creationism-britons. See also David Tyler, interview by BBC Breakfast, video, Feb. 7, 2005, http://news.bbc.co.uk/media/video/40801000/rm/_40801753_creationism_vtandlive_vi.ram; David J. Tyler, "Religious and Philosophical Inputs to Geochronology," Biblical Creation Society, Aug. 1995, www.biblicalcreation.org/scientific_issues/bcs080.html; David Tyler, " 'The Grand Design' Revisited," *The ID Update Blog*, Jan. 11, 2012, www.arn.org/blogs/index.php?author=11; Stephen Meyer, interview by David Boze, *ID the Future*; Michelangelo D'Agostino, "In the Matter of Berkeley v. Berkeley," *Berkeley Science Review*, Spring 2006, 33, http://sciencereview.berkeley.edu/articles/issue10/evolution.pdf; "William Dembski Interview," *Thebestschools.org*, Jan. 14, 2012, www.thebestschools.org/blog/2012/01/14/william-dembski-interview/.

12. "About Truth in Science," Truth in Science, accessed Feb. 4, 2012, www.truthinscience.org.uk/tis2/index.php/about-topmenu-82.html. See also "About CID," Centre for Intelligent Design; "Welcomed by Lord Mackay, Meyer Speaks in London to Centre for Intelligent Design UK," *Evolution News and Views Blog*, Nov. 29, 2011, www.evolutionnews.org/2011/11/welcomed_by_lor053561.html; "Peter Loose," British Centre for Science Education.

13. "Truth in Science," "Take Heed" Ministries newsletter, Mar. 2004, http://web.archive.org/web/20040402093215/www.takeheed.net/MARCH+2004.htm. See also

"About," Illustra Media, accessed Feb. 13, 2012, http://illustramedia.com/about/; "Resource Pack," Truth in Science, accessed July 20, 2012, www.truthinscience.org.uk/tis2/index.php/resource-pack-mainmenu-92.html; James David Williams, "Creationist Teaching in School Science: A UK Perspective," *Evolution: Education and Outreach* 1 (2008): 89–90, accessed Mar. 2, 2012, www.springerlink.com/content/4103235r16308112/fulltext.pdf, doi 10.1007/s12052-007-0006-7; Robert Crowther, "British Organization Seeks to Incorporate Teaching Scientific Criticisms of Evolution in UK," *Evolution News and Views Blog*, Sept. 21, 2006, www.evolutionnews.org/2006/09/british_organization_seeks_to002647.html.

14. "Press Release Regarding the Textbook 'Explore Evolution,'," n.d., Truth in Science, accessed July 20, 2012, http://liveweb.archive.org/www.truthinscience.org.uk/tis2/index.php/news-blog-mainmenu-63/300-press-release-regarding-the-textbook-qexplore-evolutionq.html. See also "'Explore Evolution' Exposed," British Centre for Science Education, n.d., accessed Mar. 11, 2012, http://bcseweb.blogspot.com/p/evolution-exposed.html; letter from Andrew McIntosh, Truth in Science, n.d., accessed July 20, 2012, www.humanism.org.uk/_uploads/documents/1TiSExplore_Letter.pdf; "Truth in Science—Letter to All UK Schools," British Centre for Science Education, Dec. 14, 2009, http://bcseweb.blogspot.com/2009/12/truth-in-science-letter-to-all-uk.html; Barbara Forrest, "Understanding the Intelligent Design Creationist Movement," Center for Inquiry, July 2007, 19–22, www.centerforinquiry.net/uploads/attachments/intelligent-design.pdf; "Truth in Science," British Centre for Science Education, last modified Oct. 10, 2007, www.bcseweb.org.uk/index.php/Main/TruthInScience.

15. "The Centre for Intelligent Design—Britain's Latest Creationist Organization," British Centre for Science Education, last modified Nov. 27, 2010, www.bcseweb.org.uk/index.php/Main/CentreForIntelligentDesign. See also Discovery Institute, "The Wedge"; Forrest and Gross, *Creationism's Trojan Horse*, 189–191; William Dembski, "More on ID at Justin Bierley's *Unbelievable*," *Uncommon Descent Blog*, Jan. 18, 2010, www.uncommondescent.com/intelligent-design/more-on-id-at-justin-brierleys-unbelievable; "Darwin or Design? An Evening with Michael Behe," *Unbelievable*, Premier Christian Media, Nov. 22, 2010, http://web.archive.org/web/20101009112934/, www.premier.org.uk/events/premier-events/Darwin%20or%20Design%20-%20An%20Evening%20with%20Michael%20Behe.aspx.

16. "Darwin or Design?," accessed Mar. 12, 2012, http://web.archive.org/web/20101009072316/, www.darwinordesign.org.uk/. See also Roger Stanyard, Paul Braterman, Roy Thearle, et al., "Michael Behe in Britain, 2010 Part 1," British Centre for Science Education, last modified Dec. 1, 2010, www.bcseweb.org.uk/index.php/Main/MichaelBeheInBritain; Discovery Institute, "The Wedge"; Forrest and Gross, *Creationism's Trojan Horse*, 39–42, 194, 322; Roger Stanyard, "Supporting the Show—Messrs Michael Behe, Steve Fuller, Geoff Barnard and Others," British Centre for Science Education, last modified Nov. 30, 2010, www.bcseweb.org.uk/index.php/CentreForIntelligentDesign/SupportingTheShowMessrsMichaelBeheSteveFullerAndGeoffBarnard; "'Design in Nature' Theme Proves a Big Draw at Cambridge Conference," Centre for Intelligent Design, accessed Sept. 17, 2012, www.c4id.org.uk/index.php?option=com_content&view=article&id=251:design-in-nature-theme-proves-a-big-draw-at-cambridge-conference&catid=1:latest&Itemid=28; "Design in Nature? Scientific and Philosophical Perspectives," Tyndale Fellowship, accessed Sept. 17, 2012, www.tyndalephilosophy.org.uk/Events/24500%20Design%20in%20Nature.pdf; "Mission Statement," Tyndale Fellow-

ship, accessed Sept. 17, 2012, www.tyndalephilosophy.org.uk/mission.html; "2011 Summer School on Intelligent Design, July 18–22 [2011]," Centre for Intelligent Design, accessed July 20, 2012, www.c4id.org.uk/summerschool; "2011 Intelligent Design Conference, September 9–10 [2011]," Centre for Intelligent Design, accessed Mar. 3, 2012, www.c4id.org.uk/index.php?option=com_content&view=article&id=233&Itemid=125; "2012 Intelligent Design Conference, September 28–29, Darwin or Design?," Centre for Intelligent Design, accessed Sept. 14, 2012, www.c4id.org.uk/index.php?option=com_content&view=article&id=249&Itemid=124; "People: Douglas Axe," Biologic Institute, accessed Sept. 14, 2012, www.biologicinstitute.org/people.

17. Alastair Noble, "C4ID's Inaugural Lecture 2011: 'Is There a Signature in the Cell?,'" Centre for Intelligent Design, n.d., accessed Feb. 13, 2012, www.c4id.org.uk/index.php?option=com_content&view=article&id=246:c4ids-inaugural-lecture-2011-is-there-a-signature-in-the-cell&catid=52:frontpage&Itemid=1. See also Forrest and Gross, *Creationism's Trojan Horse,* 240–242, 250–251; Focus on the Family, "State Groups," Citizenlink.com, accessed July 20, 2012, www.citizenlink.com/state-groups/; "Dr. [Stephen] Meyer: The Case for Design," *The Truth Project,* Focus on the Family, video, Sept. 23, 2007, www.focusonthefamily.com/popups/media_player.aspx?MediaId={09003731-825F-4EFD-932D-07845748A572}.

18. "Welcomed by Lord Mackay, Meyer Speaks in London to Centre for Intelligent Design UK," *Evolution News and Views Blog.* See also Mark Horne, "Law Chief Urges Scots Courts: Consult the Bible in Judgments," *Sunday Herald,* Aug. 15, 2010, www.heraldscotland.com/news/home-news/law-chief-urges-scots-courts-consult-the-bible-in-judgments-1.1048316; "Dr. Alastair Noble (Director, Centre for Intelligent Design)," British Centre for Science Education, last modified Nov. 30, 2010, www.bcseweb.org.uk/index.php/CentreForIntelligentDesign/DrAlastairNobleDirectorCentreForIntelligentDesign; "Where?," CARE, accessed July 20, 2012, www.care.org.uk/where-we-work; "Education," CARE, accessed Mar. 12, 2012, www.care.org.uk/advocacy/education; Forrest and Gross, *Creationism's Trojan Horse,* 270–271.

19. Chris Watt, "Would You Adam and Eve It? Top Scientists Tell Scottish Pupils: The Bible Is True," *Sunday Herald,* Oct. 10, 2010, www.heraldscotland.com/news/education/would-you-adam-and-eve-it-top-scientists-tell-scottish-pupils-the-bible-is-true-1.1060545. See also "An Introduction to Intelligent Design," Centre for Intelligent Design, accessed Mar. 12, 2012, www.c4id.org.uk/index.php?option=com_content&view=article&id=203&Itemid=109; "Dr. Alastair Noble (Director, Centre for Intelligent Design)," British Centre for Science Education; "Keeping Faith in Schools," CARE, accessed Feb. 19, 2012, www.care.org.uk/where-we-work/care-for-scotland/education/keeping-faith-in-schools; Alastair Noble, interview by Joshua Youngkin, *IDthefuture.com,* podcast audio, Oct. 5, 2011, http://intelligentdesign.podomatic.com/entry/2011-10-05T17_29_05-07_00; "Academic Freedom," Centre for Intelligent Design, accessed Mar. 12, 2012, www.c4id.org.uk/index.php?option=com_content&view=article&id=244:academic-freedom&catid=57:freedom&Itemid=133.

20. "Worldwide Intelligent Design Organizations." See also Kutschera, "The German Anti-Darwin Industry"; "Informationen aus der Studiengemeinschaft Wort un Wissen," Wort und Wissen, accessed Mar. 12, 2012, www.wort-und-wissen.de/index2.php?artikel=info/i10/3/i10-3.html; Kutschera, chapter 6 in this volume; Älykkään Suunnitelman, accessed July 20, 2012, www.intelligentdesign.fi/; Intelligent Design (Denmark), accessed Mar. 12, 2012, www.intelligentdesign.dk/.

21. "Another Biology Journal Publishes Article Applying Intelligent Design Theory to Scientific Research," Discovery Institute, June 8, 2005, www.discovery.org/a/2627. See also "Libri Stranieri [Foreign Books]," Progetto Cosmo, accessed Mar. 3, 2012, http://progettocosmo.altervista.org/index.php?option=content&pcontent=1&task=view&id=23&Itemid=48.

22. See ¿Darwin o Diseño Inteligente?, accessed July 20, 2012, www.darwinodi.com/. See also Organización Internacional para el Avance Científico del Diseño Inteligente, accessed Mar. 12, 2012, www.oiacdi.org/; Felipe Aizpún, interview by Casey Luskin, *IDthefuture.com*, podcast audio, Jan. 12, 2011, http://intelligentdesign.podomatic.com/player/web/2011-01-12T12_00_48-08_00.

23. See "Secular Europe and Religious America: Implications for Transatlantic Relations," Pew Research Center, Apr. 21 2005, www.pewforum.org/Politics-and-Elections/Secular-Europe-and-Religious-America-Implications-for-Transatlantic-Relations.aspx. See also John Timmer, "Attempts to Introduce Intelligent Design in Europe Spark Backlash," *Ars Technica*, June 26, 2007, http://arstechnica.com/science/news/2007/06/attempts-to-introduce-intelligent-design-in-europe-spark-backlash.ars; Parliamentary Assembly of the Council of Europe, "The Dangers of Creationism in Education," Doc. 11375, Sept. 17, 2007, http://assembly.coe.int/main.asp?link=/documents/workingdocs/doc07/edoc11375.htm; Parliamentary Assembly of the Council of Europe, "The Dangers of Creationism in Education," Resolution 1580, Oct. 4, 2007, http://assembly.coe.int/main.asp?link=/documents/adoptedtext/ta07/eres1580.htm; Casey Luskin, "Council of Europe Makes Its Dogmatism Official: Intelligent Design Poses 'a Threat to Human Rights,'" *Evolution News and Views Blog*, Oct. 22, 2007, www.evolutionnews.org/2007/10/council_of_europe_makes_its_d004378.html; Stephen Adams, "Creationism 'Banned from Free Schools,'" *Telegraph*, May 20, 2011, www.telegraph.co.uk/education/educationnews/8526161/Creationism-banned-from-free-schools.html; "Free Schools FAQs—Curriculum," Department for Education, accessed Mar. 9, 2012, www.education.gov.uk/schools/leadership/typesofschools/freeschools/freeschoolsfaqs/a0075656/free-schools-faqs-curriculum.

24. Parliamentary Assembly of the Council of Europe, Resolution 1580. See also Sophie in 't Veld, "The Rise of Europe's Religious Right," *Guardian*, Mar. 17, 2011, www.guardian.co.uk/commentisfree/belief/2011/mar/17/europe-religious-right.

25. See Frederick Clarkson, *Eternal Hostility: The Struggle between Theocracy and Democracy* (Monroe, ME: Common Courage Press, 1997). See also Damon Linker, *The Theocons: Secular America under Siege* (New York: Anchor Books, 2007); Barbara Forrest, "The Discovery Institute, the LA Family Forum, and the 'LA Science Education Act' updated," *Talk to Action Blog*, June 26, 2008, www.talk2action.org/story/2008/6/26/18920/8497; Discovery Institute, "The Wedge"; "Mission and Program," Center on Wealth, Poverty, and Morality, accessed Mar. 12, 2012, www.discovery.org/cwpm/; James Robison and Jay W. Richards, *Indivisible: Restoring Faith, Family, and Freedom Before It's Too Late* (New York: FaithWords, 2012), 154–156.

CHAPTER THIRTEEN

The Rise of Anti-creationism in Europe

PETER C. KJÆRGAARD

The increasing activism among European creationists has not gone unnoticed. Widespread media coverage of evolution debates has turned the alleged antagonistic relationship between science and religion, and the implication of a necessary choice between scientific knowledge and faith, into a standard narrative familiar to the general European public. Polls juxtaposing evolution and creation, often with a focus on human origins, support the public image of two sides with equal rights to be taken seriously in debates and policy decisions. This has worked to the benefit of various creationist groups to get more attention than they would otherwise. With routine reference to journalistic balance, creationists are habitually invited along with academics to offer their perspective on science and religion issues, creationism, intelligent design, and public opinion about evolution in U.S. newspapers, magazines, radio shows, and television programs.[1] We find the same pattern in European media. In response to the increased antievolution activism and to the media coverage giving more credit to the creationist side than scientists and science supporters would like, explicit anti-creationist reactions have become commonplace in various European countries. Although Europeans have never created organizations such as the National Center for Science Education in the United States specifically to oppose creationism, many have fought back in more limited ways. European creationism is characterized by heterogeneity with local groups acting according to their own cultural and religious agendas.[2] Anti-creationism, its political counterpart, follows the same pattern.

Early European Anti-creationism

During the nineteenth century there was a growing realization that appeals to the supernatural did not count as science. Already in 1838 the young Charles Darwin recently returned from the *Beagle* voyage noted that creation of animals with reference to a deity "is no explanation—it has not the character of a physical law & is therefore utterly useless." Darwin was by no means alone with his opinion and joined a scientific community gradually turning away from the natural theology framework where science to a certain extent was religion.[3] Arguing for "the hypothesis of progressive development" in 1856, Thomas Henry Huxley echoed Darwin's thoughts by arguing that reference to special creation passed the question of origins out of the domain of science altogether: "We have not the scientific evidence of such unconditional creative acts; nor, indeed, could we have such evidence; for if a species were to originate under one's very eyes I know of no amount of evidence which would justify one in admitting it to be a creative act independent of the whole vast chain of and effects in the universe."[4]

As one of the most outspoken critics of ideas of supernatural origins of species, Huxley set the standards for a particular form of early European anti-creationism. His direct and caustic style in conjunction with a fair amount of ridicule became a widely used weapon against nineteenth-century adherents of special creation by divine intervention. Darwin himself tried to be more cautious in the *Origin of Species* when addressing the "blindness of preconceived opinion" among certain naturalists: "These authors seem no more startled at a miraculous act of creation than at an ordinary birth. But do they really believe that at innumerable periods in the earth's history certain elemental atoms have been commanded suddenly to flash into living tissues?" Darwin's use of elemental atoms flashing into existence was instantly picked up by reviewers; among them the anatomist Richard Owen. Darwin felt, though, that his argument was misconstrued and complained about it in a letter to the Harvard botanist Asa Gray: "I ask whether creationist really believe that elemental atoms have flashed into life. He says I describe them as so believing; & this surely is a difference."[5] However little Darwin wanted to be a source of squabbles and misunderstandings, his argument advanced the differences and bolstered two camps of incompatible opinions.

The first legal battle over the status of evolutionary theory in the United States in 1925 attracted much attention and was widely reported in European newspapers, magazines, and journals. Among the many commentators, one of the most prominent was the British anthropologist Arthur Keith. The so-called Scopes trial, in which a high school teacher in the small town of Dayton, Tennessee, was accused of violating a law making it unlawful to teach human evolution in any state-funded schools, made creationism a cultural and political issue, also seen from a European perspective. Two years after the trial, as president of the British Association of the Advancement of Science, Keith warned against the threat from creationists in the United Kingdom. "Daytonism," as he called the politically motivated American creationism, "is very much alive throughout the land [the United Kingdom]." It was not just an American phenomenon, he cautioned. Yet in Britain the situation was different. Religious leaders and scientists shared the same ideals, according to Keith. Both groups wanted to understand and explain the universe and find a solution to the great riddle why we are here. He called for reconciliation between science and religion, but clearly defined on his own terms. "They [the religious leaders] have grown up in the post-Darwinian period, and no longer regard the great army of science as an enemy, but as a friendly power. They realise that religion cannot stand still, that it too must evolve."[6]

Whereas most of the debates in scientific circles were questions of the scientific legitimacy of evolutionary arguments, Keith saw a new challenge on the horizon, and not a scientific one. He was backed by the Rationalist Press Association, a membership organization specializing in publishing books by agnostic and atheistic authors fulfilling a growing demand for secular and distinctly antireligious literature in the British public. In 1925 the association lured the outspoken Canadian creationist George McCready Price to a public debate with the rationalist philosopher and ex-Franciscan priest Joseph McCabe in Queen's Hall, London. The event turned out to be a public humiliation of Price with a hostile audience of three thousand heckling. Price himself expressed disappointment in the British for being so little open to his creationist ideas. Keith was right, however. Gradually creationist ideas were taking root even in Britain. In 1932 the Evolution Protest Movement, the first explicitly creationist organization in Europe, was founded in London with the English physicist, engineer, and inventor of the diode, Ambrose Fleming, as its first president.[7]

The emerging organized creationism in Europe did not, however, generate a strong response in scientific circles. It remained a fringe discussion and was given very little consideration by pro-evolution scientists. The cultural and political response during the most part of the twentieth century came from members and partisans from the growing number of secularist organizations around Europe. Apart from Britain these were represented in countries like Belgium, the Netherlands, Germany, France, Italy, Greece, and Denmark. In 1991 the European Humanist Federation was founded as an international partnership for humanist associations now including more than fifty national societies from twenty-three European countries. The open political opposition to creationism was mostly channeled through individual activism that also included religious people such as, for example, the Dutch Calvinist geologist J. R Van de Fliert, and by members of national secularist societies, not through mainstream scientific institutions and in scientific debates.[8]

Creationism as a Political Issue

Following the rise in organized creationism in the 1970s, scientists and educators in the United States found the previous strategies to counter antievolution campaigns inadequate. The new organized opposition to creationists approached the problem as a political rather than a scientific problem, echoing the way secularist societies worked, but targeted creationism specifically. This was the point of one of the leading pro-evolution spokespersons, the biologist John A. Moore's address to the annual meeting for the American Association for the Advancement of Science in 1979. "Once one recognizes that the creationism-evolution controversy is basically political, not scientific, one also realizes how difficult it is to deal with the matter," Moore pointed out and continued: "If we accept the premise that the creationists are confronting us with a political not a scientific problem, then we must also accept that scientific arguments alone will not solve the problem."[9]

Moore was in line with the consensus of a growing community that worried that the growing funds to creationist campaigns and the ensuing legal battles would eventually succeed, at least at a local and state level. This eventually led to the formation of the Committees of Correspondence in the early 1980s, a national American network of anti-creationists organized first under the National Association of Biology Teachers and later under its own National Center for Science Education.[10] The firm belief in the convincing powers of scientific arguments alone and the idea that truth will out was weakened by numerous cases with well-prepared creationists able to convince local school boards of the justification for their cause. To win the battle, evolution defenders recognized they had to adopt the same grass-roots tactics the creationists used.[11] Consequently, American anti-creationists began a similar campaign as a political protest lobbying at a local level by circulating literature, sending letters, attending school board meetings, and confronting politicians, school officials, and committee members. Also in this period a dedicated focus on teaching evolution prompted a series of new science education initiatives and prompted new organizations of science teachers and educators.[12]

Another aspect influencing the debates, as evolution shifted from a scientific context into political and legal contexts in the media, was that it was no longer merely a topic for science journalists. Bouncing to political pages, opinion sections, and radio and television news stories, the public and journalistic frames changed. The strong scientific case for evolution was toned down in favor of a controversy narrative with two opposing sides with equal rights to be heard. While painting a false picture, it would still be politically convenient and from a journalistic perspective support the ideal of presenting a balanced story. That particular angle worked well in the hands of regular news reporters with little interest in or understanding of the science behind evolutionary theory, and became the dominant frame in coverage of evolution-creation debates from the 1980s onward.[13] Incidentally, it also served creationist grass-roots' interests by taking the discussion out of the scientists' home turf and into a political playing field. Anti-creationists on both sides of the Atlantic had no choice but, following suit, to adapt a strategy of finding the frames resonating with the general public and not just pushing ahead relying on presenting scientific facts.[14]

Two different main approaches were used, both appropriated and employed by North American and European evolution defenders as creationism gained momentum across the Atlantic. The negative anti-creationist approach built its case as a scientific and philosophical criticism of creationist arguments with the aim of demonstrating the untenable position of antievolutionists. The positive anti-creationist approach aimed instead at building a strong alternative to creationist arguments by demonstrating the power of evolutionary explanations and the weight of the scientific evidence. These approaches were not mutually exclusive, but often one or the other was emphasized as the main political approach. How and where to push them, however, depended much on the context and the overall strategy of anti-creationist initiatives in the various European countries.

Political Anti-creationism

Anti-creationism in the United States has been fairly well organized compared to the movement elsewhere in the world since the 1980s, but it remained largely within the confines of grass-roots activism and science educator associations. In Europe it took a radically different turn in the first decade of the twenty-first century as the European Council in 2007 voted for a resolution against creationism and intelligent design. The Council of Europe is an intergovernmental body founded in 1949. With its forty-seven member states representing some 820 million citizens, it is an entirely different organization from the European Union. Formed in the political aftermath of the Second World War, the council aimed to further greater unity between European countries by safeguarding and realizing the ideals and principles of their common heritage and facilitate economic and social progress. The council does not have any legislative powers. Instead, member states commit themselves to implement common policies through conventions, such as the European Convention on Human Rights, adopted in 1950.[15]

Resolution 1580, which was passed by 48 to 25, urged member states to firmly oppose the teaching of creationism in schools. In the resolution the council expressed a worry "about the possible ill-effects of the spread of creationist ideas within our education systems and about the consequences for our democracies. If we are not careful, creationism could become a threat to human rights, which are a key concern of the Council of Europe." The resolution emphasized the central role science played in the economic, technological, and social development of the European countries, and thus contributed to being a stabilizing factor in the foundation for sustaining successful democracies. Creationist groups were cast as antiscientific, which in itself was construed as "one of the most serious threats to human and civic rights." Furthermore, the "war on the theory of evolution and on its proponents" was identified as often originating in various forms of religious extremism closely linked to extreme right-wing political movements and thus directly to antidemocratic activism. This had to be taken seriously as it was claimed that the "creationist movements possess real political power."[16]

The real threat from creationist groups was documented in a series of events. Adnan Oktar's Muslim offensive, which had originated in December 2006 in Turkey and targeted various European countries, had not gone unnoticed and was of key concern to the council.[17] In France, for example, the minister of education, Gilles de Robien, called upon chief education officers to ensure that Oktar's *Atlas of Creation*, distributed for free to schools and universities around Europe, was not made available for students. The so-called Interdisciplinary University of Paris was also under accusation for harboring neo-creationist ideas with links to the Paris Natural History Museum. After anti-creationist exposure, sponsors including some prestigious companies, gradually withdrew their financial support to the institution's activities. In Switzerland there was a political reaction barring the *Atlas of Creation* from being used in schools, while the European Biblical Centre, a publisher of creationist literature, and the ProGenesis Group working toward building Genesis Land, a creationist theme park in northern Switzerland, got the council's political attention.

Following the call of Letizia Maratti, Italian minister of education and research in the Berlusconi government, in 2004 for abolishing the teaching of evolution in primary and secondary schools, a strong anti-creationist reaction mounted among scientists and journalists. This resulted in a commission report pointing out that the theory of evolution was crucial to an overall view of life and the role of science in modern culture. Furthermore, the report stressed that the teaching of Darwinian theories would counteract racism and eugenics. While some of these arguments were scientific, the report rested upon the premise that creationism was indeed a political issue needing a political answer. Also in 2004 the Serbian minister of education, Liliana Colić, ordered schools to stop teaching the theory of evolution if it was not accompanied by creationist ideas. A successful anti-creationist campaign led by the Serbian Academy of Science and supported by around forty different organizations consequently forced her to resign. In the Netherlands the minister of education, Maria van der Hoeven, spoke out for creationism and intelligent design in 2005, but was unsuccessful in promoting its acceptance owing to strong political anti-creationist opposition from both a center-left party and a liberal right-wing party. As Britain hosted a large three-day international creationist symposium in the summer of 2006, the National Union of Teachers, the biggest teaching union in the United Kingdom, called for legislation to counter the growing influence in the British education system. The National Union of Teachers earned support for this cause from the British Centre for Science Education (the only explicitly anti-creationist national organization in Europe), the Royal Society, and the archbishop of Canterbury, who spoke out against the teaching of creationism in British schools.[18]

Common for all these examples is an organized political response in individual European countries to creationist activism or creationist activities, often from within or with support from the political establishment. With the background report and the resolution, the Council of Europe wanted to make a concerted politically guided anti-creationist effort to block the documented increase in European creationist activism. Incidentally, in the following years Resolution 1580 came to serve as a reference point for increased creationist and anti-creationist organization in Europe.

Multiple Platforms

Building up to the Darwin anniversary in 2009 celebrating Charles Darwin's two-hundredth birthday and the 150 years that had passed since the publication of *On the Origin of Species*, creationists and evolutionists rallied in separate camps preparing for the biggest global discussion yet of the scientific and cultural significance of evolutionary theory.[19] In the event, pro-evolution initiatives used a strategy of multiple platforms for communication, adopting the anti-creationist strategy of lobbying locally, regionally, and nationally. This included traditional meetings, seminars, and discussions, talks at schools and public venues, appearances in newspapers and magazines, interviews on radio and television, publications, events, and exhibits. Among the most outspoken anti-creationists was the German evolutionary biologist Ulrich Kutschera, who persistently and publicly tracked creationism in Europe. But for 2009 some of the most important venues were the electronic media. Dedicated websites, blogs, opinion pieces, and comment sections in established news media were now included in web wars with creationists and anti-creationists competing for space and attention. National 2009 addresses including the words "evolution" and "Darwin" were used on both sides to oppose or support evolution. Norwegian creationists, for example, campaigned at darwin2009.no, using the national domain for Norway. This was also a pun on saying no to the Darwin 2009 celebrations. Anti-creationists at the universities of Bergen and Oslo unsuccessfully complained that it was immoral to appropriate a public space for religious propaganda and argued that it should belong to the scientific community to make a proper celebration of Darwin and evolution. Little did it help, and the site remained in creationist hands. In other countries, Darwin 2009 and evolution websites were in the hands of evolutionists with both implicit and explicit anti-creationist agendas.[20]

The efficacy of working multiple platforms to communicate evolution and to counteract creationist propaganda in contexts familiar to the specific target audiences was tested on numerous occasions and in very different national contexts during the Darwin year. Part of the strategy was to take seriously what creationist had understood long ago, that there was no way of making the evolution-creation debate merely a scientific one discussed on the premises dictated by scientists. It was necessary to take the political and cultural angle seriously and use the well-known frames and master narratives employed by the media and exploited by creationists. Anti-creationist initiatives in some countries were consequently far more successful in reaching audiences and getting their attention, including opposing the many miscomprehensions and bringing the focus back on scientific issues, such as the usefulness of evolutionary theory for modern medicine.[21] These frames worked and were effectively becoming an increasing part of public debates about evolution.

Activist Agendas

Proactive individuals and interest groups were also adopting the multiple-platform strategy, some pushing the negative anti-creationist approach trying to dismantle creationist arguments and exposing their scientific ignorance. Among many scientists and philosophers the most famous is probably the British evolutionary biologist from Oxford University, Richard Dawkins. Gradually building his arguments in his early writings from within a scientific context to eventually embracing the political and cultural scene in an explicitly atheistic and openly antireligious outlook, he understood the value of adopting the polemic lobbyist position that many creationists had worked for decades. Dawkins has done it with remarkable success by creating far more attention than anyone else on the untenable scientific position and arguments of creationists and supporters of intelligent design. The militant scientific atheism has been immensely popular and created a large community of followers, also on the social media.[22]

However, despite Dawkins's talent for getting the anti-creationist message across and especially because of the media attention he has enjoyed, not everybody in the European scientific and education communities is happy about the agenda he has set so firmly in the public mind. Dawkins's confrontational style and the way he has embraced the antagonistic narrative of science and religion, in many peoples' eyes making it a question of either-or, went against the far more cautious political strategy of the Council of Europe. In Resolution 1580 the council went to great lengths not to antagonize religious groups by stressing in the very first paragraph that it was not the aim to question or fight belief but rather to prevent belief from opposing science.

These and similar phrases were put in the resolution to get the final vote through, as the first draft, far more directly antireligious, alienated some members who threatened the entire plan of speaking in one voice against creationism in Europe from the Parliamentary Assembly. Similarly, the argument against Dawkins's antireligious campaign is that, apart from denigrating various social and religious groups, he gives resonance to the narrative of social and religious conservatives that the scientific establishment has an antireligious agenda. This helps fuel the highly popular conflict frame in the media, generating journalistic, sensationalist simplicity gliding from the false assumption of a causal link between the positions of evolution versus creation, science versus religion, and Darwin versus God. The inevitable conclusion of many religious groups is that science is indeed a threat to their religious identity, just as the creationists have been saying for decades.[23]

Dawkins has used his highly visible public profile to advance his cause further through the Richard Dawkins Foundation for Reason and Science. It was founded in 2006 as a charity organization with offices in the United Kingdom and the United States, with the mission to support scientific education, critical thinking, and evidence-based understanding of the natural world in the quest to overcome religious fundamentalism, superstition, intolerance, and human suffering. Getting tax-free status in the United Kingdom was difficult compared to religious groups, which in most cases would easily obtain it. Dawkins, in his own words, had to "jump through hoops to demonstrate that they [nonreligious organizations] benefit humanity." In the prolonged and expensive but ultimately successful attempt to obtain charitable status, the British Charity Commission sent a letter to Dawkins saying: "It is not clear how the advancement of science tends towards the mental and moral improvement of the public. Please provide us with evidence of this or explain how it is linked to the advancement of humanism and rationalism."[24] Evidently, the British Charity Commission was out of sync with the Council of Europe's recommendations and arguments for the link between science and human rights. Dawkins and followers were also not blind for the irony that an organization distributing creationist material in London received tax exemption as a charity organization and was as such supported by the British taxpayers.

Organizations with humanist or explicitly atheist agendas have played an increasingly important part in anti-creationist activism. In 2009 the British Humanist Association joined forces with the Richard Dawkins Foundation in the Atheist Bus Campaign sponsoring an advertising campaign with buses carrying the slogan: "There's probably no god. Now stop worrying and enjoy your life." This inspired similar campaigns in other European countries, including Ireland, Italy, Spain, Germany, Sweden, and Finland. Russian atheists were also planning an event but were banned from doing so by government officials. The campaigns are a result of a growing atheist awareness and organization.[25] But it is also linked directly to an anti-creationist agenda. The British Humanist Association has been sponsoring an annual Darwin Day celebrating the evolutionary view of life, and the Rational Association's magazine *New Humanist* has published numerous anti-creationist articles explaining and exposing creationists at work in Europe and the rest of the world.

At an inter-European level in an attempt to change the Romanian government's decision to remove evolution from the national school syllabus, the European Humanist Federation wrote a strongly worded letter to the Romanian minister of education, Cristian Mihai Adomnitei, in 2008. The letter had no effect on the policies of the Romanian government, although it was widely circulated. It did, however, prompt a harsh reaction from a conservative group called the Alliance of Romania's Families, allegedly representing 650,000 adult Romanians. The Humanist Federation was accused of trying to force upon the free Romanian people the very evolutionist ideology that had deprived them of their freedom and human rights under the totalitarian rule of the communist

Nicolae Ceaușescu. Insisting upon human rights as the central concern of the Humanist Federation and that evolution as a scientific theory to understand life on earth was not a political ideology was fruitless.[26] At a general level, trying to set the agenda gave anti-creationist activism more attention and more influence at a local and at a national level. It did not imply, however, that every call for action succeeded.

Institutional Initiatives

Scientific and academic institutions have also been home for anti-creationist initiatives, both taking the negative approach directly challenging creationist ideas and the positive approach building a strong defense for evolutionary theory to communicate positively about the science. At an international level, numerous European scientific academies co-signed a statement on the teaching of evolution from the Interacademy Panel, a global network of science academies. The explicit purpose of the statement was to counter creationist activities around the world by providing a clear support of the basic facts of evolutionary theory, including the age of the earth, natural causes for the changes in earth's geology and life, and the converging evidence for evolution from multiple scientific disciplines.[27]

At the Darwin Festival at the University of Cambridge in 2009, the Canadian rapper, Baba Brinkman, created a pro-evolution and anti-creationist happening as he had more than eight hundred specially invited guests singing along with him in the Cambridge Botanical Gardens, including the vice chancellor of the university. The university was responsible for the festival but not the specific acts. During the Darwin year, similar events took place in Europe. Some of these were connected to special exhibits at libraries, universities, or museums. Various approaches were used, some avoiding creationism altogether, some circumventing creationist arguments by addressing the challenges but without explicit reference to creationism, and some by bringing the discussions in and dealing with the issues directly in institutional settings.

In the Scandinavian countries, for example, three different approaches were employed at the major national natural history museums, which had a direct influence on the public debates on evolution in 2009. The celebration at the Natural History Museum in Stockholm was limited to a professional academic meeting. In consequence, the museum had little impact on the overall limited media coverage of evolution and creation in Sweden during the Darwin year. The situation in Norway and in Denmark was different. The natural history museums in Oslo and in Copenhagen were both engaged in building major exhibitions.

In Oslo the common creationist political frame of moral implications of evolutionary theory was used to confront and provoke visitors to think critically under the headline: "Can we forgive Darwin?" Both the Oslo and the Copenhagen museums are university museums, which allows for close collaboration with university researchers. In Oslo this was partly done through a special exhibit on a spectacularly well-preserved Eocene primate fossil, *Darwinius masillae*, receiving unprecedented international attention.[28] At the museum in Copenhagen, a collaboration between the two major Danish universities resulted in a national research-based internet encyclopedia, evolution.dk, to communicate evolution, including public debates and misconceptions, to accompany the exhibition. Both the exhibit and the internet encyclopedia were permanent. The exhibit in Oslo was temporary. The Copenhagen exhibit did not address creationism explicitly but presented the positive approach, communicating the science of evolution, while addressing politically, culturally, and religiously motivated objections implicitly. Comparing the substantial media coverage in Norway and Denmark, the Norwegian press had more opinion pieces and less articles directly referring to the museum and the exhibit, while the Danish press was dominated by articles referring to or using the exhibit and the internet encyclopedia.[29]

Following the rise of creationism in Europe, the number of individual and organized anti-creationist events has also increased. Like European creationists, the groups and initiatives are varied and widely spread, mostly tied to local or national contexts or as responses to specific events. In that respect European anti-creationism is very much similar to its organized or semi-organized North American counterpart. Some inter-European initiatives and responses to creationist challenges have made an impact, especially the Council of Europe's Resolution 1580. During the Darwin year in 2009, scientists and science educators were exploring new ways of meeting the creationist challenge by moving grassroots activities more and more to the electronic and social media. On that point, many anti-creationists have finally caught up to well-organized and well-funded creationist groups in their adoption of political lobbying strategies that can meet their target audience on multiple platforms.

NOTES

1. Jason Rosenhouse and Glenn Branch, "Media Coverage of 'Intelligent Design,'" *BioScience* 56:3 (2006): 247–252.

2. Stefaan Blancke, Hans Henrik Hjermitslev, Johan Braeckman, and Peter C. Kjærgaard, "Creationism in Europe: Facts, Gaps, and Prospects," *Journal of the American Academy of Religion* 81:4 (2013): 996–1028.

3. Charles Darwin, "Macculloch. Attrib of Deity [Essay on Theology and Natural Selection]" (1838), 5, Manuscript ID: CUL-DAR71.53-59, University Library, Cambridge, in *The Complete Work of Charles Darwin Online*, ed. John van Wyhe (2002–), darwin-online.org.uk. See also Ronald L. Numbers, *Science and Christianity in Pulpit and Pew* (New York: Oxford University Press, 2007), 52, and Janet Browne, *Charles Darwin*, vol. 1: *Voyaging* (London: Jonathan Cape 1995), 129.

4. Thomas Henry Huxley, "Lectures on General Natural History," *Medical Times & Gazette* 17 (May 1856): 482.

5. Charles Darwin, *On the Origin of Species by Means of Natural Selection, or the Preservation of Favoured Races in the Struggle for Life* (London: John Murray, 1859), 483; Charles Darwin to Asa Gray, May 22, 1860, in *The Correspondence of Charles Darwin*, vol. 8, ed. Frederick Burkhardt et al. (Cambridge: Cambridge University Press, 1993), 223. See also Ronald L. Numbers, *Darwinism Comes to America* (Cambridge, MA: Harvard University Press, 1998), for the American appropriation of Darwin's argument with atoms flashing into elephants.

6. Arthur Keith, *Concerning Man's Origin* (London: Watts, 1927), vi–vii.

7. Ronald L. Numbers, *The Creationists: From Scientific Creationism to Intelligent Design*, expanded ed. (Cambridge, MA: Harvard University Press, 2006), 141–144, 162–166.

8. Abraham C. Flipse, "The Origins of Creationism in the Netherlands: The Evolution Debate among Twentieth-Century Dutch Neo-Calvinists," *Church History* 81:1 (2012): 104–147, 142–145; Jim Herrick, *Humanism: An Introduction* (Amherst, NY: Prometheus Books, 2005).

9. John A. Moore, "Dealing with Controversy: A Challenge to the Universities," *American Biology Teacher* 41:9 (1979): 544–547, 545–546.

10. Hee-Joo Park, "The Politics of Anti-Creationism: The Committees of Correspondence," *Journal of the History of Biology* 33:2 (2000): 349–370; Wayne A. Moyer, "The Problem Won't Go Away," *American Biology Teacher* 42:4 (1980): 234.

11. Hee-Joo Park, "The Creation-Evolution Debate: Carving Creationism in the Public Mind," *Public Understanding of Science* 10:2 (2001): 173–186.

12. Brian J. Alters and Craig E. Nelson, "Perspective: Teaching Evolution in Higher Education," *Evolution* 56:10 (2002): 1891–1901.

13. Chris Mooney and Matthew C. Nisbet, "Undoing Darwin," *Columbia Journalism Review* 44:3 (2005): 30–39.

14. Matthew C. Nisbet and Dietram A. Scheufele, "What's Next for Science Communication? Promising Directions and Lingering Distractions," *American Journal of Botany* 96:10 (2009): 1767–1778, 1772–1773.

15. Martyn Bond, *The Council of Europe: Structure, History and Issues in European Politics*, Global Institutions (Oxford: Routledge, 2011).

16. Parliamentary Assembly of the Council of Europe, "The Dangers of Creationism in Education" Resolution 1580, Oct. 4, 2007, §§2, 12–13, http://assembly.coe.int/main.asp?link=/documents/adoptedtext/ta07/eres1580.htm.

17. Peter C. Kjærgaard, "Western Front," *New Humanist* 123:3 (2008): 39–41. See also Hjermitslev and Kjærgaard, chapter 5 in this volume.

18. Parliamentary Assembly of the Council of Europe, "The Dangers of Creationism in Education," Doc 11375, Sept. 17, 2007, http://assembly.coe.int/main.asp?link=/documents/workingdocs/doc07/edoc11375.htm.

19. James A. Secord, "Focus: Darwin as a Cultural Icon. Introduction," *Isis* 100 (2009): 537–541; Steven Shapin, "The Darwin Show," *London Review of Books* 32 (2010): 3–9; Peter C. Kjærgaard, "After the Storm: Parties and Partisans in the Darwin Year," *Viewpoint: Newsletter of the British Society for the History of Science* 90 (2009): 1–3.

20. Peter C. Kjærgaard, "The Darwin Enterprise: From Scientific Icon to Global Product," *History of Science* 48:1 (2010): 105–122, 114–116.

21. Casper Andersen, Jakob Bek-Thomsen, Mathias Clasen, Stine Slot Grumsen, Hans Henrik Hjermitslev, and Peter C. Kjærgaard, "Evolution 2.0. The Unexpected Learning Experience of Making a Digital Archive," *Science & Education* 22:3 (2013): 657–675, 662–663.

22. Laura van Eperen and Francesco M. Marincola, "How Scientists Use Social Media to Communicate Their Research," *Journal of Translational Medicine* 9:1 (2011): 1–3; Brian Trench, "Towards an Analytical Framework of Science Communication Models," in *Communicating Science in Social Contexts*, ed. Donghong Cheng, Michel Claessens, Toss Gascoigne, Jenni Metcalfe, Bernard Schiele, and Shunke Shi (Houten: Springer Netherlands, 2008), 119–135.

23. Matthew C. Nisbet and Dietram A. Scheufele, "What's Next for Science Communication? Promising Directions and Lingering Distractions," *American Journal of Botany* 96:10 (2009): 1773.

24. Richard Dawkins, *The Greatest Show on Earth: The Evidence for Evolution* (London: Bantam Press, 2009), 436.

25. Tatjana Schnell and William J. F. Keenan, "Meaning-Making in an Atheist World," *Archive for the Psychology of Religion / Archiv für Religionspychologie* 33:1 (2011): 55–78.

26. Letters: Pollock to Adomnitei (Mar. 9, 2008); Costea to Pollock (Mar. 12, 2008); Pollock to Costea (Mar. 15, 2008). The correspondence is published by the European Humanist Federation on its website to disclose what it finds outrageous argumentation by creationists.

27. The Interacademy Panel,"IAP Statement on the Teaching of Evolution," 2006.

28. Peter C. Kjærgaard, "Ida and Ardi: The Fossil Cover Girls of 2009," *Evolutionary Review* 2 (2011): 1–9.

29. Hanne Strager and Peter C. Kjærgaard, "Is Darwin Dangerous? Museums, Media, and Public Understanding of Evolution," *Nordic Museology* 24 (2013): 98–115.

Reclaiming Science for Creationism

NICOLAAS A. RUPKE

In bringing this collection of essays on creationism in Europe to a close, I propose to focus on the question, Why did creationists do so well in the decades following World War II? What is the reason that creationist theory has been given the wide acceptance we have witnessed over the past half century or more, not just in North America, but also in Europe? How are we to understand the fact that advocates of creation belief so confidently speak of "scientific creationism" and have confronted Darwinian and neo-Darwinian evolution theory, meeting eyeball-to-eyeball with their evolutionist adversaries about issues of biology and geology?

In the historical literature, the evolution part of the binomial "creation versus evolution" has received an enormous amount of attention. A veritable "Darwin industry" has sprung up, especially during and in the wake of the 1959 centennial of *On the Origin of Species*. Over the past two decades or so, also the history of "creation," the other part of the "creation versus evolution" antithesis, has begun to receive serious scholarly attention. The classic study is Numbers's *The Creationists* (1992), which brought the considerable sophistication of Darwin historiography to bear on the development of creationism, focusing on America. In the second, much expanded edition of 2006, Numbers added a substantial chapter on "Creationism Goes Global," drawing attention to the existence of creationists and creationist institutions around the globe, especially also in Europe.

The present volume of essays significantly expands upon Numbers's pathbreaking work, showing the surprising spread of creationism in Europe and providing the national and "local" contexts and conditions, especially for the period that followed World War II. In full agreement with David N. Livingstone, Numbers, and others who have emphasized the importance of "location," the authors convincingly show that we should think of the European phenomenon of creationism not as a monolithic "European creationism" but as a multifaceted, variegated "creationism in Europe" that has to be understood in terms of local, regional, and national contexts of sociopolitical purposes, of different histories and personalities, and, last but not least, of different religious traditions.

This volume, moreover, brings to our attention the remarkable fact that the history of creationism in recent decades is a success story, not only in the United States, where many respectable academic institutions till this very day have remained openly and unashamedly Christian-religious in their avowed purposes, but also in Europe, in spite of the fact that just about all top-notch academic institutions have long adopted the secular philosophy of Darwinian theory. The question presents itself: How come that creationism has also done well in Europe in the decades following World War II?

Creationism Reclaims Its Scientific Credentials

It may, perhaps, be enlightening to place the noticeable resurgence of creationism during the second half of the twentieth century and into the beginning decades of the twenty-first century in a broader context than that of its own, historic moments. There is of course nothing wrong with pointing out the seminal importance of the twelve-volume set, *The Fundamentals*, of 1910, in the United States, and of such public spectacles as the Scopes trial of 1925. What's more, it is entirely legitimate to acknowledge the traction that creationism acquired as a result of the classic *The Genesis Flood* (1961) by the conservative evangelical Old Testament scholar John C. Whitcomb and the Southern Baptist professor of hydraulic engineering Henry M. Morris.

However, the creationism that followed in the wake of *The Fundamentals* and as it manifested itself in the Scopes trial objected to science, which was spurned as a dangerous form of anti-biblical modernism. By contrast, the later creationism of Morris reclaimed science, in some sense going back to the times that mainstream science itself was creationist, in early modern times. With Morris, the creationist movement broke out of its antiscientific shell to reclaim its early modern scientific identity. Admittedly, in the early twentieth century, creationists such as the Canadian Seventh-day Adventist George McCready Price, in his textbook *The New Geology* (1923) and similar publications, already had kept the tradition of scientific creationism alive, a tradition that before him, during much of the nineteenth century, had been established by the scriptural geologists in the United Kingdom, among whom were Granville Penn, George Bugg, Andrew Ure, George Fairholme, John Murray, George Young, and William Rhind; but Price's success was limited. So, why did Morris with his *The Genesis Flood* succeed? Why did his claim to be scientific have such a remarkable effect?[1]

The answer, I believe, is that Morris and his brand of creationism were part of the larger phenomenon of neo-catastrophism that emerged after World War II. It was the synergy that the new creationism experienced with various forms of neo-catastrophism, both maverick and mainstream scientific, that provided the opportunity and credibility to put itself forward as "scientific creationism." The neo-catastrophist revival emboldened the creationist movement to reclaim its early modern scientific identity.

Perhaps it should be added that this synergy with neo-catastrophism was only one of the factors that contributed to the post-World War II success of scientific creationism in Europe. There were several other circumstances, each and all of them addressed in the various chapters of this book. First, there were cognitive developments in theology that favored biblical literalism and moved away from the hermeneutics of an exegetical accommodation of the Bible to the secular earth and life sciences. A directly related, second feature was the formation of new, conservative church factions and breakaway churches with their spinoff institutions of denominational high schools, teacher training colleges, and theological seminaries that functioned as greenhouses in which biblical literalism luxuriated. Moreover, a third factor was that these new groups proved savvy in making effective use of modern electronic media for spreading their message. Still, here we focus on how creationism, in particular young-earth creationism, regained its original scientific self.

During early modern times and well into the nineteenth century, apocalyptic catastrophe thinking was an integral part of the natural philosophy of the earth and, with it, of creationist theory, either young earth or, later on, old earth. To most early modern natural philosophers, the history of the earth covered the same stretch of time as the history of humankind—they are coeval—and the purpose of the former is that it provides the stage on which the latter can play out. In many and varied ways, earth history was considered part of human history, starting at the creation of the world, passing through the universal deluge, and soon to end—many believed—in a final conflagration. *Erdgeschichte* was seen in terms of *Heilsgeschichte*, that is, of the historical process of human salvation as taught by the Bible—from Genesis to Revelations. Catastrophic events from the past, such as the biblical deluge, as known from written records, were interpreted as human catastrophes, ultimately caused by humanity's sinful ways, and they might be portents of things to come—milestones in a goal-directed, teleological history with an *Endzeit*, a final, apocalyptic fire. Thus catastrophist doomsday thought was an integral and incontrovertible part of earth and human history combined.

From early Victorian days on, however, catastrophism was replaced with a model explaining the origins of geologic and biological phenomena in terms of gradual and uniform change. This gradualist model was championed by the

Scottish geologist Charles Lyell and later incorporated by Darwin into his theory of evolution, thus becoming the ruling paradigm in the earth and life sciences.

The Neo-Catastrophist Turn

Following World War II, a spectacular revival of catastrophist thinking in "proper" science took place. In cosmology, the big bang theory, already enunciated in 1927 by the Jesuit priest Georges Lemaître, astronomer and professor of physics at the Catholic University of Louvain, now gained ascendancy over the steady-state model and was explicitly connected with the biblical story of creation by the American astronomer and cosmologist Robert Jastrow. Through the fifties, sixties, and seventies, in planetary science, the notion of cometary impact reentered, most famously when Nobel laureate and Columbia University cosmochemist Harold Urey published his paper "Cometary Collisions and Geological Periods." Questions of past global annihilation and possible imminent, future disaster recaptured center stage. Even in geology, still tightly holding on to traditional Lyellian gradualism, "discontinuity" and "the rare event" became acceptable catchwords, taking on new meanings.[2]

In 1963 Columbia University geologist Norman D. Newell, who also was curator of invertebrate paleontology at the American Museum of Natural History in New York, published a by now classic paper "Crises in the History of Life," returning to the old Cuvierian observation that whole groups of animals have simultaneously died out, reintroducing the suggestion that cataclysmal events may have been responsible. Newell trained several students who went on to become leading paleontologists in their own right, among them Niles Eldredge, who, like his mentor, worked as curator of invertebrates at the American Museum of Natural History, and Harvard University's Stephen Jay Gould. In 1977, Eldredge and Gould put forward the discontinuous "fits and starts" model of evolution, known as the punctuated equilibria theory.[3]

This nongradualist model of earth history received a boost from an unlikely ally, namely mathematics. In the late 1960s and early 1970s, the French topologist René Thom gained popularity for his mathematical catastrophe theory. This is a branch of dynamical systems theory that describes phenomena in which small changes lead to sudden and major shifts. An example involves turbidity currents, when gradual, piecemeal sediment accumulations at the edge of the continental slope may lead to sediment overload and slope instability, triggering a catastrophic event of massive sediment redistribution by density currents.[4]

At this time, too, in the 1960s and early 1970s, as part of the Apollo program, the study of the moon's impact craters added to the search of such scars on earth.

The notion that the moon's craters are due to volcanic eruptions—lunar vulcanism—had been defended as late as 1944 by the American geologist Josiah Edward Spurr, who dismissed lunar impact on the basis that the earth lacks an impact record—a view that persisted well into the 1960s; and, indeed, on our earth only remnants of the impact record remain. When, however, the Caltech astrogeologist Eugene Shoemaker conducted a classic study of Meteor Crater in Arizona and pioneered the global search for crater phenomena, unexpectedly many of them were found. "Shatter cones" and other diagnostic impact features were added to the vocabulary of catastrophist geology. By 1990, more than one hundred structures on earth had been identified as impact craters, in the United States and Canada, in Germany, in South Africa, and indeed across the globe.[5]

Impact phenomena and mass extinctions were sensationally linked by Luis Alvarez, a Nobel Laureate in physics from Berkeley, and his son, the geologist Walter Alvarez, professor of earth and planetary sciences at Berkeley, who put forward the hypothesis, published in *Science* in 1980, that an asteroid struck the earth 65 million years ago, bringing about a mass extinction and the close of the Mesozoic era. The hypothesis appeared corroborated by the later discovery of a major impact crater of the right age on the Yucatan Peninsula in Mexico, the Chicxulub Crater. Walter Alvarez's book, *T. Rex and the Crater of Doom* (1997), contributed to the popular spread of the notion that the riddle of the extinction of the dinosaurs is now solved, having been brought about by a giant meteorite impact.[6]

With a rapidly growing enthusiasm for asteroid studies in the earth and planetary sciences, the geologic establishment began giving up its Lyellian opposition to catastrophe thinking. Already in 1965, Gould openly asked: "Is uniformitarianism necessary?" answering his question in the negative and suggesting that "substantive uniformitarianism" should be abandoned. In the meantime, doomsday science has progressed to a further stage, adding to the identification of possible causes of global havoc several initiatives toward their prevention. After all, these events should not be regarded as unavoidable "acts of God." From the early 1980s on, impact hazard and prevention gained the attention of NASA and bodies concerned with national and international security. Surveys of near earth objects (NEOs) were conducted and space missions organized with, among other things, the objective to devise strategies of NEO defense, deflection, or interception. The word *tsunami*, previously restricted to the technical jargon of marine geologists and oceanographers, entered public discourse, as impact hazards were analyzed and the 2004 tsunami, caused by a deep ocean earthquake that devastated coastal regions around the Indian Ocean, provided in miniature an image of what an asteroid impact wave might cause around the globe.[7]

In recent years, scientific doomsday literature has surged. Remarkable about this trend is that doomsday warnings no longer come from just popular and pseudoscientific sources but also from mainstream science. A representative instance is *Our Final Hour,* which carries the subtitle: *A Scientist's Warning: How Terror, Error, and Environmental Disaster Threaten Humankind's Future in this Century—on Earth and Beyond* (2003). It is written by none other than Martin Rees, the onetime Plumian professor of astronomy and experimental philosophy at Cambridge University, and president of the Royal Society of London. Today's scientific establishment shares the doomsday anxieties that used to be the domain of religion and science fiction. "The end" has entered scientific discourse as a serious subject of research and science policy.[8]

In 2010, *Scientific American* devoted a special issue to our fascination with humanity's "final hour." Considering the odds of modern civilization coming to an apocalyptic end, this issue lists eight possible doomsday scenarios. Ranging in their effects from "moderate devastation" to "total extermination," they are a solar superstorm, runaway global warming, a killer pandemic, the eruption of a supervolcano, nuclear war, a nearby gamma-ray burst, a giant asteroid impact, and—extremely unlikely—a bubble nucleation "whereby our universe flips into a new state with different fundamental forces." The bubble scenario should not keep anyone awake at night; nor should a cosmic gamma-ray blast. The other threats are considered of sufficiently high probability, however, to require the urgent attention of national and international policy makers. Especially the interconnected phenomena of global warming, climate change, and sea-level rise have goaded scientists and politicians into action, for the most part in the form of alarm bell reports (rather than binding international agreements).[9]

Creationism and the Neo-Catastrophist Turn

Unlike geology, Victorian planetary science never abandoned catastrophist doomsday thinking altogether and, for many decades after the demise of diluvialism, provided shelter to global catastrophe scenarios, past and future. The English astronomer George Howard Darwin postulated in 1879 that the moon had originated as a chunk of the earth, ripped away when in its early days the earth was believed to have rotated at a velocity of about one rotation in five hours. Others additionally connected the catastrophic birth of the moon to continental drift, which flood geologists and young-earth creationists attributed to the after effects of the deluge. By and large, however, the doomsday discourse survived primarily in popular literature on astronomy.[10]

Other maverick catastrophists, after World War II, drew for their doomsday scenarios on astronomy and astrophysics. The Russian-born Jewish psychiatrist Immanuel Velikovsky saw in erratic satellites and planets the cause of global, catastrophic havoc and presented evidence for geologic catastrophes in his *Earth in Upheaval* (1955). All of these authors yearned for acceptance by mainstream science, and Velikovsky took pride in intellectual support given to him by Albert Einstein. Opposition by the scientific mainstream led to the so-called Velikovsky Affair when a cultlike following, especially among the counterculture young of the 1960s and 1970s, turned him into an antiestablishment hero.[11]

For a number of years, I myself took part in these developments, when in 1961, as an aspiring young-earth creationist, but still in high school, I read *Earth in Upheaval.* I learned about Price and corresponded with him; learned about Morris and read *The Genesis Flood;* then, as an undergraduate student, worked on the history of biblical catastrophism by reading John Ray, John Woodward, Jean Andre Deluc, even George Fairholme; and reintroduced, together with new, current examples, the old argument for cataclysmal sedimentation in the form of polystrate dendrolites. At the same time, I enthusiastically took part in mainstream neo-catastrophist sedimentology, just then gaining ground in such forms as the turbidity current theory put forward by my mentor Philip Henry Kuenen. In many ways, neo-catastrophist science seemed to reconfirm a Judeo-Christian picture of the history of the world. It has given succor and support to creationism—young-earth, old-earth, also the more recent intelligent design movement.[12]

NOTES

1. The scriptural geologists are discussed in detail by Terry Mortenson, *The Great Turning Point: The Church's Catastrophic Mistake on Geology—Before Darwin* (Green Forest, AR: Master Books, 2004).

2. For further details, see Nicolaas Rupke, "The Doomsday Discourse in the Earth- and Planetary Sciences, 1700-Present," in *Historical Disasters in Context: Science, Religion and Politics,* ed. Andrea Janku et al. (New York: Routledge, 2012), 115–139; Robert Jastrow, *God and the Astronomers* (New York: Norton, 1978); Harold C. Urey, "Cometary Collisions and Geological Periods," *Nature* 242 (1973): 32–33.

3. Norman D. Newell, "Crises in the History of Life," *Scientific American* 208 (1963):

76–92. See also his earlier "Catastrophism and the Fossil Record," *Evolution* 10 (1956): 97–101; Stephen Jay Gould and Niles Eldredge, "Punctuated Equilibria: The Tempo and Mode of Evolution Reconsidered," *Paleobiology* 3 (1977): 115–151. For an early, neo-catastrophist reinterpretation of the stratigraphic record, see Derek V. Ager, *The Nature of the Stratigraphic Record* (London: Macmillan, 1973).

4. See David Aubin, "Forms of Explanations in the Catastrophe Theory of René Thom: Topology, Morphogenesis, and Structuralism," in *Growing Explanations: Historical Perspective on the Sciences of Complexity*, ed. M. N. Wise (Durham: Duke University Press, 2004), 95–130.

5. Eugene M. Shoemaker, "Impact Mechanics at Meteor Crater, Arizona," in *The Moon, Meteorites and Comets*, ed. B. M Middlehurst and G. P. Kuiper (Chicago: University of Chicago Press, 1963), 301–336; "Asteroid and Comet Bombardment of the Earth," *Annual Review of Earth & Planetary Sciences* 11 (1983): 461–494. For a general history, see Richard J. Huggett, *Catastrophism: Asteroids, Comets, and Other Dynamic Events in Earth History* (London: Verso, 1997).

6. L. W. Alvarez, W. Alvarez, F. Asaro, and H. V. Michel, "Extraterrestrial Cause of the Cretaceous-Tertiary Extinction," *Science* 208 (1980): 1095–1108.

7. Stephen Jay Gould, "Is Uniformitarianism Necessary?" *American Journal of Science* 263 (1965): 223–228.

8. In other academic fields, too, reputable scholars produce titles such as John Leslie, *The End of the World: The Science and Ethics of Human Extinction* (London: Routledge, 1996), and John Polkinghorne and Michael Welker, eds., *The End of the World and the Ends of God* (Harrisburg, PA: Trinity Press International, 2000).

9. "The End: Or Is It? The Eternal Fascinations—and Surprising Upsides—of Endings," *Scientific American*, Sept. 2010, 83.

10. Nicolaas A. Rupke, "Eurocentric Ideology of Continental Drift," *History of Science* 34 (1996): 251–272.

11. Immanuel Velikovsky, *Earth in Upheaval* (New York: Doubleday, 1955). Other notable maverick catastrophists include the American college professor of history Charles H. Hapgood (*Earth's Shifting Crust: A Key to Some Basic Problems of Earth Science* [New York: Pantheon, 1958], which featured a foreword by Albert Einstein) and the American geographer Donald W. Patten (*The Biblical Flood and the Ice Epoch* [Seattle: Pacific Meridian, 1966]). Hapgood became entangled in spirit communication (purportedly with the spirits of Albert Einstein, Mark Twain, Christ, and Vishnu) and in the Acámbaro figurines hoax, depicting the contemporaneity of humans with dinosaurs. Patten, for his part, has been active in the Christian fundamentalist search for Noah's Ark in Armenia. See, e.g., *The Velikovsky Affair—Scientism versus Science*, ed. Alfred de Grazia, Ralph E. Juergens, and L. C. Stecchini, 2nd ed. (Princeton, NJ: Metron Publications, 1978).

12. Nicolaas A. Rupke, "Prolegomena to a Study of Cataclysmal Sedimentation," *Creation Research Society Annual* 3 (1966): 16–37; reprinted in Walter E. Lammerts, ed., *Why Not Creation?* (Presbyterian and Reformed Publishing, 1970), 141–179. Perhaps it should be added that at a later stage I left the creationist movement.

A Note on Sources

STEFAAN BLANCKE

The main reason for publishing this book was that we found that the literature lacked a much-needed coherent study of creationism in Europe. This does not mean, however, that the topic has been completely neglected. In recent years, the group of scholars studying or discussing the issue has grown gradually larger, resulting in more publications, both in local or national venues and in international peer-reviewed journals. The fact that this group had reached a critical mass provided yet another reason why we thought it appropriate to instigate this project. The following note on sources should be regarded as a guide to, and not as an exhaustive list of, this literature. As such, this discussion includes what we consider to be the most significant works on creationism, divided according to different topics and perspectives.

Whereas creationism in Europe has only recently drawn academic attention, American creationism has been the subject of a number of detailed and engaging historical studies. Indispensable is Ronald L. Numbers's *The Creationists: From Scientific Creationism to Intelligent Design* (Cambridge, MA: Harvard University Press), originally published in 1992 and reissued as an expanded edition in 2006. This book offers an autoritative and detailed study of the history of creationism around the world but primarily focuses on North America. Edward J. Larson has written two books on the history of American creationism, *Summer for the Gods: The Scopes Trial and America's Continuing Debate over Science and Religion* (Cambridge: Basic Books, 2006 [1997]), which focuses on the Scopes trial in 1925, its impact on American culture, and its role in the American popular imagination, and *Trial and Error: The American Controversy over Creation and Evolution* (New York: Oxford University Press, 2003 [1985]), which highlights the trials that marked the creation/evolution debates in the United States. Several recent books cover early creationism from several different angles: *God—or Gorilla: Images of Evolution in the Jazz Age* (Baltimore: Johns Hopkins University Press, 2008) by Constance Areson Clark looks at popular images; the political scientist Michael Lienesch published *In the Beginning: Fundamentalism, the Scopes Trial, and the Making of the Antievolution Movement* (Chapel Hill: University of North Carolina Press, 2009); the attitudes of African Americans and women constitute the main topic of *American Genesis: The Antievolution Controversies from Scopes to Creation Science* (New York: Oxford University Press, 2012) by Jeffrey P. Moran; and, finally, Adam R. Shapiro's *Trying Biology: The Scopes Trial, Textbooks, and the Antievolution Movement in American Schools* (Chicago: University of Chicago Press, 2013) discusses the battle over evolution in public schools. Scientific creationism is discussed by James Moore in "The Creationist Cosmos of Protestant Fundamentalism," in *Fundamentalisms and Society: Reclaiming the Sciences, the Family, and Education*, vol. 2 of *The Fundamentalism Project*, ed. Martin E. Marty and R. Scott Appleby (Chicago: University of Chicago Press, 1993), pp. 42–72, and by Christopher P. Toumey in *God's Own Scientists: Creationists in a Secular Age* (New Brunswick, NJ: Rutgers University Press, 1994). The history of the intelligent design movement has been described in detail by Barbara Forrest and Paul R. Gross in *Creationism's Trojan Horse: The Wedge of Intelligent Design* (New York: Oxford University Press, 2007 [2004]). Other indispensable volumes on intelligent design are Robert T. Pennock's *Tower of Babel: The Evidence against the New Creationism* (Cambridge, MA: MIT Press, 1999) and *The Evolution-Creation Struggle* (Cambridge, MA: Harvard University Press, 2005) by Michael Ruse.

Darwin and the Bible: The Cultural Confrontation (Boston: Penguin, 2009), edited by Richard H. Robbins and Mark N. Cohen, collects essays on the relationship between evolution and Christianity with the focus primarily on North America. Peter J. Bowler's *Monkey Trials and Gorilla Sermons: Evolution and Christianity from Darwin to Intelligent Design* (Cambridge, MA: Harvard University Press, 2007) discusses the relationship between evolutionary theory and Christianity in Great Britain and the United States since the publication of Charles Darwin's *On the Origin of Species* in 1859 until today. For those who are interested in consulting original sources, a good place to start is *Evolution and Creationism: A Documentary and Reference Guide*, edited by Christian C. Young and Mark A. Largent (Westport, CT: Greenwood Press, 2007), a compilation of relevant historical documents regarding the American creation/evolution debates. For analyses of the concerns of creationist individuals in America and Britain we suggest Jason Rosenhouse's *Among the Creationists: Dispatches from the Anti-evolutionist Frontline* (New York: Oxford University Press, 2012) and *Doubting Darwin: Creationism and Evolution Skepticism in Britain Today*, by Robin Pharoah et al. (London: Theos/ESRO, 2009).

For an overview of studies of creationism outside the United States, see the relevant chapters of Ronald L. Numbers's *The Creationists* and his chapter "Myth

24: That Creationism Is a Uniquely American Phenomenon," in *Galileo Goes to Jail and Other Myths about Science and Religion* (Cambridge, MA: Harvard University Press, 2009), edited by Ronald L. Numbers. Simon Coleman and Leslie Carlin broadened the scope with an edited volume, *The Cultures of Creationism: Anti-evolutionism in English-Speaking Countries* (Aldershot: Ashgate, 2004), a collection of essays on creationism in Britain, North America, Australia, New Zealand, and Kenya. *Science and Religion: New Historical Perspectives* (Cambridge: Cambridge University Press, 2010), edited by Thomas Dixon, Geoffrey Cantor, and Stephen Pumfrey, contains solid essays on the relationship between science and religion. This compilation includes contributions by Salman Hameed on evolution and creationism in the Islamic world and by Bronislaw Szerszynski on differences between creationist movements and antievolutionary sympathies in Europe and North America.

As to the historical position of evolutionary theory in the different European countries, one can consult the two volumes of *The Reception of Charles Darwin in Europe* (London: Continuum, 2008) edited by Eve-Marie Engels and Thomas F. Glick, and the companion volumes *The Literary and Cultural Reception of Charles Darwin in Europe* (London: Bloomsbury, 2014), edited by Thomas F. Glick and Elinor Shaffer. *Negotiating Darwin: The Vatican Confronts Evolution, 1877–1902*, by Mariano Artigas, Thomas F. Glick, and Rafael A. Martínez (Baltimore: Johns Hopkins University Press, 2006), discusses the early reactions of the Vatican to evolution. Short reviews of recent creationist activities in Europe can be found in articles by Athel Cornish-Bowden and María Luz Cárdenas ("The Threat from Creationism to the Rational Teaching of Biology," *Biological Research* 40:2 [2007]: 113–122), Almut Graebsch and Quirin Schiermeier ("Antievolutionists Raise Their Profile in Europe," *Nature* 444:7121 [2006]: 803), Peter C. Kjærgaard ("Western Front," *New Humanist* 123:3 [2009]: 39–41), and Ulrich Kutschera ("Darwinism and Intelligent Design: The New Anti-evolutionism Spreads in Europe," *Reports of the National Center for Science Education* 23:5–6 [2003]: 5–6). An extensive review essay that highlights particular findings of this volume is "Creationism in Europe: Facts, Gaps and Prospects" by Stefaan Blancke, Hans Henrik Hjermitslev, Johan Braeckman, and Peter C. Kjærgaard, published in the *Journal of the American Academy of Religion* 81:4 (2013): 996–1028.

Also available are historical studies discussing creationism in a particular country. Kutschera has published several short reports on creationism in Germany (e.g., "The German Anti-Darwin Industry: Twentieth Anniversary of a Multi-media Business," *Reports of the National Center for Science Education* 28:1 [2008]: 12–13), and Polish creationism is discussed by Bartosz Borczyk ("Creationism and the Teaching of Evolution in Poland," *Evolution: Education and Outreach* 3:4 [2010]: 614–620). Creationism in Turkey has been covered in articles by Taner Edis ("Islamic Creationism in Turkey," *Creation/Evolution* 14:1 [1994]: 1–12, and "Cloning Creationism in Turkey," *Reports of the National Center for Science Education* 19:6 [1999]: 30–35) and by Ümit Sayin and Aykut Kence ("Islamic Scientific Creationism," *Reports of the National Center for Science Education* 19:6 [1999]: 18–20, 25–29). Moreover, Martin Riexinger has written two articles on Islamic opposition to evolution ("Propagating Islamic Creationism on the Internet," *Masaryk University Journal of Law and Technology* 2:2 [2008]: 99–112; "Islamic Opposition to the Darwinian Theory of Evolution," in *Handbook of Religion and the Authority of Science*, ed. Olav Hammer and James R. Lewis [Leiden: Brill, 2011], 484–509). Both Abraham C. Flipse and Stefaan Blancke have written articles on creationism in the Netherlands, each covering a different period ("The Origins of Creationism in the Netherlands: The Evolution Debate among Twentieth-Century Dutch Neo-Calvinists," *Church History* 81:1 [2012]: 104–147, and "Creationism in the Netherlands," *Zygon. Journal of Religion and Science* 45:4 [2010]: 791–816, respectively). From a sociological perspective, focusing primarily on how the creation-evolution controversy is represented in the media, Joachim Allgaier has written three articles on recent creationist incidents in the United Kingdom (with Richard Holliman, "The Emergence of the Controversy around the Theory of Evolution and Creationism in UK Newspaper Reports," *Curriculum Journal* 17:3 [2006]: 263–279; "Scientific Experts and the Controversy about Teaching Creation/Evolution in the UK Press," *Science & Education* 19:6 [2010]: 797–819; and "Networking Expertise: Discursive Coalitions and Collaborative Networks of Experts in a Public Creationism Controversy in the UK," *Public Understanding of Science* 21:3 [2012]: 299–313).

In several European countries works in the native languages have been published that discuss creationism, either in general or in that particular country. Ulrich Kutschera has written two books on creationism in Germany, *Kreationismus in Deutschland. Fakten und Analysen* (Berlin: LIT Verlag, 2007) and *Streikpunkt Evolution. Darwinismus und Intelligent Design* (Berlin: LIT Verlag, 2007). Also in Germany, Dittmar Graf has edited a volume which tackles creationism in Europe (*Evolutionstheorie—Akzeptanz und Vermittlung im Europäischen Vergleich* [Berlin and Heidelberg: Springer, 2012]). For France, several books discuss creationism in general, including Jacques Arnould's *Les Créationnistes* (Paris: Cerf/Fides, 1996) and *Dieu versus Darwin. Les Créattionistes Vont-ils Triompher de la Science?* (Paris: Albin Michel, 2007) and Thomas Lepeltier's *Darwin Hérétique. L'éternel Retour du Créationnisme* (Paris: Seuil, 2007).

Quantitative surveys also gauge the popularity of creationist beliefs in Europe. In 2010, Angus Reid Public Opinion found that 51 percent of Americans, 22 percent of Canadians, and 17 percent of Britons think God created human beings in their present form within the past ten thousand years ("Origin of Humans: Americans Are Creationists; Britons and Canadians Side with Evolution," www.visioncritical.com/wp-content/uploads/2010/07/2010.07.15_Origin.pdf, released July 15, 2010). Two years later, Angus Reid Public Opinion more or less confirmed these numbers (47 percent of Americans, 24 percent of Canadians, and 16 percent of Britons) ("Britons and Canadians More Likely to Endorse Evolution than Americans," www.angus-reid.com/wp-content/uploads/2012/09/2012.09.05_CreEvo.pdf, released Sept. 5, 2012). The European Commission's 2005 Eurobarometer survey *Social Values, Science and Technology*, Special Eurobarometer 225, Wave 63.1, constitutes a comprehensive survey of Europeans' views on science and technology, including the acceptance of human evolution in thirty-two European countries (Brussels: European Commission, 2005; http://ec.europa.eu/public_opinion/archives/ebs/ebs_225_report_en.pdf, accessed Oct. 22, 2012). The data of this survey have been incorporated in the study by Jon D. Miller, Eugenie C. Scott, and Shinji Okamoto ("Public Acceptance of Evolution," *Science* 313 [2006]: 765–766). This often-cited review article spans several surveys on the acceptance of (human) evolution in thirty-four countries. The Ipsos Global @dvisory, a comprehensive online survey of religious views, including views on evolutionism and creationism, in twenty-four countries found that while 51 percent of the citizens surveyed believed in a supreme being and 51 percent believed in an afterlife, only 28 percent identified as creationists ("Supreme Being(s), the Afterlife and Evolution," www.ipsos-na.com/news-polls/pressrelease.aspx?id=5217, released Apr. 25, 2011). A discussion of the problems concerning polling creationist beliefs can be found in the previously mentioned articles by Szerszynski and by Stefaan Blancke, Hans Henrik Hjermitslev, Johan Braeckman, and Peter C. Kjærgaard.

Contributors

JOACHIM ALLGAIER is a sociologist and media and communications researcher. He is a senior scientist at the Institute of Science, Technology and Society Studies at the Alpen-Adria-Universität Klagenfurt in Austria. He was Honorary Fellow at the School of Journalism and Mass Communication at the University of Wisconsin at Madison, and was employed at the Research Center Jülich in Germany, the University of Vienna in Austria, and the Open University in the United Kingdom, where he was awarded a PhD in sociology in 2008. He studied sociology at LMU Munich in Germany, and Science and Technology Studies at Maastricht University in the Netherlands. His research interests are science and technology studies, scientific controversies, public communication of science, technology and medicine, and (new) media, communication, and popular culture. Recent articles appeared in *BioScience, Public Understanding of Science, Science Communication, Science and Education,* and *Journalism Practice.*

STEFAAN BLANCKE is a postdoctoral fellow in the Department of Philosophy and Moral Sciences at Ghent University, Belgium. He obtained his PhD in 2011 with an interdisciplinary study of creationism in Europe. He has published articles on the topic of creationism and the relation between science and religion in several academic journals, including *Zygon: Journal of Religion and Science, Biology and Philosophy, Journal of Biological Education, Science & Education, Foundations of Science, Journal of Religious History,* and *Journal of the American Academy of Religion.* His main interests are creationism, science and religion, outreach and education of evolutionary theory, public perceptions of biotechnology, and cognitive and evolutionary approaches to religion, pseudoscience, and the philosophy and history of biology.

BARTOSZ BORCZYK is an associate professor in the Department of Evolutionary Biology and Conservation of Vertebrates at the University of Wroclaw. His research interests involve mechanisms of morphological evolution, vertebrate comparative anatomy, and herpetology. He is also interested in the evolution-creationism struggle and the popularization of science.

JOHAN BRAECKMAN is a professor of philosophy at Ghent University, where he teaches courses on the history of philosophy, the history of biology, philosophical anthropology, and critical thinking. For five years, he also taught courses on bioethics and the philosophy of science at Amsterdam University, the Netherlands. He studied philosophy at Ghent University (Belgium), human ecology at the Free University of Brussels, and environmental history at the University of California, Santa Barbara. His doctoral thesis is on the influence of the theory of evolution on philosophy. Braeckman has published articles and books on Darwin and evolutionary biology, cloning, the history of philosophy, bioethics, and critical thinking. He is codirector of the research group The Moral Brain, which consists of ten collaborators with backgrounds in moral sciences, philosophy, psychology, biology, medicine, and other disciplines. For more details, see www.johanbraeckman.be.

JESÚS I. CATALÁ-GORGUES is associate professor of history of science at the University CEU Cardenal Herrera in Valencia, Spain. He studies the history of natural history, geologic surveys, and evolutionism in Spain and Portugal during the nineteenth and twentieth centuries. He is joint editor of *Darwin. El seu temps, la seua obra, la seua influència* (with Víctor Navarro, 2010), and joint author of *Els nostres naturalistes* (with Josep Maria Camarasa, 2008). He is currently working on the ideological framework of societies of natural history in twentieth-century Spain.

ABRAHAM C. FLIPSE is a historian of science at VU University Amsterdam. He is especially interested in the historical relationship between science and religion, the history of physics, and the history of (confessional) universities. His publications include books and articles on the history of the originally Calvinist "Free University" in Amsterdam, religious scientists in the early twentieth century, the debate about science and religion among Dutch Calvinists and Roman Catholics, and the origins and present state of creationism in the Netherlands.

BARBARA FORREST is professor of philosophy at Southeastern Louisiana University. With Paul R. Gross, she coauthored *Creationism's Trojan Horse: The Wedge of Intelligent Design*, 2nd ed. (2007 [2004]), detailing the Discovery Institute's strategy to promote intelligent design creationism. She has published in both scientific and legal journals and the popular media concerning the intelligent design movement in the United States. In 2005 she testified for the plaintiffs in *Kitzmiller et al. v. Dover Area School District*. She serves on the board of directors of the National Center for Science Education and the board of trustees of Americans United for Separation of Church and State.

She cofounded the Louisiana Coalition for Science, which works to protect science education in Louisiana.

THOMAS F. GLICK is professor of history at Boston University. He has written extensively on the reception of Darwinism, relativity, and psychoanalysis in Latin countries and is the editor of *The Reception of Charles Darwin in Europe* (with Eve-Marie Engels, 2008) and *What about Darwin?* (2010).

HANS HENRIK HJERMITSLEV is assistant professor of social science at University College South Denmark in Aabenraa. In 2010 he received his PhD in the history of science from the Department of Science Studies at Aarhus University. Hjermitslev's research interests include the reception of Darwinism in Denmark, the history of creationism in Scandinavia, and the popularization of science in late nineteenth- and early twentieth-century Denmark. He has published articles in *Journal of the History of Ideas, Annals of Science, Centaurus, Journal of the American Academy of Religion,* and *Science & Education* and contributed to *The Reception of Charles Darwin in Europe,* vols. 1 and 3 (ed. T. F. Glick et al., 2008 and 2014), and *Science in Denmark: A Thousand-Year History* (ed. P.C. Kjærgaard et al., 2008). He is currently investigating appropriations of Herbert Spencer in Scandinavia and the circulation of scientific knowledge in rural Denmark in the decades around 1900.

UWE HOSSFELD studied biology and the history of science as an undergraduate at the Friedrich-Schiller-University of Jena, Germany. He received his Magister Artium in 1994, his PhD in 1996, and his Habilitation in 2003, also from the University of Jena. Between 1993 and 1997 he taught at the University of Tübingen and at the University of Göttingen. Between 1998 and 2006 he was an assistant professor at the University of Jena (Ernst Haeckel House and Department of History). Since 2006 Hoßfeld has been the head of the working group "Didactics of Biology" at the University of Jena, and in 2009 he accepted a full professorship for didactics of biology. His fields of interest are the history of biology (evolutionary biology, morphology, embryology/evo-devo, "race movement" during the Third Reich), didactics of biology, and the history of the university.

PETER C. KJÆRGAARD is professor of evolutionary studies and director of the Centre for Biocultural History at Aarhus University. Originally trained in history of science and ideas, he obtained his PhD after completing a dissertation on Victorian debates about science. He has published widely in general history of science. Currently his research focuses on all aspects of evolution, with a particular interest in human evolution. He has held fellowships at a number of prestigious universities around the world, including the

University of Cambridge, University of Oxford, École Normale Supérieure, Paris, Harvard University, and the University of California, Los Angeles. He is editor-in-chief of the large-scale public engagement projects evolution.dk and darwinarkivet.dk, and scientific consultant on the human evolution exhibition at Moesgaard Museum.

ULRICH KUTSCHERA is a professor of plant physiology and evolutionary biology at the University of Kassel (Germany) and a visiting investigator in the Department of Plant Biology, Carnegie Institution for Science, Stanford, California. He is best known for his work on the mechanism of phytohormone-mediated cell expansion, his contributions to the taxonomy and molecular phylogeny of aquatic annelids, and his publications on evolution versus creationism from a historical-philosophical perspective. He has published more than two hundred scientific papers, nine books, and several YouTube-Science Videos. (For details, see *Wikipedia*, Ulrich Kutschera, in German.)

THOMAS LEPELTIER is a French independent scholar working on the history and philosophy of science. His research interests include the history of creationist and evolutionary theories and the history of modern cosmology. He has published various articles and books on these subjects, among them *Darwin hérétique. L'éternel retour du créationnisme* (Seuil, 2007), *Univers parallèles* (Seuil, 2010), and *La face cachée de l'Univers. Une autre histoire de la cosmologie* (Seuil, 2014). He has also started to investigate animal ethics and published his first book on this now hotly debated subject: *La révolution végétarienne* (Éditions Sciences Humaines, 2013).

GEORGY S. LEVIT (University ITMO, St. Petersburg) studied history and history of philosophy at the Herzen State Pedagogical University of Russia, St. Petersburg and received his PhD from the University of Oldenburg (Germany), where his research was on the history of the biosphere theory and biogeochemistry. Since 2002 he served as a research scientist investigating the history and philosophy of the life sciences. In 2008 Levit joined the University of King's College (Canada). He has published on a wide range of topics, including the history of evolutionary theory in Russia and Germany, global issues, and the application of Darwinism in the social sciences. There is a common thread, however, running through these diverse topics—the interaction of science with the social-cultural environment.

INGA LEVIT studied history and sociology at the Herzen State Pedagogical University of Russia, St. Petersburg (MA 1992), and completed her PhD in 1999 at the same university. She was assistant professor at the Department of Religious Studies at Herzen University (2002–9), after that becoming a postdoctoral fellow at the Department of Philosophy of Religion and Religious Studies at the St. Petersburg State University, Russia. In recent years as a DAAD grant holder she closely cooperated with the Working Group Biodidactics at the Friedrich-Schiller Jena University, Germany. Levit's research interests are in German evangelicalism, the role of religion in the collective self-awareness, science and religion in Russia, folk religions, and medical aspects of religious beliefs.

RAFAEL A. MARTÍNEZ is professor of philosophy of science at the Pontifical University of the Holy Cross (Rome). Educated as a physicist and philosopher, he is interested in the historical and epistemological aspects of scientific concepts and their relation to religion. He is author of *Immagini del dinamismo fisico. Causa e tempo nella storia della fisica* (Images of physical dynamism: Causation and time in the history of physics) (1996), coauthor, with Mariano Artigas and Thomas F. Glick, of *Negotiating Darwin: The Vatican Confronts Evolution, 1877–1902* (2006), and coeditor, with Gennaro Auletta and Marc Leclerc, of *Biological Evolution: Facts and Theories; A Critical Appraisal 150 Years after "The Origin of Species"* (2011).

EFTHYMIOS NICOLAIDIS is director of the History, Philosophy, and Didactics of Science and Technology Program at the Institute of Historical Research of the National Hellenic Research Foundation (www.hpdts.gr). He obtained his Baccalauréat in France and studied physics at Paris XI—Orsay and history of science at the École des Hautes Etudes en Sciences Sociales, Paris. After serving as a collaborator of the National Observatory of Athens (1979–84), he joined the program of history of science of the Institute of Historical Research of the National Hellenic Research Foundation in 1984. Since 2003 he has been director of this program, president of the International Union of the History and Philosophy of Science / Division of History of Science and Technology (2013–17), and member of the Council of the International Academy of History of Science. He is coeditor of *Almagest* (Brepols publishers) and of the *Newsletter for the History of Science in Southeastern Europe*. He coordinates the program *Hephaestus* (E.U., FP7, Capacities), *Dacalbo* (Byzantine and Modern Greek alchemy), and NARSES (Science and Religion in Southeastern Europe and Eastern Mediterranean). His main publications are on the history of science in Byzantium, the Ottoman Empire, and the modern Greek state; the spreading of Modern European science; and the history of the relation between science and religion. His latest book is *Science and Eastern Orthodoxy* (Baltimore: Johns Hopkins University Press, 2011).

RONALD L. NUMBERS is Hilldale Professor Emeritus of the History of Science and Medicine at the University of Wisconsin–Madison, where he taught for nearly four decades. He has written or edited more than two dozen books, including, most recently, *Galileo Goes to Jail and Other Myths about Science and Religion* (2009), *Biology and Ideology from Descartes to Dawkins* (with Denis Alexander, 2010), *Science and Religion around the World* (with John Hedley Brooke, 2011), and *Wrestling with Nature: From Omens to Science* (with Peter Harrison and Michael H. Shank, 2011). He is general editor, with David C. Lindberg, of the eight-volume *Cambridge History of Science* (2003–), as well as a past president the History of Science Society, the American Society of Church History, and the International Union of History and Philosophy of Science.

LENNART OLSSON was born into a countryside family of farmers and workers, where no one had received a college education. He first became a dental technician (1980) and later studied biology, chemistry, and history of ideas at Stockholm University, Sweden, obtaining a MSc in zoology in 1988. Olsson received his PhD at Uppsala University, Sweden, in 1993. He then taught as a lecturer at Uppsala University and did postdoctoral research in the United States and Australia. As of November 2000, Olsson is a tenured professor of comparative zoology at the Friedrich-Schiller-Universität in Jena, Germany. He has spent sabbaticals at the Konrad-Lorenz-Institute for Evolution and Cognition in Vienna, at Harvard University, in Sydney, and at NESCent in Durham, North Carolina. Currently, most of the research in his group is focused on the evolution and development of the vertebrate head. Olsson also has an interest in the history and philosophy of biology and is working on the history of evo-devo and comparative morphology.

MARTIN RIEXINGER is associate professor of Arabic and Islamic Studies at Aarhus University in Denmark. He took courses in Islamic studies at Tübingen University (MA 1997). During his work on his PhD thesis *Ṣanāʾullāh Amritsarī (1868–1948) und die Ahl-i Ḥadīṯ im Punjab unter britischer Herrschaft*, at Freiburg University (Drphil 2002), his attention was drawn to the debates about the reception of modern astronomy in the Islamic world during the nineteenth and early twentieth centuries. He further pursued his research interest in Muslim attitudes toward modern science in his Habilitationsschrift *Die verinnerlichte Schöpfungsordnung: Weltbild und normative Konzepte in den Schriften Said Nursis (gest. 1960) und der Nur Cemaati* (2009), which deals with the Turkish Islamic movement behind the emergence of Islamic creationism.

NICOLAAS A. RUPKE is Johnson Professor of History at Washington and Lee University in Lexington, Virginia, having recently retired from a Lower Saxony Research Professorship of the History of Science at Göttingen University. Educated at Groningen (BSc, 1968) and Princeton (PhD, 1972), he held research fellowships at the Smithsonian Institution, Oxford University, Tübingen University, the Netherlands Institute for Advanced Studies, the Wellcome Institute for the History of Medicine, the National Humanities Center, and the Institute of Advanced Studies in Canberra. Rupke was the inaugural holder of the Nelson O. Tyrone Jr. Chair of the History of Medicine at Vanderbilt. He is the author of several scientific biographies, including a study of the Oxford geologist William Buckland, the London biologist Richard Owen, and the German scientist and explorer Alexander von Humboldt. Rupke currently works on the non-Darwinian tradition in evolutionary biology. He is a fellow of the German Academy of Sciences Leopoldina and the Göttingen Academy of Sciences.

Index

Aalders, G. Ch., 69
Abdülhamit II, 181, 184
Acworth, Bernard, 53
Adam (and Eve), creation of, 71, 78, 111, 112, 113, 114, 167, 182, 184, 186, 201, 203, 205
Adam of Aap? (tv series), 71
Adomnitei, Cristian Mihai, 237
Affair of Max Planck that Never Existed, The (Lönnig), 119
Agassiz, Louis, 183
Aguirre, Emiliano, 33–34
Aizupún, Felípe, 221
Akyol, Mustafa, 191
Alexy II, 168, 173
Alferov, Zhores, 169
al-Jisr al-Tarabulusi, Husayn, 182, 185
Alliance of Romania's Families, 237
Al-Nasr Trust, 59
Althaus, Dieter, 115
Alvarez, Luis, 246
Alvarez, Walter, 246
American Association for the Advancement of Science, 21
American creationism, 2–5, 243
Andersson, N. J., 87
And God Created Darwin's World (Hemminger), 118
Andrews, Edgar Harold, 91
Answers in Genesis, 55, 95, 136, 216
anthroposophy, 79, 87, 90
anti-creationism, 2; activist agendas for, 235–38; early European, 228–31; France, 21, 26–28; institutional initiatives, 238–39; multiple platforms, 234–35; political, 231–34; Russia, 177; Scandinavia, 239–40; UK, 10, 229–30, 235–37, 238
anti-Darwinism, 5–6; Catholicism and, 201; France, 15, 21–28; Germany, 107, 110; Greece, 157; Netherlands, 72–73; Poland, 129; Portugal, 43; Russia, 173, 174; Scandinavia, 11, 87–90, 92–95, 96, 97; Spain, 32, 41; Turkey, 183, 186, 188, 189, 191, 192
antievolutionism: Belgium, 79; Catholicism and, 199–210; France, 15; Germany, 107–10; global spread of, 5–6; Greece, 12, 146, 150, 155–56, 159–60; in media, 6; Netherlands, 75–76; North America, 2–5; Oktar's *Atlas of Creation*, 6, 16, 59–60, 78, 97, 109, 233; Poland, 125, 131–36; Russia, 12, 172–74; Scandinavia, 10–11, 87, 91–95; Spain, 9, 33, 39; Turkey, 181, 182–84; UK, 53–54, 61
Aral, Seda, 97
Arnould, Jacques, 26
Artigas, Mariano, 36–37
Ashirov, Nafigullah, 169
Atatürk, Mustafa Kemal, 180, 184
atheism, 6; France, 17; Germany, 106, 112, 118–19, 121; Greece, 145, 147, 148, 149, 152–54; Poland, 127, 132; Russia, 12, 163, 165, 169, 172, 237; Scandinavia, 87, 93; Spain, 9, 31, 34, 37, 42; Turkey, 187; UK, 52, 56, 60, 61, 230, 231, 237
Atheist Bus Campaign, 237
Atkins, Peter, 219
Atlas of Creation (Oktar), 6, 16, 59–60, 78, 97, 109, 190–91, 233
Attic Calendar, 146
Au commencement, 19, 20
Australia, 6, 54, 55, 61
Austria, 105, 108, 125, 126, 128
Avatar (film), 24
Avgoustinus of Florina, 151
Axe, Douglas, 219
Ayala, Francisco J., 208, 209

Babuna, Oktar, 59
Back, Knud Aage, 90, 94, 96
Baker, Sylvia, 133
Barnard, Geoff, 133
Barnes, Thomas G., 37
Barr, Stephen M., 208, 209
Basic Types of Life theory, 92, 105–6, 110–15, 117, 118, 120
Baudouin, Cyrille, 15
Beck, Horst W., 107, 110, 117
Begzos, Marios, 157
Behe, Michael, 21, 39, 72, 73, 74, 93, 114, 115, 117, 134, 208, 209, 219
Belarus, 12, 162, 163–64, 175
Belgium, 7, 8, 10, 71, 75, 77; Catholicism and evolution, 10, 65, 77–78, 79; creationism, 78–79
Belyaev, Demyan, 163
Benedict XVI (pope), 116, 209
Bennet, Robert, 207
Berger, Leszek, 126
Berthault, Guy, 18, 177, 206, 207, 208
Beyaz, Zekeriya, 186
Beyond Darwin (Staune), 23
Bible and Science (Van Delden), 71
Biblical Creationism Society, 133–34, 135
Biblical Creation Ministries, 54
Biblical Creation Society, 54
Bieganski, Wladyslaw, 138
big bang theory, 245
Bijbel en Onderwijs, 75
Bijbel en Wetenschap (journal), 71
Bilmen, Ömer Nasuhi, 183–84
BIO-Complexity (journal), 216, 217–18
Biologic Institute, 216, 217, 219
Biology Teacher Discovers Creation, A (Scheven), 109
Blair, Tony, 56
Blechschmidt, Erich, 117
Boehner, John, 219
Bradley, Walter, 74
Branchaninov, Ignatius (saint), 172, 173
Brandes, Georg, 87, 89
Brandi, Salvatore, 205
Brasseur, Anne, 208
Breaking the Spell (Dennett), 138
Breitung, Amand, 88
Brinkman, Baba, 238
Britain. *See* United Kingdom
British Association for the Advancement of Science, 229
British Centre for Science Education (BCSE), 52, 54, 218, 219, 220, 234
British Council survey, 52, 53
British Humanist Association, 59, 237
Brosseau, Olivier, 15
Bryan, William Jennings, 2
Büchner, Ludwig, 149, 181
Buckser, Andrew, 85
Buddhism, 127, 163
Bufeeff, Konstantin (archpriest), 170, 173–74, 177
Bugg, George, 243
Bulgaria, 144, 145
Buonpensiere, Enrico, 204–5
Burgess, Stuart, 216, 217, 218
Burn, John, 55

Calvinists, Dutch, 65–71
Cameron, James, 24–25
Cameron, Nigel M. de S., 54
Campbell, John G., 206
Canada, 54, 246
capitalism, 171, 177, 187, 222
Carreira das Neves, Joaquim, 43
Caterini, Pietro, 205
Catholicism, 1, 2; antievolutionist and creationist groups, 199, 206–8; Belarus, 164; Belgium, 10, 65, 77–78, 79; Darwinian, 25–28; early reaction to evolution, 200–203; fighting against evolution, 203–6; France, 16, 17–19, 25–28; Germany, 106; *Humani Generis*, 33, 205; Netherlands, 65, 66; Poland, 11, 127–28, 129–30, 135–36; Portugal, 9, 31, 43; present tendencies vs intelligent design, 199, 208–10; Russia, 163; Scandinavia, 86; Spain, 9, 31, 32–37, 42
Caverni, Raffaello, 202, 205
Ceausescu, Nicolae, 238
Center for (the Renewal of) Science and Culture, 4, 72, 74, 75, 214, 216, 222
Centre d'Études et de Prospective sur la science (CEP), 18–19
Centre for Intelligent Design, 59, 217–20
CEP, Le (journal), 19
Cercle d'Études Scientifique et Historique (CESHE), 18, 206, 207

Cevdet, Abdullah, 181
Chaberek, Michal, 135–36
Chapman, Geoff, 54
Chauvin, Rémy, 17
Christian Action Research and Education (CARE), 220
Christian Center for Science and Apologetics, Crimea, 175–76
Christian Union of Scientists, 149
Christodoulos (archbishop), 152–53, 154
Christou, Panagiotis, 150
Chrysopigi, 151, 153, 154
Church of England, 10, 60, 61
Cicognani, Marcolino, 203
Civiltà Cattolica, La (journal), 205
Colic, Liliana, 234
Collins, Francis, 42, 73
Committees of Correspondence, 231
Commonwealth of Independent States, 162
communism, 6, 93, 188, 192, 193; East Germany, 119; Greece, 12, 146, 147, 148, 149, 150, 151; Poland, 11, 126, 128, 129, 137; Romania, 237; Russia, 12
Comte, Auguste, 129
Copernicus, Nicolaus, 130
Council of Europe Resolution 1580, 1, 95, 99, 221, 232–33, 234, 236, 237, 239
Counter Evolution Group, 206
Coyne, George, 209
Creabel, 79
Creatio-Biblical Creationism (vom Stein), 117
Creation as Holy Communion (Zizioulas), 157
creationism, European, 9–13; opinion polls, 6–8, 42, 43. *See also specific countries*
creationist movement, 2–6, 242–43; global, 5–6, 242–43; reclaiming of scientific credentials, 243–45; US, 2–5, 243
Creationists, The (Arnould), 26
Creationists, The (Numbers), 199, 242
Creationists, The: A Threat to French Society? (Baudouin and Brosseau), 15
Creation Ministries International, 6, 136
"Creation Museum," 190
Creation Research, 54–55
Creation Research Society, 70, 90
Creation Science Foundation, 55
Creation Science Movement, 5, 54, 79. *See also* scientific creationism
Creations Resources Trust, 54
Creation through Evolution (Maldamé), 26
Creative Evolution (Goswami), 39
Cremo, Michael, 96
Crick, Francis, 3
Croatia, 216
Crombette, Fernand, 18, 206
Crumbling Theory of Evolution, The (Johnson), 131
Crusafont, Miquel, 33–34
Cruz, Antonio, 41
Cursed Legend of the Twentieth Century, The (Dambricourt-Malassé), 21
Cuvier, Georges, 145, 168
Czech Republic, 216
Czy istniały dinozaury?, 137–38

Damascene, 173
Damaskin, 177
Dambricourt-Malassé, Anne, 21–22, 28
Danish Society for Intelligent Design, 96
Danneels, Godfried, 77
Darwin, Charles, 2, 3, 16, 50, 65, 87, 93, 107, 117, 125, 134, 146, 164, 181, 200, 208–9, 228–29, 234, 242, 245
Darwin, George Howard, 247
Darwin and Christianity (Euvé), 26
Darwinian Catholicism, 25–28
Darwin industry, 242
Darwinism, responses to: Catholic, 201, 207; in Germany, 110, 113, 115; in Greece, 145–46, 149; in Low Countries, 72, 75–77; in Poland, 129–30, 134, 139; in Russia, 173, 174; in Scandinavia, 87–90, 92, 93, 95; in Spain, 32, 34–35; in Turkey, 188–89; in UK, 50–53, 237. *See also* anti-Darwinism
Darwin no mató a Dios (Cruz), 41
Darwin on Trial (Johnson), 21, 37, 113, 134, 139
Darwin's Black Box (Behe), 21, 39, 72, 74, 134, 139, 208
Darwin Strikes Back (Woodward), 41
Darwin year, 6, 234–35, 238–39; Belgium, 77; Catholicism and, 199, 207; Germany, 106, 117; Greece, 157; institutional anti-creationist initiatives, 238; multiple anti-creationist platforms, 234–35; Netherlands, 10, 65, 75–77; Poland, 139; Scandinavia, 92, 238; Spain, 39; Turkey, 191; UK, 238

Data on the Teaching of Evolution in Biology Classes (Scheven), 11, 108
Daugaard, Holger, 94
Davidheiser, Bolton, 37
Dawkins, Richard, 36, 42, 56, 60, 138, 190, 235–37
day-age theory, 3
Daylight (newsletter), 206, 207
Daylight Origins Society, 206
Daytonism, 5, 229
Declaration of the Christian Union of Scientists, 149
De Deo Creante, 201, 204
de Dorlodot, Henry, 205
de Felipe, Pablo, 40–41
definition of creationism, 3, 4–5
Dekker, Cees, 72–76
Delmouzos, Alexander, 148
Deluc, Jean Andre, 248
de Mattei, Roberto, 207
Dembski, William, 39, 73, 74, 93, 114, 216, 217, 219
Demirel, Süleyman, 187
Denmark, 7, 10–11; anti-creationism, 230, 238–39; early Protestant responses to Darwinism, 87–90; free churches fighting Darwin, 95; intelligent design, 95–96, 217, 221; Islamic creationism, 97–99; organizing antievolutionism, 91–95; religious background, 85–87; scientific creationism, 90–91
Dennett, Daniel, 138
Denton, Michael, 21, 72
de Prada, Juan Manuel, 39–40
de Robien, Gilles, 233
Descent of Man, The (Darwin), 87, 129, 146, 164
de Vries, Hugo, 89
Dewaer, Douglas, 33
Dichman, Ole, 92
Did God Create via Evolution? (Gitt), 118
Dinçerler, Vehbi, 187
Dinosaurerne og syndfloden (Hoffmann), 91
Discovery Institute (DI), 4, 21, 41, 43, 58, 59, 72, 92, 115, 189–90, 191, 214–22; continental allies, 220–21; European focus, 215–17; UK ties, 217–19; Wedge strategy, 214–15
DNA structure, 3
Dobzhansky, Theodosius, 112, 150, 157, 158
Dogma of Evolution, The (More), 91
Dogmet om evolusjonen (Sæbö), 90–91
doomsday warnings, 171, 243, 246–48
Dorenbos, Bert, 74
Dosios, Leandros, 146
Ducrotay de Blainville, Henri, 145
Dutch Reformed Church, 65, 66, 67, 69, 71
Dybowski, Benedykt, 128–30

Earth in Upheaval (Velikovsky), 247
Eastern Orthodox Church, 86, 127, 128, 144, 150
Ecevit, Bülent, 187
Edwards v. Aguillard, 4, 214
Eggen, André, 19
Einstein, Albert, 248
Eldredge, Niles, 245
Elim Pentecostal Church, 216, 219
Emmanuel College, 55–58
Emmanuel School Foundation, 58
England. *See* United Kingdom
Enoch, Hannington, 90
Erbakan, Necmettin, 181
Erdogan, Recep Tayyip, 181
Erroneous Way of Darwinism, The (Nachtwey), 107
Escuain, Santiago, 37–38, 41
Eskov, Kirill, 177
Estonia, 162
European Commission's Eurobarometer survey, 7, 42, 43, 85, 193
European Court of Human Rights, 128, 168
European Creationist Congress, 1, 38, 71, 94
European Evangelical Alliance, 1
European Humanist Federation, 230, 237–38
European Parliament, 1, 207, 222
European Union (EU), 1, 50, 162, 222, 232
Euvé, Francois, 26–27
evangelicalism, 53, 199; Belarus and Ukraine, 175; Belgium, 79; Germany, 11, 105, 106, 109, 110, 115, 116–17, 120, 121; intelligent design and, 219–20; Netherlands, 10, 70–71, 72, 75; Poland, 133; Scandinavia, 10–11, 85, 86–95, 99; Spain, 9, 37–38, 40, 41, 44; UK, 53; US, 2–5
Evangelische Kirche in Deutschland, 106
Evolución, La (Crusafont, Aguirre, and Mélendez), 34
Evolution: A Challenge for Science, a Danger to the Faith (Tassot), 19

Evolution: A Critical Textbook (Junker and Scherer), 77, 114–16, 120
Evolution and Dogma (Zahm), 203, 205
evolutionary theory: vs intelligent design, 2, 4; opinion polls of creationism and, 6–8; religious opposition, 5; US science education, 2–5
Evolution: A Theory in Crisis (Denton), 21, 72
Evolution, Creation and Science (Marsh), 112
Evolution: Fact or Belief? (video), 131, 137
Evolution News and Views, 220
Evolution or Creation (Enoch), 90
Evolution Protest Movement, 5, 33, 53–54, 90, 230
evolution restreinte aux espèces organiques, L' (Leroy), 203
Evren, Kenan, 187
Expelled: No Intelligence Allowed (documentary), 6, 137, 219
Explore Evolution (DI), 59, 218
extinctions, mass, 246

Factum (journal), 107
Fairholme, George, 243, 248
fascism, 6, 55, 93, 148, 171, 187, 188–89; Spain and Portugal, 9, 31, 43
Fascism: The Bloody Ideology of Darwinism (Oktar), 189
Fenni, Ismail, 183, 184
Feuerbach, Ludwig, 173
Finland, 94, 216, 217, 221, 237
Fleming, Ambrose, 230
flood geology, 68, 70, 90, 99, 109, 248
Flood in the Light of the Bible, Geology and Archaeology, The (Rehwinkel), 70
Focus on the Family, 219
Fontana, Ernesto (bishop), 204
Forbidden Archeology (Cremo and Thompson), 96
Förening for Biblisk Skapelsestro, 91, 94
fossil record, 33, 37, 69, 109, 135, 146, 149, 183, 186, 207, 239
France, 7, 8, 9, 24; Biblical creationism, 17–19, 28; creationism and education, 15–16, 28; Darwinian Catholicism, 25–28; historical background, 16–17; intelligent design, 9, 21; teaching creationism, 15–16; UIP and criticism of Darwinism, 20–23, 25, 27, 28
Franco, Francisco, 33, 34, 37, 43, 150
From Nothing to Nature (Andrews), 91
fronteras del evolucionismo, Las (Artigas), 36
fundamentalism, 199, 207, 236; France, 15; Germany, 106, 115, 116, 117, 120, 121; Greece, 151–52, 153, 154–55; Netherlands, 70; Poland, 135; Scandinavia, 86, 89, 91, 99; Spain, 39; Turkey, 192; UK, 55, 58; US, 2–5
Fundamentals, The, 243
Fursenko, Andrei, 166

Gaia theory, 174
gap theory, 3
Gärdeborn, Anders, 94
Garner, Paul, 54, 133
Garrigues Díaz-Cañabate, Antonio, 35
Gaudry, Albert, 146
Geelkerken, J. G., 67
General Biology (Mamontov), 166
General Zoological Tables (Blainville), 145–46
Genesis (biblical) account of creation, 3, 27, 55, 67, 69, 108, 111–12, 136, 137, 154
Genesis, Creation and Early Man (Rose), 156
Genesis Expo, 54
Genesis Flood, The (Whitcomb and Morris), 3, 37, 90, 137, 243, 248
Genesis (journal), 94
Génesis (journal), 38
Genesis Land, 233
Genesis (organization), 94, 95
Genesis Problems (journal), 132–33, 134, 137–39
genetics, Mendelian, 3, 16
Genocide of the White Race, The (Slezin), 168
Georgia, 162
Germany, 7, 8, 11; Basic Types of Life theory, 105–6, 110–15, 117, 118, 120; creationism and religious education, 119–20; creationism in schools, 116–18; *Evolution: A Critical Textbook*, 114–16; intelligent creator, 117–19; intelligent design, 105, 110, 115, 117–19, 220–21; religious landscape, 106–7, 120; rise of modern creationism, 107–10; Study Community Word and Knowledge, 11, 105, 107–8, 109, 110, 113, 115, 117, 118, 120, 220; teaching creationism, 105, 108–10, 115–17, 120–21
Giertych, Maciej, 1, 131, 134, 137, 139, 207–8
Giertych, Roman, 134

Gilbert, Scott, 174
Gilson, Etiènne, 206
Ginzburg, Vitaly, 166, 169
Gish, Duane T., 37, 91, 154, 186, 190
Gitt, Werner, 118, 133
Glick, Thomas F., 32
global creationism, 5–6, 242–43. *See also specific countries*
Glubokovsky, N. N., 172
God Delusion, The (Dawkins), 42, 138
God Is Not Great (Hitchens), 138
God versus Darwin (Arnould), 26
Golovin, Sergei, 175
González, Guillermo, 41
Goswami, Amit, 39
Gould, Stephen Jay, 34, 36, 245, 246
Grassé, Pierre-Paul, 17, 23, 27
Gray, Asa, 229
Greece, 12; change in 1974 and new trends in orthodoxy, 150–52; creationism and evolution in schoolbooks, 158–60; Darwin, Haeckel, and Greek Orthodox Church, 145–48; Holy Synod and creationism, 152–53; moderate views and New Orthodox, 155–56; Orthodox scientists, political regime, and materialism, 148–50; Orthodox supporters of evolutionary theories, 156–57; religious history, 144–45; right-wing views on creationism, 153–55; teaching creationism, 150, 158–60
Greek Orthodox Church, 12, 144–57
Grundtvig, N. F. S., 88
Gülen, Fethullah, 185–86, 187
Gündogdu, Cihat, 59, 97

Haeckel, Ernst, 67, 129, 146, 149, 165, 174, 181
Ham, Ken, 55, 95, 136
Hardy, Randall, 55
Hare Krishna, 87, 96
Hasan, Usama, 60
Has Our Existence a Meaning? (Staune), 23
Hasouros, Georgios, 155–56
Haught, John, 208
Heams, Thomas, 24–25
Hemminger, Hansjörg, 118
Hepp, Valentijn, 68, 69
Heribert-Nilsson, Nils, 149
Hilsberg, Thomas, 110
Hinduism, 5, 50, 97, 170
History of the Human Race, The (Stavrianos), 158
Hitchens, Christopher, 138
Hitler, Adolf, 114, 189
Hoffmann, Helge, 94
Hoffmann, Kirsten, 91
Hoffmann, Poul, 91, 94
Hollevoet, Chris, 79
Hovind, Kent, 136
Huibers, Johan, 75
Humani Generis, 33, 205
Huxley, Thomas, 128, 165, 229

Iberia, 7, 8, 9, 31–44. *See also* Portugal; Spain
Icons of Evolution (Wells), 92
Ieronymos I (archbishop), 150, 151
Ieronymos II (archbishop), 151, 154–55
Ierotheos of Nafpaktos, 155
Imming, Peter, 113
India, 90, 99
Indivisible (Robison and Richards), 222
Inner Mission, 86, 88, 91, 94
Inönü, Ismet, 180
Institute for Creation Research, 55, 71, 91, 94, 136, 189
intelligent design, 214; Catholicism and, 199, 208–10; Discovery Institute in Europe, 215–17; France, 9, 21; Germany, 105, 110, 115, 117–19, 220–21; Italy, 221; Netherlands, 72–75; North America, 2, 4, 5; Poland, 11, 133, 134; prospects for advancement in Europe, 221–22; Scandinavia, 11, 85, 93, 95–96, 99; Spain, 38–39, 42, 221; UK, 51, 58–59, 60, 217–20; US, 217
Intelligent Design (Dembski), 39
Intelligent Design 101, 139
Interacademy Panel, 238
In the Beginning, 19
Ipsos Mori poll, 8, 42, 51, 52, 217
Islam Denounces Terrorism (Oktar), 189
Islamic creationism. *See* Muslim creationism
Ismail, Hekimoglu, 192
Israel, 90, 94, 214
Is the Bible Right? (film), 114
Italy, 8, 205, 217, 221, 230, 233, 237

Jacobsen, Jens Peter, 87
Jastrow, Robert, 245

Jehovah's Witnesses, 18; Germany, 107, 115; Poland, 127, 131, 135; Scandinavia, 87, 95; Ukraine, 163
Jensen, Johannes V., 89
Jensen, Leif Asmark, 96
Jewish creationism, 5, 10, 78
Jews, 127, 163
Jodkowski, Kazimierz, 134, 138
Johannsen, Wilhelm, 89
John of Kronstadt (saint), 173
John Paul II (pope), 128, 200, 205, 209
Johnson, J. W. G., 131
Johnson, Phillip E., 21, 37, 61, 72, 73, 75, 113, 134, 214, 216, 217
Jones, John E., III, 214, 220
Jones, Steve, 59
Jordan, James B., 133
Journal of Creation, 216
Junker, Reinhard, 77, 110–18, 120–21

Kakolewski, Leon, 128
Kallinikos of Piraeus, 153–54
Kantiotis of Florina, 159
Karlsmose, Marianne, 93
Karoussos, Konstantinos, 153
Keith, Arthur, 5, 229–30
Kemal, Mustafa, 181, 184
Kemalism, 185, 187, 190, 192
Kenyon, Dean H., 208
Kepler, Johannes, 130
Kierkegaard, Søren, 88
Kitzmiller v. Dover Area School District, 4, 208, 214, 217, 218, 220
Knevel, Andries, 72, 74–76
Kolbe Center for the Study of Creation, 207, 208
Konashev, Mikhail, 177
Kopeikin, Kirill (archpriest), 170, 171
Kópru (journal), 186
Koszteyn, Jolanta, 137
Kouznetsov, Dmitri, 166
Kowalevsky, A. O., 165
Krimbas, Costas, 158
Kuenen, Philip Henry, 248
Kuhn, Thomas, 35
Kuraev, Andrei, 170, 171
Kutschera, Ulrich, 235
Kuyper, Abraham, 66

Lamarck, Jean-Baptiste, 16, 149, 200, 201
Language of God, The (Collins), 42, 73
Lassalle, Ferdinand, 146
Latvia, 162
Laun, Andreas (bishop), 207
Layfield, Stephen, 56, 58
Leben. Deutsches Schöpfungsmagazin (journal), 109
Lebendige Vorwelt, 109
Le Bon, Gustave, 183
Lecointre, Guillaume, 20, 22
Leisola, Matti, 216, 221
Lemaître, Georges, 245
Lenartowicz, Piotr, 137
Lengagne, Guy, 208
Leo XIII (pope), 202
Leroy, Dalmace, 201, 203, 204
Lever, Jan, 69, 70
Lie, The: Evolution (Ham), 136, 139
Lithuania, 162
Livets uppkomst (Molén), 94
Livingstone, David N., 242
Lloyd, Steve, 54
Lönnig, Wolf-Ekkehard, 114, 119
Loose, Peter, 216, 218
Lorencez, Isaac, 41
Lovelock, James, 174
Low Countries, 7, 8, 10, 65–80
Lyell, Charles, 244
Lysenkoism, 126

Machado, Jónatas, 44
Mackay, John, 54, 55
Mackay of Clashfern, Lord, 220
Major, John, 220
Maldamé, Jean-Michel, 26
Mamontov, Sergei, 166
Man from Monkey? (Kallinikos), 153
Man's Place in Nature (Huxley), 165
Maratti, Letizia, 233
Marcozzi, Vittorio, 205
Margulis, Lynn, 174
Marsh, Frank L., 92, 111–12, 113
Martínez, Antonio, 41
Marx, Karl, 146
Marxism, 153–54, 165, 185, 187, 190
Mastropaolo, Joseph, 208
Matsoukas, Nikos, 157

Mayr, Ernst, 112
Mazzella, Camillo, 202–3, 204
McCabe, Joseph, 230
McIntosh, Andrew, 56, 133, 218
McLean v. Arkansas, 4
McQuoid, Nigel, 55
Meester, Ronald, 72–73
Megasuccessions and Climax in the Tertiary (Scheven), 109
Mélendez, Bermudo, 34
Menderes, Adnan, 184
Merino, Santiago, 42
Messenger, Ernst, 205
Met andere ogen (van den Beukel), 72
Metaxas, Ioannis, 148
methodological naturalism, 20, 27
Metschinikov, I. I., 165
Meyer, Stephen, 216, 219–20
Miller, Kenneth R., 208
Mitterrand, François, 39
Mivart, St. George Jackson, 138, 201
Moczydlowski, Eugeniusz, 133, 135
modern synthesis, 3, 16, 90
Molén, Mats, 94
Monrad, D. G., 87
Moore, John A., 231
More, Louis T., 91
Moreland, James Porter, 134
Morris, Henry M., 3, 37, 54, 70, 90, 107, 137, 243, 248
Morris, John, 190
Most Recent Phase of Materialism, The: Darwinism and Its Lack of Foundation (Sougras), 146
Mota, Enrique, 38
Murray, John, 243
Muslim creationism, 5; Belgium, 10, 78; France, 9, 29n10; Scandinavia, 11, 97–99; Turkey, 6, 12–13, 181, 182, 185–92, 233; UK, 51, 52, 59–60. *See also* Oktar, Adnan
Muslims: Belgium, 65; former Soviet republics, 162, 163, 164, 169; France, 17, 29n10; Greece, 145; Netherlands, 66; Poland, 127; Scandinavia, 87; Turkey, 180; UK, 50

Nachtwey, Robert, 107
National Center for Science Education, 218, 228
National Defense Education Act, 3
NATO (North Atlantic Treaty Organization), 50, 162, 180, 192
Natural History of Creation (Haeckel), 165
natural selection, viewpoints on, 3; Catholicism and, 209; France, 16, 17, 21, 24–26; Germany, 118; Greece, 149, 155, 158; Low Countries, 72, 77, 78; Poland, 129; Russia, 167; Scandinavia, 87, 89, 92; Spain, 35, 36
Nature's Destiny: How the Laws of Biology Reveal Purpose in the Universe (Denton), 21
Navarro, Arcadi, 42
Nelson, Byron C., 90
Nelson, Paul, 216
neo-catastrophism, 243–48; creationism and, 247–48
neo-Lamarckianism, 16–17, 27
Neruda, Pablo, 39
Netherlands, 7, 10, 65; Calvinists and evolution science, 66–67; creationism institutionalized, 70–72; Darwin year, 10, 65, 75–77; first wave of creationism, 67–69; intelligent design, 72–75; pillarization of society, 66, 69, 71, 80; religious background, 65–66, 79; second wave of creationism, 69–70; teaching creationism, 67; wider acceptance of evolutionary theory, 69
Nevard, Anthony, 206
Nevin, Norman, 218
Newell, Norman D., 245
New Geology, The (Price), 243
New Humanist (journal), 237
Newton, Isaac, 130
New Zealand, 6, 54
Nielsen, Rasmus, 88
95 Theses against Evolution (pamphlet), 118
Nisiotis, Nikos, 156, 157
Nissen, Karsten, 91
Noah's ark, 75, 192
Noah's Ark Zoo Farm, 54
Noble, Alastair, 59, 219, 220
North American creationism, 2–5, 243
Northern Ireland, 50, 52, 53, 59, 216, 237
Norway, 7, 10–11, 96; anti-creationism, 235, 238–39; early Protestant responses to Darwinism, 87–90; Islamic creationism, 97–99; organizing antievolutionism, 91–95; religious background, 85–87; scientific creationism, 90–91

Nowodworski, Michal, 129
Numbers, Ronald L., 199, 242
Núñez, Diego, 32
Nurbaki, Haluk, 186
Nurcus in Turkey, 185–87, 191–93
Nursi, Said, 184–85
Nussbaum, Martha, 39

Of Pandas and People (Kenyon), 4, 208
Øhrstrøm, Peter, 91, 93
Oktar, Adnan, 6, 13, 16, 59–60, 97, 99, 188–91, 233. *See also* Yahya, Harun
old-earth creationism, 3, 5, 54, 97
Olkhovsky, Vladislav, 175
On the Origin of Species (Darwin), 2, 41, 50, 65, 87, 107, 117, 125, 128–29, 140, 146, 164, 200, 229, 234, 242
Origin and History of the Organisms (Junker and Scherer), 110
Origins (journal), 54
Origo (journal), 10, 92–94, 96
Orzechowski, Miroslaw, 134, 139
Østergaard, Kristian Bánkuti, 93, 94
Oude Wereld, 76–77
Our Final Hour (Rees), 247
Ouweneel, Willem, 37, 70, 71, 72, 75, 76, 107
Owen, Hugh, 177, 207
Owen, Richard, 229
Özal, Turgut, 187

Paisios, 155
Pajewski, Mieczyslaw, 131–33, 137
Papandreou, Andreas, 158
Parker, Gary E., 94, 137
Peet, John, 133
Penn, Granville, 243
Pentecostals, 216, 218, 219; Low Countries, 75, 79, 86; Poland, 131; Scandinavia, 92, 93, 94, 95, 99; UK, 52, 53; Ukraine, 163
Perinçek, Dogu, 190
Perrier, Pierre, 21
Phenomenon of Man, The (Teilhard de Chardin), 26
Philippaerts, Jos, 79
Physicians and Surgeons for Scientific Integrity, 41, 42
Pietrzak, Zbigniew, 135
Piotrowski, Robert, 134, 138
Piraeus Association of Scientists, 154
Pius XII (pope), 33, 164, 205
Poland, 7, 8, 11; Catholic Church and creationism, 135–36; creationism after 1990, 131–35, 140; creationism between 1945 and 1989, 130–31, 140; creationism in public domain, 138–40; historical background, 125–27; intelligent design, 11, 133, 134; international creationist organizations, 136–38; reaction to publication of *On the Origin of Species*, 128–30, 140; religious landscape, 127–28; teaching creationism, 136
Polish Anti-Macroevolution Society, 131–33, 134, 137
political anti-creationism, 231–34
polling creationism, 6–8
Poppenberg, Fritz, 114
Portugal, 7, 9, 31, 43–44
Price, George McCready, 68–69, 70, 230, 243, 248
Problemy Genezy (journal), 132–33, 134, 137–39
ProGenesis society, 76, 118, 233
Prolegomena in Theological Theory of Knowledge (Nisiotis), 157
Protestantism, 1; Belarus, 164, 175; Belgium, 65; France, 15, 19; Germany, 106; Greece, 151, 157; Netherlands, 10, 65–66, 67, 75, 79–80; Poland, 11, 127–28, 133, 135; Russia, 12, 162, 163, 165, 166, 170, 175, 176–77; Scandinavia, 11, 85, 86, 87–90, 92, 99; Spain and Portugal, 9, 31, 37–38, 40–41, 43–44; Turkey, 192; Ukraine, 163, 175–76
Pujiula, Jaume, 32
punctuated equilibria theory, 245
Putin, Vladimir, 169

Queen University Belfast Creation Society, 52
Qur'an, 182, 183, 188

Rachinsky, S. A., 164
Rahner, Karl, 206
Ramfos, Stelios, 156
Rationalist Press Association, 230
Ratzinger, Joseph, 116
Ray, John, 248
Read, David, 59
Rees, Martin, 247
Reeves, Colin, 216

Reflections on Natural History (Gould), 34
Rehwinkel, Alfred M., 70, 90
Reiss, Michael, 60–61
Renard, Krister, 95
Reynolds, John Mark, 134
Rhind, William, 243
Richard Dawkins Foundation for Reason and Science, 236–37
Richards, Jay, 219, 222
Riddle of the Universe, The (Haeckel), 165
Rivista Biologia (journal), 221
Rode, Helge, 89
Rodríguez de la Fuente, Félix, 34
Roldán, Alejandro, 33
Romania, 237–38
Romanov, Konstantin, 166
Rose, Seraphim (Eugene Dennis), 156, 177
Rosevear, David, 79
Royal Society of London, 59, 60–61, 234, 247
Ruiz de la Peña, Luis Juan, 35–36, 37
Rupke, Nicolaas A., 70
Russia, 2, 8, 12, 162–77; creationism in post-Soviet, 165–69; creationism in tsarist, 164–65; radical Orthodox creationism, 170–72; religious background, 162–64; teaching creationism, 166–68; theological discussions about evolutionary theory, 172–75
Russian Orthodox Church (ROC), 12, 112, 144, 156, 162, 163, 164, 166–73, 175, 176–77

Sadun, Ali, 59
Sæbö, Odd, 90–91
Sagan, Carl, 36
Sagan, Dariusz, 134, 138
Sala, Ignacio, 33
Salazar, António de Oliveira, 43, 150
Salis Seewis, Francesco, 205
Santorum, Rick, 219, 220
Sayın, Ümit, 190
Scandinavia, 7, 8, 10–11; anti-creationism, 238–39; early Protestant responses to Darwinism, 87–90; free churches fighting Darwin, 95; intelligent design, 11, 85, 93, 95–96, 99; Islamic creationism, 11, 87, 97–99; organizing antievolutionism, 91–95; religious background, 85–87; scientific creationism, 90–91, 99. *See also* Denmark; Norway; Sweden
Scheeben, Matthias J., 202
Scherer, Siegfried, 77, 92, 110–16, 118, 120–21, 177
Scheven, Esther, 109
Scheven, Joachim, 11, 108–10, 120
Schitterend ongeluk of sporen van ontwerp (Dekker and Meester), 73, 74
Schönborn, Christoph (cardinal), 42, 199, 209–10
Schreiber, Kirill, 166, 167
Schreiber, Maria, 166, 167, 176
Science, Philosophy and Theology in the Hexaemeron of Saint Basil (Matsoukas), 157
Science Research Foundation (BAV), 1, 189–91
scientific creationism, 2, 3, 236, 242; Germany, 117; neo-catastrophism and, 243–44; Poland, 131, 134; Russia, 165, 171, 176; Scandinavia, 90–91, 99; Spain, 36–38; UK, 54
Scientists Speak about God (Kallinikos), 153
Scopes trial, 67, 229, 243
Scotland, 50, 53, 56, 59, 216, 217–20
Seifert, Joseph, 207
separation of church and state, 3, 4, 9, 15, 16; absence in UK, 217, 220
Sequeiros, Leandro, 42
Serafeim (archbishop), 151
Serbia, 144, 145, 234
Sermonti, Giuseppe, 221
Seventh-day Adventists, 3, 68, 86, 92, 94, 99, 111, 131, 139, 243
Shestodnev society, 173, 177
Shoemaker, Eugene, 246
Silberstein, Marc, 27–28
Simmons, Geoffrey, 41
Simpson, George G., 33
Siotis, Marcos, 150
Sızıntı (journal), 186
Skaltsounis, Ioannis, 146–47
Slezin, V. B., 168
Slusher, Harold S., 37
Snelling, Andrew, 61, 216
socialism, 12, 66, 129, 145, 147, 148, 151, 154, 158, 165
Society for the Study of Creation, the Deluge, and Related Science, 111

Society without God (Zuckerman), 85
Soler, Manuel, 42
Sôter, 151
Sougras, Spyridon, 146
South Africa, 54, 246
Spaemann, Robert, 206
Spain, 7, 8, 9, 31–44; and centenary of Darwin's death, 34–37; influences of Catholicism, 31, 32–37, 42; intelligent design, 38–39, 42, 221; polarizations around evolution, 32–33; teaching creationism, 42; 21st century, 38–42; US-style creationism, 37–38, 41
Spanish Evangelical Alliance, 40
Spanish Society for Evolutionary Biology, 41–42
Spärck, Ragner, 90
Spencer, Herbert, 67
Spurr, Josiah Edward, 246
Staune, Jean, 20–25, 28
Stavrianos, Lefteris, 158
Steiner, Rudolf, 90
Sternberg, Richard, 216
Stichting tot Bevordering van Bijbelgetrouwe Wetenschap, 70–71
Studium Integrale Journal, 108
Study Community Word and Knowledge, 11, 105, 107–8, 109, 110, 113, 115, 117, 118, 120, 220–21
Stworzenie czy Ewolucja (Pajewski), 137
Sugenis, Robert, 207
Sweden, 2, 7, 8, 10–11, 96; anti-creationism, 237, 238; early Protestant responses to Darwinism, 87–90; free churches fighting Darwin, 95; Islamic creationism, 97–99; organizing antievolutionism, 91–95; religious background, 85–87; scientific creationism, 90–91
Swedish Evangelical Mission, 86
Switzerland, 105, 107, 108, 118, 233
Syndfloden (Thorngreen), 90
Sysoev, Daniil, 170–71
Szujski, Jósef, 129

Taouil, Nordine, 78
Tassot, Dominique, 15, 19, 207
Tatlı, Ädem, 186, 187
teaching creationism: Council of Europe resolution, 1, 95, 99, 221, 233, 234, 236, 239; France, 15–16; Germany, 105, 108–10, 115–17, 120–21; Greece, 150, 158–60; Netherlands, 67; Poland, 136; Portugal, 44; Russia, 166–68; Scandinavia, 93–94, 95; Spain, 42; UK, 51, 55–58, 60–61, 217, 218, 221, 234; US, 2–5, 214–15
Teilhard de Chardin, Pierre, 10, 26, 33, 78, 157, 170, 206
Tevfik, Baha, 181
Thatcher, Margaret, 220
theistic evolutionism, 10, 52, 70, 73–74, 75, 78, 106, 119, 121, 132, 136, 206
Theophan the Recluse (saint), 172–73
theory of Basic Types, 105–6
Theotokis, Alexander, 145
Third International Conference on Creationism, 113
Thom, René, 245
Thompson, Richard, 96
Thorngreen, Frode, 90
Thorvaldsen, Steinar, 91, 92
Three Views on Creation and Evolution (Moreland and Reynolds), 134
Timofeev, Alexander, 170
Tønnsen, Aminah, 99
Traces of God in His Creation? (Junker), 118
Trends in Ecology and Evoluiton (journal), 114
Tripepi, Luigi, 204
Truth in Science, 58, 217–18
turbidity current theory, 248
Turkey, 2, 7, 8, 12–13, 94; evolution in late Ottoman and early Republic Period, 181–82; global impact of Turkish Islamic creationism in 1990s, 188–91; historical background, 180–81; Nurcu movement, 12; rejection of evolution and internalization of Islamic norms, 192–93; religious opposition to evolution theory, 182–84; rise of Islamic creationism in 1970s–1980s, 185–88; role of Nursi, 184–85; US Christian and Turkish Islamic creationists, 191–92
Turkmenistan, 162
Twilight of Evolution, The (Morris), 70
Tyler, David, 217
Tyndale Fellowship, 219
Tyvand, Peder A., 92, 94

Ukraine, 12, 162, 163, 175–76
United Kingdom, 2, 5, 7, 8, 10; anti-creationism, 10, 229–30, 235–37, 238; creationist organizations, 53–55; demographics, 50; Emmanuel College controversy, 55–58; fragmented creationist movement, 60–61; intelligent design, 51, 58–59, 60, 217–20; Muslim creationism, 51, 52, 59–60; teaching creationism, 51, 55–58, 60–61, 217, 218, 221, 234; views on evolution and creation, 51–53
United States Supreme Court rulings, 3–4
Université Interdisciplinaire de Paris (UIP), 19–23, 25, 27, 28, 29n18
Unlocking the Mystery of Life (DVD), 137
Ure, Andrew, 243
Urey, Harold, 245
Uzbekistan, 162

van de Fliert, Jan R., 69, 70, 230
Vandel, Albert, 17
van Delden, J. A., 70, 71, 74
van den Beukel, Arie, 72, 73, 74
van der Hoeven, Maria, 73, 208, 234
van Helden, Kees, 75–76
van Woudenberg, René, 72–73
Vardy, Peter, 56, 57
Vardy Foundation, 56, 58
Variation and Fixity in Nature (Marsh), 112
Vatican, 1, 26, 157, 164, 200, 203, 205, 207. *See also* Catholicism
Vaz, Armindo dos Santos, 43
Velikovsky, Immanuel, 248
Vernadsky, Vladimir I., 173–74
Vertjanov, Sergej, 166–67, 177
Victoria Institute, 53
Vidal, César, 40
Vogel, Bent, 91, 94
Vogt, Carl, 164, 181
vom Stein, Alexander, 117, 118
Voyage That Shook the World, The (documentary), 6

Wales, 50, 56, 59
Walesa, Lech, 126
Ware, Kallistos, 153
Warming, Eugen, 88
Wartenberg, Feliks, 129
Was Darwin Wrong?, 139
Watson, James, 3
Weet Magazine, 76
Wells, Jonathan, 92, 93, 221
Weyel, Ulrich, 110
What Is Creation Science? (Parker), 94, 137
What Now, Mr. Darwin? (vom Stein), 118
Whitcomb, John C., 3, 37, 54, 70, 90, 137, 243
Wiecek, Gary, 55
Wieth-Knudsen, Knud Asbjørn, 89
Wilberforce, Samuel, 128
Wilder-Smith, Arthur E., 71, 91, 93, 107, 120, 216
Wojtyla, Karol, 128
Wolf, Jakob, 95
Wolff, Karin, 116, 121
Wolpert, Lewis, 219
Woodward, John, 248
Woodward, Thomas, 41
Woolley, Paul, 52
Word of Faith movement, 95, 99

Yahya, Harun, 6, 59–60, 78, 97, 109, 188–89, 190. *See also* Oktar, Adnan
Young, George, 243
young-earth creationism, 3, 5, 6, 19, 248; Catholicism and, 208; Germany, 107, 108, 109, 110–11, 115, 120; intelligent design and, 216, 217, 218; Netherlands, 68, 70, 72, 74, 75, 76; Scandinavia, 93, 94; UK, 52, 54, 56

Zafer (journal), 186
Zagadka Biblijnego Potopu (Pasiud), 137
Zahm, John A., 203, 204, 205
Zigliara, Tommaso, 202
Zillmer, Hans, 208
Zizioulas, Ioannis, 156, 157
Zochios, Ioannis, 147
Zoë, 148, 150, 151
Zuckerman, Phil, 85